Minerals and Rocks

10

John T. Wasson

Meteorites

Classification and Properties

With 70 Figures

Springer-Verlag Berlin Heidelberg New York 1974

Professor Dr. *John T. Wasson*
Department of Chemistry
and Institute of Geophysics
and Planetary Physics
University of California
Los Angeles, CA 90024/USA

Volumes 1 to 9 in this series appeared under the title *Minerals, Rocks and Inorganic Materials*

ISBN-13: 978-3-642-65865-5 e-ISBN-13: 978-3-642-65863-1
DOI: 10.1007/978-3-642-65863-1

Softcover reprint of the hardcover 1st edition 1974

Preface

My goal in writing this book was to provide an introduction to meteorite science and a handbook on meteorite classification. Insofar as I succeeded it should prove useful both to the practicing professional and to university students at the upper-division and graduate levels.

I originally intended the book to be nearly twice as long. The second half was to be a review of properties relating to the origin of each group of meteorites. Chapter XVIII is an example of how these later chapters would have looked, although most would not have been as interpretative. These chapters would have been useful chiefly to meteorite researchers looking for a quick summary of group properties; they were not written because of lack of time. Perhaps I will start to prepare this "second volume" in a year or so when my family and I have recovered from the preparation of the present volume.

Although some parts of the classification portion are mildly iconoclastic, I have attempted either to avoid the inclusion of speculative interpretations or to flag them with a caveat to the reader. I have relaxed these principles somewhat in Chapter XVIII to conserve space, but even there the discussion of alternative speculations should give the reader a feeling for the degree of uncertainty attached.

The book was started in the summer of 1971 and completed in the summer of 1973. This was a good period to attempt to summarize the literature on meteorites, since the production rate of quantitative meteorite studies declined during 1970—1973 as a result of the diversion of meteorite researchers' efforts to the study of lunar samples. The rate of meteorite research activity is now increasing as researchers apply to meteorites techniques refined for their lunar research.

I am indebted to numerous persons for advice rendered during the past decade regarding problems of meteorite origin and classification. A complete list would include more than one hundred names. Here I will cite only eleven who have kindly given me many hours of consultation and help: V. F. Buchwald, W. R. Van Schmus, H. Wänke, G. W. Wetherill, and the postdoctoral scholars I have had the pleasure to work with, P. A. Baedecker, C.-L. Chou, O. Müller, R. Schaudy, E. R. D. Scott, S. N. Tandon, and C. M. Wai.

The quality of this book has greatly benefited from constructive criticism by a number of persons. E. R. D. Scott and P. A. Baedecker read the entire text, and C.-L. Chou, M. Fuller, W. M. Kaula, L. Schultz, W. R. Van Schmus, and G. W. Wetherill read one or more chapters each. E. R. D. Scott supervised the addition of material from recent publications into the data bank and assisted in the preparation of the subject index. H. Kruse carried out the programming necessary to produce Appendix II. Other persons too numerous to mention provided advice on individual aspects of the book.

A number of photos were kindly provided by persons acknowledged in the figure captions. As cited in the text, several drawings were taken from published papers; I thank the authors and the publishers of the journals and books for their friendly cooperation. Most original drawings were prepared by J. Sells and D. Heine. Prints of a number of these were prepared by P. Deibele.

The entire manuscript was typed by J. F. Kaufman, who also suggested numerous stylistic improvements. Her cheerful diligence during the final months made it possible to complete the manuscript as quickly as we did.

Financial support during the early phases was provided by the J. S. Guggenheim Memorial Foundation and the Max-Planck-Gesellschaft zur Förderung der Wissenschaften. The University of California, Los Angeles, provided facilities and a steady source of income.

I owe the previously mentioned individuals and institutions an enormous debt, and I extend them my warmest thanks. Most of all I thank my wife Gudrun whose patience and confidence made the effort possible and worthwhile, and I dedicate this book to her.

Los Angeles, Summer 1974 J. T. Wasson

Contents

I. Introduction: Meteorites as Probes of Processes Occurring Very Early in the History of the Solar System

The solar system consists of the sun, nine planets, 32 known satellites (and a ring system), several thousand telescopically visible asteroids, $>10^5$ comets, meteoroids (see Glossary), interplanetary dust, and a tenuous and transitory plasma (the solar wind). A primary purpose of meteorite research is to learn more about the origin and evolution of the solar system.

Meteorites carry information about a wide variety of solar-system processes: 1. those which occurred in the solar nebula (see Glossary) prior to the formation of planets; 2. those which occurred in planet-like bodies, and which were similar to processes occurring in the interiors of the earth and other planets; 3. those resulting from collisional interaction between interplanetary objects, i.e., shock and fragmentation processes; and 4. those produced by interaction with solar and galactic cosmic rays.

Until the Apollo-11 mission to the moon in 1969, meteorites were the only extraterrestrial samples of the solar system available for laboratory study. Many meteorites are relatively undifferentiated objects which have not been altered by igneous processes. The lunar samples from six Apollo missions and from Luna 16 and 20 are products of igneous processes, and have experienced additional alteration by shock and reheating as a result of the bombardment of the lunar surface by comets, asteroids and meteoroids during the period since these igneous events occurred. Thus, although the investigation of returned lunar samples has provided invaluable information bearing on the early history of the moon (and to some degree, the earth), the relatively undifferentiated meteorites continue to be unique sources of information about processes occurring in the solar nebula.

In order to put meteorite research in perspective, it is useful to review briefly the present properties of the solar system and to follow this with a discussion of some models which attempt to describe its origin.

In Table I-1 are listed some data on the largest members of the solar system. Most of the mass of the solar system is concentrated in the sun, a typical medium-sized, middle-aged star. Most of the mass of the planetary system resides in Jupiter, which also carries the bulk of the

Table I-1. Data regarding the **sun**, planets and *major satellites*, after KAULA (1968, 1971), MORRISON (1973), NEWBURN and GULKIS (1973); 1-atm densities from UREY (1966)

Body	Orbital semimajor axis (cm)	Orbital semimajor axis (AU)	Titius-Bode rule (AU)	Mass (g)	Equatorial radius (cm)	Density ($g\ cm^{-3}$)	Density at 1 atm, 25° C ($g\ cm^{-3}$)	Angular momentum ($g\ cm^2\ sec^{-1}$)
Sun	—	—	—	$1.99 \cdot 10^{33}$	$6.96 \cdot 10^{10}$	1.4	—	$1.7 \cdot 10^{48}$
Mercury	$5.79 \cdot 10^{12}$	0.387	0.4	$3.35 \cdot 10^{26}$	$2.44 \cdot 10^{8}$	5.5	5.2	$9.3 \cdot 10^{45}$
Venus	$1.08 \cdot 10^{13}$	0.723	0.7	$4.87 \cdot 10^{27}$	$6.05 \cdot 10^{8}$	5.2	4.0	$1.8 \cdot 10^{47}$
Earth	$1.50 \cdot 10^{13}$	1.000	1.0	$5.98 \cdot 10^{27}$	$6.38 \cdot 10^{8}$	5.5	4.0	$2.7 \cdot 10^{47}$
Moon	$3.84 \cdot 10^{10}$	—	—	$7.35 \cdot 10^{25}$	$1.74 \cdot 10^{8}$	3.3	3.4	—
Mars	$2.28 \cdot 10^{13}$	1.524	1.6	$6.42 \cdot 10^{26}$	$3.39 \cdot 10^{8}$	4.0	3.6	$3.5 \cdot 10^{46}$
(asteroids)	$\sim 4.2 \cdot 10^{13}$	~2.8	2.8	$\sim 4 \cdot 10^{24}$	—	—	—	—
Jupiter	$7.78 \cdot 10^{13}$	5.204	5.2	$1.90 \cdot 10^{30}$	$6.99 \cdot 10^{9}$	1.31	—	$1.9 \cdot 10^{50}$
Io	$4.22 \cdot 10^{9}$	—	—	$7.23 \cdot 10^{25}$	$1.82 \cdot 10^{8}$	2.9	—	—
Europa	$6.71 \cdot 10^{9}$	—	—	$4.71 \cdot 10^{25}$	$1.55 \cdot 10^{8}$	3.0	—	—
Ganymede	$1.07 \cdot 10^{10}$	—	—	$1.55 \cdot 10^{26}$	$2.64 \cdot 10^{8}$	2.0	—	—
Callisto	$1.88 \cdot 10^{10}$	—	—	$9.58 \cdot 10^{25}$	$2.50 \cdot 10^{8}$	1.5	—	—
Saturn	$1.43 \cdot 10^{14}$	9.580	10.0	$5.69 \cdot 10^{29}$	$5.85 \cdot 10^{9}$	0.70	—	$7.8 \cdot 10^{49}$
Titan	$1.22 \cdot 10^{11}$	—	—	$1.38 \cdot 10^{26}$	$2.43 \cdot 10^{8}$	2.3	—	—
Uranus	$2.86 \cdot 10^{14}$	19.14	19.6	$8.73 \cdot 10^{28}$	$2.54 \cdot 10^{9}$	1.3	—	$1.7 \cdot 10^{49}$
Neptune	$4.52 \cdot 10^{14}$	30.19	38.8	$1.03 \cdot 10^{29}$	$2.13 \cdot 10^{9}$	1.7	—	$2.5 \cdot 10^{49}$
Triton	$3.56 \cdot 10^{10}$	—	—	$1.4 \cdot 10^{26}$	$1.88 \cdot 10^{8}$	4.9	—	—
Pluto	$5.90 \cdot 10^{14}$	39.44	—	$6.6 \cdot 10^{26}$	$3.2 \cdot 10^{8}$	4.8	—	$1.8 \cdot 10^{47}$

solar system's angular momentum. One problem which must be solved by any satisfactory model for the origin of the solar system is why the planets, with only 0.1% of the mass, possess 99.5% of the angular momentum of the solar system.

The orbits of the planets are very regular. With the exception of Pluto, whose orbit undergoes strong alteration by Neptune, all have inclinations (see Chapter XII) within 7° of the ecliptic. Except for Mercury and Pluto, the orbits are nearly circular, with eccentricities of <0.1. Neglecting Pluto, their semimajor axes increase in regular fashion according to the approximate relationship called the Titius-Bode rule (or, sometimes, "law"):

$$r_n = 0.4 + 0.3 \times 2^n$$

where r_n is the semimajor axis, and $n = -\infty$ for Mercury, 0 for Venus, 1 for earth, etc. The values calculated by this relationship (Table I-1) give a remarkable fit for all planets except Neptune, providing that the asteroids are considered to occupy the $n=3$ position. Although the Titius-Bode rule is surely related to the sweeping out of given areas of the early system by protoplanets, it has not been explained quantitatively, and may in fact have more mnemonic than theoretical importance (ALFVÉN and ARRHENIUS, 1970a). The approximate relationship

$$r_n = 0.354\,(1.733)^n$$

where $n=0$ for Mercury, 1 for Venus, etc., is more tractable theoretically. It reproduces the orbital radii to within about 10% on the average, the worst fit being a value 20% too high for Mars.

The planets fall into two natural groupings: the terrestrial planets Mercury, Venus, earth and Mars; and the major planets Jupiter, Saturn, Uranus and Neptune. The moon and the asteroids are probably members of the first group; Pluto can be placed in a tentative third division together with the major satellites of the major planets.

For comparison with meteorite samples it is of interest to use equation-of-state relationships to convert the densities of the terrestrial planets to those which would obtain at 1-atm pressure and 25° C. Such values calculated by UREY (1966) are listed in Table I-1. The adjusted densities of Mercury, Venus and earth are higher than those observed for any class of chondrites (Table XVII-1), which is generally interpreted to indicate that the Fe/Si atom ratios in these planets are greater than the chondritic value of about unity (UREY, 1952, 1966). The adjusted densities of the moon and Mars are essentially chondritic. The bulk of Jupiter and Saturn appears to be H and He; and of Uranus and Neptune to be C, N, and O. As a result of the high content of volatile elements, the amount of non-volatile elements in the major planets is unknown and cannot be compared with meteoritic values. It is widely assumed that Jupiter has an essentially solar composition; in this case its non-volatile fraction (i.e., silicates and other non-volatile oxides) should amount to about 6.5×10^{27} g, very similar to the mass of the earth. The satellites of the major planets appear to be intermediate in composition between the terrestrial and major planets. The masses of Pluto and Triton are highly uncertain (NEWBURN and GULKIS, 1973) and their calculated densities may be erroneously high. There is a remarkable regularity in the orbital spacing of the satellites of Jupiter, Saturn and Uranus (ALFVÉN and ARRHENIUS, 1970a) which is probably the result of tidal friction and gravitational coupling. There appears to be a systematic decrease in the density of the large (Galilean) satellites of Jupiter which suggests that these have formed by processes similar to those responsible for the planetary system.

Reviews of models for the origin of the solar system are given by SAFRONOV (1969). TER HAAR and CAMERON (1963). WILLIAMS and CREMIN (1968) and WOOLFSON (1969); the book edited by REEVES (1972a) gives scenarios by a number of writers which together provide a good overview of more recent theoretical speculations. BODENHEIMER (1972) reviewed models and observational evidence regarding star formation.

Solar-system models can be considered to fall into two general classes; 1. cataclysmic and 2. nebular. The cataclysmic theories are based on the somewhat improbable occurrence of a collision or near collision of one or more stars with the sun, such that material was removed from one or both of these bodies to form the planetary portion of the solar system. Such models can be traced back to BUFFON (1745); during the present century variations on this theme have been proposed by CHAMBERLIN (1901), MOULTON (1905), JEFFREYS (1929), JEANS (1931), WOOLFSON (1964) and others. There are substantial difficulties with this class of models, including: 1. the low probability of a grazing encounter between the sun and another star; 2. the fact that very hot material from the sun would undergo excessive expansion before dissipating its heat (SPTZER, 1939); and 3. the very eccentric planetary orbits resulting from such models unless very special conditions are assumed (NÖLKE, 1930; RUSSELL, 1935). As a result, cataclysmic models currently have very few advocates. Some of these ideas remain germane, however, since stars may form as close clusters (see LARSON, 1972).

Nebular models can be divided into two subtypes: 1. those which assume an accretionary origin of the planetary part of the solar system after the formation of the sun; and 2. those which assume that the origin of the extrasolar portion of the solar system occurred as part of the same processes which resulted in the formation of the sun. Accretionary models have been proposed by ALFVÉN (1942, 1954), ALFVÉN and

ARRHENIUS (1970a, b), and by SCHMIDT (1944). The latter envisioned the capture of a nebula during the passage of the sun through an interstellar cloud; if this capture occurred on a short time scale, there need be little difference between this model and those which consider the sun and planets to have formed from the same solar nebula. Capture on a slow time scale would not yield the heating necessary to produce the nebular-type fractionations observed in chondritic meteorites (see Chapter XVIII).

The accretionary model of ALFVEN (1954) does lead to fractionations. He proposed that the early sun had a substantial magnetic field. Atoms from a surrounding interstellar gas cloud underwent gravitational acceleration near the earth; when these collided they became ionized and trapped in the magnetic field. The distance at which a particular element became trapped in the solar magnetic field depended on its ionization potential; the lower the ionization potential, the greater its equilibrium distance from the sun. Thus, easily ionized elements such as the alkalies should be enriched and the inert gases depleted in a planet formed far from the sun. Gradually this plasma condensed to particles, at which point further accumulation proceeded by gravitational processes. A detailed discussion of several aspects of this model is given by ALFVÉN and ARRHENIUS (1970a, b, 1973).

Most recent models of the origin of the solar system consider the sun, planets and other extrasolar material to have formed from the same batch of material at approximately the same time and by the same or complementary processes. Observational evidence (see, e.g., SPITZER, 1963; HERBIG, 1970; BODENHEIMER, 1972) shows that young stars are generally immersed in dense interstellar clouds of gas and dust. Examples are found in the Orion nebula and in dust clouds in the Taurus region. One of the latter stars, T-Tauri, has supplied the name for a whole class of recently formed stars (see HERBIG, 1962, for a review). These stars are generally irregularly variable in luminosity, and their strongest emission lines are blue-shifted, indicating that the lines arise in material moving away from the star. KUHI (1964, 1966) estimates rates of mass loss of the order of 10^{-7} solar masses per year, about 10^7 times greater than current mass loss from the sun via the solar wind. The T-Tauri stars have high Li contents, which is evidence of their youth, since Li is easily burned in stellar interiors. T-Tauri stars are generally found in clusters, and their spatial density is proportional to the apparent density of dust in their formation regions (HERBIG and PEIMBERT, 1966). Thus, their youth and genetic connection with the surrounding interstellar clouds appear to be established.

Important aids in the classification of stars are plots of the Hertzspring-Russell type (Fig. I-1), in which log luminosity is plotted

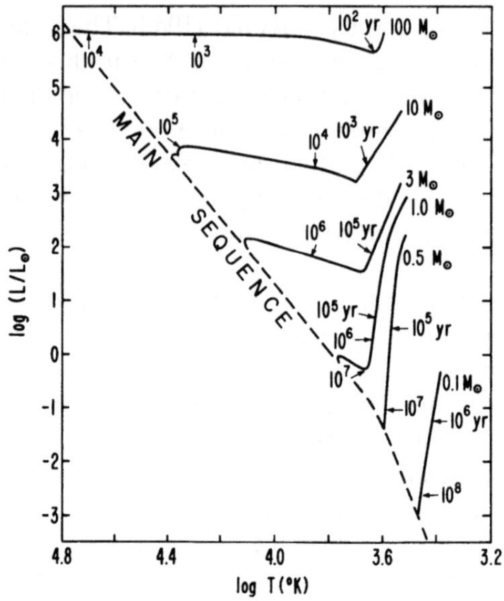

Fig. I-1. Hertzsprung-Russell diagram. The log of a star's luminosity normalized to the luminosity of the sun, $L_{\odot}$ is plotted against log temperature (decreasing to the right). The dashed line labelled *main sequence* shows the loci of stars which are in their hydrogen-burning phase. Theoretical tracks show the paths by which new stars of differing mass (expressed in units of the sun's mass, $M_{\odot}$) evolve onto the main sequence. During this latter, highly convective, Hayashi phase, the luminosity of small and medium-sized stars can be much greater than the equilibrium value achieved when they reach the main sequence. The time elapsed since the onset of the Hayashi-type convective phase is shown

against color, or more recently, log temperature (with temperature decreasing to the right). Most stars plot on or near a diagonal track called the main sequence on such a diagram. A star remains on the main sequence as long as it is in the hydrogen-burning phase, which is most of its life as a visible object. New stars evolve onto the main sequence toward its lower right end, the exact position depending on the mass of the star. Just prior to its joining the main sequence a typical star passes through a convective stage during which heat is liberated very rapidly and the luminosity reaches a value many times higher than the main-sequence value (HAYASHI, 1961). Some typical evolutionary Hayashi-type tracks onto the main sequence are shown in Fig. I-1 (after BODENHEIMER, 1972). These calculations are based on somewhat idealized conditions (cloud densities of about 10^{-9} g cm^{-3}, spherical symmetry, no angular momentum, no magnetic field), and there is much current discussion as to whether such highly luminous stages exist in early stellar evolution. LARSON (1972) finds a lower maximum luminosity if he uses an initial cloud density of 10^{-19} g cm^{-3}.

Observational evidence and theoretical models suggest that several distinct phases occur during the formation of a star (CAMERON, 1962; SPITZER, 1963; HERBIG, 1970; BODENHEIMER, 1972):

1. Gravitational collapse of a dense cloud; as the density increases, fragmentation occurs with the final fragmentation resulting in the isolation of the solar nebula.
2. Rapid collapse of the nebula.
3. With increasing density the opacity becomes great enough to trap thermal energy; the interior temperature rises, and the rate of contraction decreases.
4. When the interior temperature reaches 1800 °K, H_2 dissociates. This provides a new energy sink; the collapse again becomes rapid and continues so until H_2 is entirely dissociated and substantial fractions of H and He are ionized.
5. A slow contraction phase follows during which the star is highly luminous, and the central temperature rises from about 10^5 to 10^6 °K.
6. Fusion of H to He begins, contraction ceases, and the star becomes a member of a main sequence.

Unfortunately, actual observations of young stars (such as T-Tauri stars) have generally been made when they were on or near the main sequence. Two visual observations of rapid increases in luminosity may involve stars during their Hayashi phase (see BODENHEIMER, 1972). During recent years a number of low-temperature objects have been discovered by infrared observations (NEUGEBAUER et al., 1971; NEUGEBAUER and BECKLIN, 1973). Some of these (e.g., one associated with the Orion nebula) appear to be protostars in the third phase listed above, but conclusive evidence is lacking. The development of high-resolution interferometers will allow the investigation of very dense interstellar clouds on the basis of visible and ultraviolet molecular absorption bands (LEQUEUX, 1972). The difficulties associated with telescopic observation of star formation emphasize the importance of meteorite studies as a source of experimental evidence relating to such phenomena.

The earliest models in which the planetary system is considered to originate as a direct consequence of the formation of the sun are those of KANT (1755) and LAPLACE (1796). Among the more detailed recent models of this type are those of HOYLE (1960, 1963; HOYLE and WICKRAMASINGHE, 1968), MCCREA (1960), CAMERON (1962, 1963a, 1973; CAMERON and PINE, 1973), and SCHATZMAN (1962, 1967). SAFRONOV'S (1969) detailed calculations can be cited here although he assumes the solar nebula as an initial condition.

HOYLE (1960, 1963) proposed the first model directly linking the formation of the sun and the planetary system. He suggested that the solar system formed from an interstellar cloud fragment with a mass of 1.01 $M_{\odot}$ ($M_{\odot}$ = one solar mass – 0.01 $M_{\odot}$ is the minimal mass which can account for the non-volatile portion of the planets) which had the same angular momentum density as the galaxy as a whole. He assumed that the level of ionization was low in the fragment, and that magnetic effects played no role in the initial collapse phase. The angular momentum halted the collapse of the protosun when its radius was 4×10^{12} cm, about 0.7 times the present orbit of Mercury. At this point the protosun began shedding rings of material. Magnetic coupling of the rings with the protosun resulted in the transfer of angular momentum to the ring, increasing the orbital radius of the ring and allowing the continued contraction of the sun. SCHATZMAN (1962) proposed that the angular momentum transfer was effected by ejection of large amounts of high-energy "solar-flare" particles and their associated magnetic fields from the surface of the contracting protosun. The total angular momentum of the solar system is now 3.2×10^{50} g cm^2 sec^{-1}, 15 times smaller than the "galactic" amount assumed by HOYLE (1960); he assumes that the missing angular momentum was transferred magnetically to the interstellar medium in the form of H and He lost from the region of Uranus and Neptune.

CAMERON (1962, 1963b) proposed that the angular momentum density in the solar system today is the same as that present in the protosolar interstellar cloud fragment and that the level of ionization of the matter was such that the original magnetic field remained trapped during the collapse phase. He suggested that the equatorial shedding of material from the contracting protosun commenced at about 100 AU. Like HOYLE (1960) he invoked magnetic transfer of angular momentum to allow further contraction of the protosun and to transfer to the planets their observed large share of the final angular momentum. He took an initial mass of 2—4 $M_{\odot}$ to allow for inefficient accretion by the planets, and additional mass loss during a T-Tauri phase.

In recent papers CAMERON (1973; CAMERON and PINE, 1973) considerably changed and quantified his model. Magnetic fields are no longer important, the nebular mass is now taken to be 2 $M_{\odot}$. Calculations are carried out for two types of initial density distributions in rotating spheres: 1. uniform, i.e., constant throughout; and 2. linear, i.e., varying linearly from an assumed value at the center to zero at the surface. Figure I-2 shows the pressure along the median plane of the nebula for these two cases, and the maximum adiabatic (compressional-heating) temperature achieved at these locations calculated from LARSON's (1969) relationship

$$T = 5 \times 10^{8} \varrho^{2/3}$$

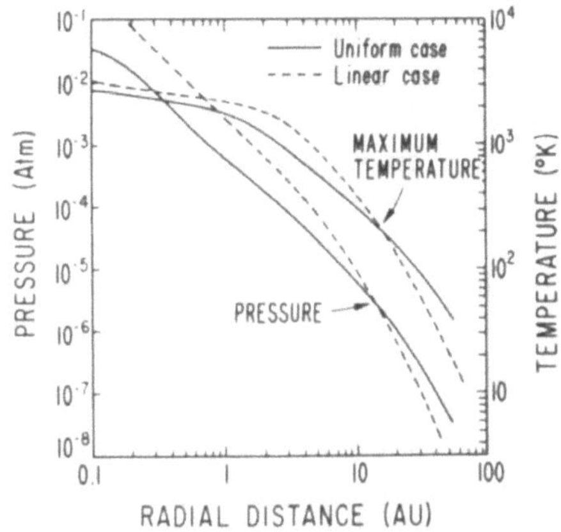

Fig. I-2. Pressure and maximum adiabatic temperature resulting from the collapse of an interstellar cloud fragment (mass = $2M_{\odot}$) to form the solar nebula (CAMERON and PINE, 1973)

where ϱ is the density. The maximum temperature is that achieved in an opaque nebula; the value for a partially transparent nebula will be lower.

Careful research on chondritic meteorites and others which may have escaped extensive post-nebular alteration can provide important boundary conditions for models such as those cited. This point was first stressed by UREY (e.g., 1951, 1952, 1954), and more recently by ANDERS (1964, 1968, 1971a), SUESS (1965) and others. For example, LARIMER (1973) and GROSSMAN and CLARK (1973) have interpreted, respectively, ordinary chondrite and CV chondrite data to indicate that these objects formed at nebular pressures of 10^{-3}—10^{-6} atm (ordinary chondrites) and $<2.10^{-3}$ atm (CV chondrites) (see Chapter II for classification information). Thus, CAMERON'S (1973) pressure distribution is consistent with these values at radial distances above ~0.8 AU. On the other hand, the fractionation observed in all chondrite groups except the CI and possibly CM chondrites indicates maximum nebular temperatures in excess of 1800—2000 °K (LARIMER and ANDERS, 1967; ANDERS, 1971a; WASSON, 1972a; GROSSMAN and LARIMER, 1974). According to CAMERON'S (1973) model such high temperatures obtain at heliocentric distances of ≤ 1 AU for the linear case, or ≤ 0.4 AU for the uniform case. Since it is likely that most of these classes of chondrites formed at distances > 2 AU, it would appear that CAMERON'S temperature profile needs adjustment.

The compositions and structures of chondritic or other primitive meteorites record a wealth of detail about mechanical fractionations as well. Solid-solid fractionations probably accounted for some of the observed large fractionations in the Fe/Si ratio; this fractionation may have resulted from a greater ability of the metal particles to stick together because of magnetic (HARRIS and TOZER, 1967; BANERJEE, 1967) or ductility (OROWAN, 1969) effects. One of the most dramatic metal-silicate fractionations is that recorded in the enstatite chondrites (LARIMER and ANDERS, 1970; BAEDECKER and WASSON, 1974).

Perhaps the most fascinating and persistent problem in research on primitive meteorites is the question of the origin of chondrules. These are spheroids generally 0.2—2 mm in diameter, which are present in members of all chondrite groups except the most primitive CI chondrites. Their origin has been attributed to equilibrium condensation (WOOD, 1963a), disequilibrium condensation (BLANDER and KATZ, 1967), lightening (WHIPPLE, 1966), impact melting on the surface of a parent body (e.g., FREDRIKSSON, 1963) and impact melting on small chunks of material (WASSON, 1972a, CAMERON, 1972, 1973). The issue remains unresolved, although the discovery of chondrule-like bodies in lunar surface breccias (KING et al., 1972; KURAT et al., 1972; NELEN et al., 1972) has shifted the emphasis in the direction of impact processes.

The differentiated meteorites provide information about fractionation processes in planetary interiors. In most groups processes occurred and the material cooled about 4.5 Gyr ago, within about 100 Myr of the formation of the oldest primitive meteorites (see Chapter X). The largest achondrite class, the eucrites, is basically basaltic in composition, and presumably formed by processes similar to those which formed terrestrial or lunar basalts, although distinct compositional differences are observed. Most iron-meteorite groups show evidence of fractionation by fractional crystallization, as one would expect in a convecting core (SCOTT, 1972). Reported differences in cooling rates between members of a single group weigh against such an origin (FRICKER et al., 1970), but these cooling-rate differences need better substantiation (WASSON, 1972b). Numerous authors have suggested that pallasites formed at a core-mantle interface in a parent body. Thus, some meteorites may represent very deep layers in planet-like bodies.

There is a remarkable number of varieties of meteorites. In Table II-8 are listed 26 groups with at least five members, and there are an additional 40—80 types with fewer members. There is no direct evidence that different categories were formed (or stored) in the same parent body. An important challenge to meteorite researchers is to search for such relationships. Knowledge of these genetic relationships is needed to facilitate the development of parent-body models. These models should be developed to as sophisticated a degree as possible before the first sampling mission to an asteroid or comet, currently projected for the late 1980's. Because of their high cost very few of these missions will be flown, and it is important that the targets be chosen so as to yield maximal information about nebular and planetary processes.

II. Classification of Meteorites

A. Introduction

The purpose of a classification is to group together related objects in order to facilitate comparative investigations. A classification should consist of as few categories as possible without forcing together objects for which the evidence of a relationship is inadequate. The subdivision of categories is good insofar as it is justified by sound data. Although it is impossible to avoid the occasional placement of potentially related objects in separate categories, one should attempt to avoid implying that differences between such objects are greater than they really are.

The most commonly used general classification scheme is that of MASON (1962a), which is based on that of PRIOR (1920). This scheme, revised according to MASON (1967a), is shown in Table II-1. The only major weakness with this system is in the classification of iron meteorites, in which the classes are defined entirely in terms of a geometrical series of kamacite bandwidths, rather than on the basis of evidence of genetic relationships between the different irons. This reflects the fact that adequate data for the classification of the iron meteorites were gathered only during the past few years.

Two additional problems with the classification are: 1. It accords separate entries to a few unique meteorites (e.g., the only siderophyre is STEINBACH), while leaving other unique meteorites (e.g., SOROTI) completely out of the classification system. A separate entry for each unique object is, in principle, a correct solution to the problem of classification, but the number of such objects is so great (among the iron meteorites alone about 50 such cases are now known) that these objects (and those cases where there are only 2—4 related objects) are best lumped together in broad "anomalous" categories in the basic classification, and then listed separately in supplementary tables. 2. By dividing the meteorites into four broad general categories of *chondrites*, *achondrites*, *irons* and *stony-irons*, the classification separates some groups for which there is good evidence of a genetic relationship, while leaving together others which are distantly related, if at all. For example, the achondrite groups of aubrites and howardites show little evidence of a genetic relationship, whereas a series of different properties suggests a relationship between the howardites and the stony-iron group of

Table II-1. Classification of meteorites according to MASON (1962a, 1967a)

I. Chondrites
  A. Enstatite chondrites
  B. Olivine-bronzite chondrites
  C. Olivine-hypersthene chondrites
  D. Carbonaceous chondrites

II. Achondrites
  A. Calcium-poor achondrites
    1. Enstatite achondrites (aubrites)
    2. Hypersthene achondrites (diogenites)
    3. Olivine achondrites (chassignites)
    4. Olivine-pigeonite achondrites (ureilites)
  B. Calcium-rich achondrites
    1. Augite achondrites (angrites)
    2. Diopside-olivine achondrites (nakhlites)
    3. Pyroxene-plagioclase achondrites
      a) Eucrites
      b) Howardites

III. Stony-irons
  A. Olivine stony-irons (pallasites)
  B. Bronzite-tridymite stony-irons (siderophyres)
  C. Bronzite-olivine stony-irons (lodranites)
  D. Pyroxene-plagioclase stony-irons (mesosiderites)

IV. Irons
  A. Hexahedrites
  B. Octahedrites
    1. Coarsest octahedrites
    2. Coarse octahedrites
    3. Medium octahedrites
    4. Fine octahedrites
    5. Finest octahedrites
  C. Nickel-rich ataxites

mesosiderites. WASSON and WAI (1970) have argued that the enstatite chondrites, the aubrites, the anomalous stony-iron Mount Egerton, and the anomalous iron Horse Creek are closely related. There are numerous examples of meteorites (e.g., mesosiderites, iron-meteorites-with-silicate-inclusions) which are not easily fitted into one of the four broad categories.

The solution to the latter problem seems to be to eliminate the four broad general categories from basic meteorite classifications. The words *chondrite*, *stony-iron*, etc., are themselves very useful, and their continued use is reasonable, provided they are used unambiguously.

In reviewing the evidence on the classification of the meteorites I have relied on the following two principles for determining whether certain meteorites were sufficiently related to one another to be classified as a group.

1. Meteorites belonging to the same group should be compositionally very similar. On plots of one compositional parameter versus another, such meteorites should form a distinct cluster, or they should lie on or near a smooth curve. Although such comparisons can normally be made from whole-rock data, a generally more stringent test is the comparison of the composition of individual phases.

2. Closely related meteorites should have very similar structures insofar as this is allowed by the bulk composition, and insofar as the original structure has not been altered by later processes such as a shock event. Structure mainly reflects composition and the pressure and temperature history during formation. It is likely that all of these factors will have been similar for meteorites of a single group.

Most of the meteorites which fall are *chondrites*, so named because they contain variable amounts of chondrules, i.e., spheroidal or ellipsoidal bodies usually a few tenths of a millimeter to a few millimeters in diameter. The mineralogy of the chondrules is essentially the same as that of the matrix in which they are embedded (KEIDEL, 1965; KEIL, 1969a; see also Chapter VI). Chondrules often account for 70% or more of the mass of a given chondrite. Chondrule-like bodies are rare or absent in terrestrial rocks.

Chondrule-containing meteorites were shown by NORDENSKJÖLD (1878) to be very similar to one another in their major element contents, and more recent research has shown that this similarity also holds for minor and trace elements as well, with the exception of some highly volatile elements. The relative amounts of non-volatile elements in chondrites are remarkably similar to those found in the solar atmosphere and in the atmospheres of stars similar to the sun.

A number of other meteorites are known in which chondrules are absent, but in which the non-volatile elemental content is very similar to that of the chondrule-rich meteorites. Such meteorites are also called chondrites. Because of the compositional similarity of chondritic meteorites to the sun, and thus, probably to the undifferentiated primitive material from which the solar system formed, one sometimes refers to all other types of meteorites as *differentiated meteorites*. Such a designation is probably better than the use of the separate categories of achondrite, stony-iron and iron for these meteorites, but it should be borne in mind that some of the differentiated meteorites may never have experienced igneous-type differentiation on a planet-like body.

Logically, the name *achondrite* should be applied to all classes of differentiated meteorites. In fact, its use is traditionally restricted to those differentiated objects containing small ($\leq 1\%$) amounts of reduced metal. A histogram showing the frequency of differentiated meteorites as a function of their metal contents is trimodal, with peaks at about 1, 50, and 99% metal. This distribution has led to the historical division of these meteorites into achondrites, stony-irons and irons, respectively.

The great abundance of meteorites with roughly equal amounts of metal and silicates results from the fact that two relatively large groups, the mesosiderites and pallasites, have such compositions. Despite this similarity, these two groups show little evidence of being closely related. Some pallasites (e.g., Brenham and Glorieta Mountain) have large areas of silicate-free metal, and appear to be closely related to the iron meteorites, whereas mesosiderite metal is generally present as fine grains and occasionally as cm-sized nodules. The mesosiderite silicates are distinctly different from the pallasitic silicates in both major-element composition and oxidation state. In general, it appears better to use the name of the group rather than "stony-iron" when discussing these meteorites.

Following the usage of VAN SCHMUS and WOOD (1967), I shall use the word *group* to designate a well-defined cluster of meteorites which appear, on the basis of available evidence, to be genetically related. A subdivision within a group is called a *type*. The word *class* is a broader term which can be used as a synonym for group, but can also be used for anomalous meteorites.

Current techniques for the classification of silicate-rich meteorites primarily involve chemical and mineralogical studies of the major elements and the phases which they form, whereas those used to classify the metal-rich meteorites primarily involve microscopy and the chemical study of Ni and various trace elements. For this reason it is convenient to discuss the metal-rich and silicate-rich meteorites separately. Some meteorites will bear discussion in both sections.

In many cases, an experienced meteorite researcher can confirm the meteoritic character of an unknown sample and assign it to a group on the basis of its macroscopic petrographic texture alone. Even in these cases, however, detailed chemical and mineralogic studies are necessary, both to confirm the identification and to increase the amount of available data on that particular group of meteorites. Logical first steps in such a study are to obtain an accurate chemical analysis (preferably by an analyst who has analyzed several other meteorites) and to study the meteorite petrographically and petrochemically, with particular emphasis on obtaining accurate phase-composition data.

The classification of the meteorites which are admixtures of apparently unrelated materials (as in Cumberland Falls, Fig. VI-7) presents a special problem. Group assignments of such meteorites should be made for the most abundant, petrographically distinct component. When complete petrographic and chemical studies are available, other major components should also be assigned to groups. Some typical meteorite breccias are discussed in Section VI F.

It is beyond the scope of this chapter to provide all the details necessary to definitively describe each group of meteorites. More details will be found in later chapters and in the literature cited there.

B. Classification of Silicate-Rich Meteorites

Figure II-1 illustrates differences in bulk composition among three groups of chondrites and three groups of differentiated meteorites. All data are for observed falls, and are taken from a single source, VON

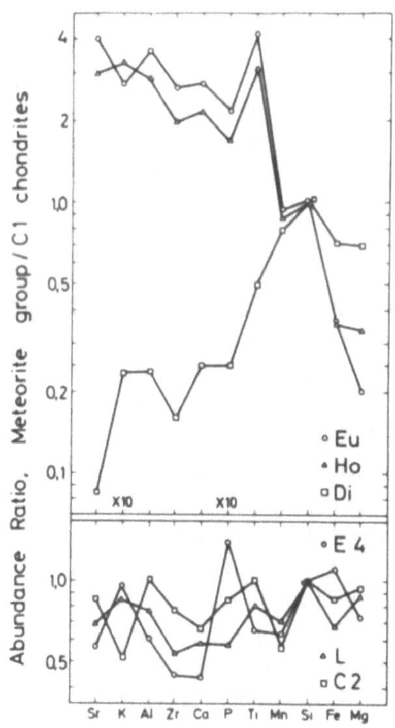

Fig. II-1. Meteorite group/CI abundance ratios (normalized at Si ≡ 1) for three groups of achondrites (differentiated silicate-rich meteorites) and three groups of chondrites. Elements arranged roughly in order of decreasing tendency to concentrate in low-melting fraction during igneous differentiation. Chondrite abundances of these elements are the same as mean solar-system abundances (estimated from CI chondrite data) to within a factor of 2, whereas much larger fractionations are found in the achondrites. (All data from VON MICHAELIS et al., 1969)

MICHAELIS et al. (1969), in order to avoid interlaboratory biases. The average abundances of 11 elements relative to their abundances in CI chondrites are plotted for two eucrites, one howardite, two diogenites, two E4 chondrites, 15L chondrites and three CM2 chondrites. The elements are arranged left to right roughly in order of decreasing tendency to concentrate in the low-melting fraction during igneous differentiation. Note that there is a tenfold difference in the scale used to plot the ratios of K and P in the upper part of Fig. II-1 as compared to the other elements.

The patterns for the achondrite groups in Fig. II-1 are strikingly different from those of the chondrite groups. The elemental ratios for eucrites and howardites decrease, those for diogenites increase as one proceeds to the right across the graph. The total range from the highest to lowest ratio within a group is greater than a factor of 10 for each of these groups. In contrast, all but four of the 33 ratios for the chondrite groups fall within the range 0.5—1.0, and the maximum range is a factor of 4 (observed in the E4 chondrites). The chondritic ratios fluctuate more or less randomly as one proceeds from left to right across the plot. If igneous fractionation has been primarily responsible for the trends among the achondrite groups, it appears that some other process has resulted in the differences among the chondrites.

1. Chondritic Meteorites

Classification of a meteorite as a chondrite is generally on the basis of its having abundances of non-volatile elements similar to those found in the sun. Since spectrographic determinations of solar abundances include rather large errors, the actual comparisons are in fact made to the very primitive CI chondrites, which have accurately determined compositions very similar to those of the sun (see ANDERS, 1971b). Thus, any meteorite showing elemental abundances in the same general range as that observed for the E4, L, and CM2 chondrites in Fig. II-1 would be considered chondritic. General discussions of chondritic classification are given by VAN SCHMUS (1969a) and KEIL (1969a).

Ratios of certain refractory elements to Si can be used to divide the chondrites into three categories, as shown in Fig. II-2. Here are plotted Al/Si, Mg/Si and Ca/Si ratios based on the data of VON MICHAELIS et al. (1969). Data are plotted only for observed falls which have been assigned to a petrologic type (defined below). For the carbonaceous chondrites, the second letter of the group symbol is shown rather than the type. Note that the chondrites divide naturally into three clusters on the basis of any of these ratios. The lowest ratios are found in the enstatite chondrites, the highest in the carbonaceous chondrites, and the three groups of ordinary chondrites (H, L, and LL) plot together in the centers of the histograms.

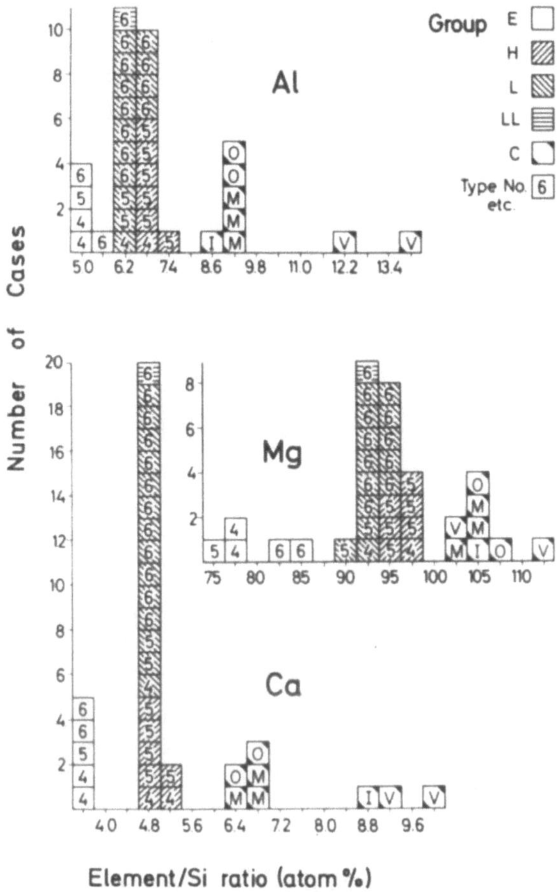

Fig. II-2. Histograms showing Al/Si, Mg/Si and Ca/Si atom ratios in the chondrite groups. The ratios of these refractory elements to Si are useful classificatory parameters for chondrites. Arabic numbers give petrologic type, letters give group designation of carbonaceous chondrites. (Data from VON MICHAELIS et al., 1969)

The abundance of Fe and its distribution between phases offers another means for distinguishing between chondritic groups. Figure II-3 is a diagram of the sort first used by UREY and CRAIG (1953), and shows the abundance of Fe present as metal and troilite plotted against that present in silicates and other oxidized forms. I have followed the usage of YAVNEL (1968) and plotted abundance rather than wt %, as used by most other authors. On this plot, a line of -1 slope corresponds to the loci of points having the same bulk Fe/Si ratios, but differing in their contents of oxidized Fe. Lines corresponding to bulk Fe/Si ratios of 0.6 and 0.8 are shown for reference purposes. All available data obtained

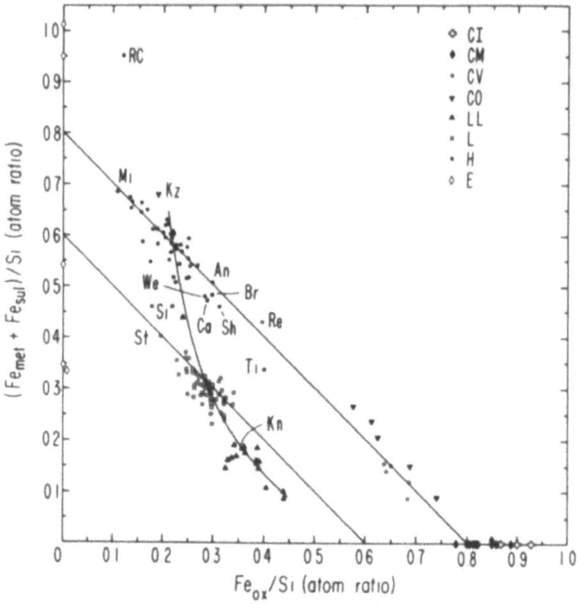

Fig. II-3. Distribution of Fe between "reduced" metal and troilite phases and the more oxidized phases of chondritic meteorites. The solid lines show the loci of points having bulk Fe/Si ratios of 0.6 and 0.8. The curve through the three ordinary chondrite groups (H, L, and LL) is a possible fractionation trend connecting these closely related groups. All analyses of chondrite falls published through 1972 by the analysts DYAKONOVA, JAROSEWICH, KHARITONOVA and WIIK are included. Some anomalous points designated by abbreviations are discussed in the text

from analyses by M. I. DYAKONOVA, E. JAROSEWICH, V. Y. KHARITONOVA, and H. B. WIIK, of chondritic observed falls are plotted. Where two analyses of the same meteorite were available, an average value is plotted.

The manner in which the plotted values are obtained deserves amplification. Metallic Fe is determined by selective dissolution. Troilitic Fe is calculated on the basis of an S determination, and on the assumption that all S is present as FeS, except in the CI and CM groups, where the assumption is made that none of the S is bound to Fe. Both these assumptions are only partially correct. Oxidized Fe is calculated by subtracting metallic and troilitic Fe from the total Fe value. Since total Fe is the most accurately determined value, errors in the determination of metallic Fe or the estimation of troilitic Fe have the effect of scattering points along the lines of constant bulk Fe/Si ratio. For example, the actual range in Fe_{ox} content in the H group is probably much smaller than indicated. The locations of several of the exceptional points designated by symbols in Fig. II-3 may reflect unusually large analytical errors.[1]

The highly reduced enstatite chondrites lie along the vertical axis in this plot. Their bulk Fe/Si ratios vary by almost a factor of 2, a much larger range than is observed in any of the other groups. The Fe/Si ratio in the CI chondrites is distinctly higher than that in other carbonaceous chondrites. The CO and CV chondrites generally contain quite small amounts of reduced Fe; Kainsaz and Renazzo are exceptions.

The ordinary chondrites plot in three distinct fields in Fig. II-3. The H (i.e., high-Fe) group is the most reduced and has bulk Fe/Si ratios between 0.67 and 0.93. The L (low-Fe) group is somewhat more oxidized and has bulk Fe/Si ratios which lie between 0.50 and 0.65. Although suspected to exist for several decades, the first clear proof of the existence of a separate LL group was provided by the careful electron-microprobe study of KEIL and FREDRIKSSON (1964). That its members were originally considered to be part of the L group is reflected in its name, LL, for low-Fe, low-metal. It has bulk Fe/Si ratios ranging from 0.48 to 0.58 and is the most oxidized of the ordinary chondrite groups. There is considerable evidence indicating that the three groups of ordinary chondrites are closely related. The broken curve in Fig. II-3 is a suggested fractionation path linking the three groups (WASSON, 1972a; see Section XVIII H).

A few words about the nomenclature of chondrites are in order. The enstatite chondrites are named after their most abundant minerals, the Mg-rich end members of the orthopyroxene or clinopyroxene solid-

---

[1] Since essentially all of the Mg and the oxidized Fe is in the ferromagnesian minerals (olivine and orthopyroxene), the Fe/(Fe+Mg) ratios in these minerals provide a control on the Fe_{ox} values reported in bulk analyses. Data listed in Appendix II B show that Ankober (An), Miller, Arkansas (Mi), Stavropol (St) and Weston (We) have olivine compositions which are typical for members of their groups; the unusual Fe_{ox} values in Table II-3 probably indicate sizable errors in these values. The Knyakinya (Kn) analysis appears to be correct; the meteorite appears to be an anomalous member of the L group with a very low metal content (see MÜLLER et al., 1971). Four of the H-group chondrites with high Fe_{ox} values—Bremervörde (Br), Castalia (Ca), Sharps (Sh) and Tieschitz (Ti)—are unequilibrated, and thus their mean Fe/(Fe+Mg) ratios are not well known. Concentrations of siderophilic elements are uniformly low in Bremervörde and especially so in Tieschitz (C.-L. CHOU, private communication, 1973). It appears that the bulk contents of Fe reported for these meteorites may be erroneously high. Concentrations of siderophilic elements in Castalia and Sharps are typical for the H group, and the high Fe_{ox} content is unexplained and possibly a real effect. The "Siena" ("Si") analysis falls directly between the H and L groups and cannot have been made on authentic Siena material, which is LL. Since it plots in the hiatus, it would be very interesting to identify the analyzed material. The metal in Rose City (RC) has been redistributed due to shock, and the analyzed specimen was probably not typical of the material as accreted. The two carbonaceous chondrites Kainsaz (Kz) and Renazzo (Re) are anomalous members of their groups.

solution series (see Chapter V), which are nearly free of Fe in these meteorites. This nomenclature has remained unchanged since the work of PRIOR (1920).

The ordinary chondrites are so named because they constitute the three largest groups of meteorites recovered following observed falls. The H, L, and LL groups account for 32, 39, and 7% of such recoveries, respectively. PRIOR (1920) assigned names to these groups: H–bronzite-olivine chondrites; Baroti-type and Soka-Banja-type hypersthene-olivine chondrites for the latter two. MASON and WIIK (1964) proposed that the name hypersthene-olivine chondrite be reserved for the L group and that amphoterite chondrite be used for the LL group. The nomenclature of PRIOR is awkward, and was shortened by many users to that of the pyroxene mineral, i.e., bronzite and hypersthene chondrites. Even these names are confusing, since they do not correctly describe the composition of the orthopyroxene minerals found in these meteorites. Current petrological usage defines the composition of bronzite as 12—30 mole % Fs. According to this definition the orthopyroxene in all three groups of ordinary chondrites is bronzite (see Section III D). Most authors currently use the nomenclature H, L, and LL (defined above) for the ordinary chondrite groups, and this is the practice which I recommend.

The name carbonaceous chondrite was applied by TSCHERMAK (1883) to a group of meteorites, most of which are still given this classification. PRIOR (1916, 1920) failed to recognize their different properties and included them together with his Soka-Banja-type hypersthene-olivine chondrites. The first quantitative definition of their different character was given by WIIK (1956). Today the main problem is not nomenclature, but more the basic difficulty of deciding which of these meteorites are sufficiently related to be assigned to the same group.

WIIK (1956) reported bulk-chemical data on carbonaceous chondrites, which he divided, on the basis of the C and H_2O contents, into categories called types I, II, and III, in order of decreasing content of these substances. YAVNEL (1963) and ANDERS (1964) divided the enstatite chondrites into types I and II, and KEIL (1968) added an intermediate type between these two. As with the carbonaceous chondrites, increasing type number in the E chondrites is accompanied by decreasing content of volatiles and increasing degree of recrystallization (see GLOSSARY). These trends apparently reflect degrees of heating which increase with increasing type number, although not necessarily in the sense of differing degrees of thermal metamorphism of the same starting material. DODD and VAN SCHMUS (1965; DODD et al., 1967) have called attention to similar trends which can be observed among the ordinary chondrites.

VAN SCHMUS and WOOD (1967) attempted to incorporate these additional facts into a new, two-parameter classification system, in which the same chemical groups are retained but are combined with an additional parameter which they call petrologic type. They arbitrarily divided the chondrites into six types, with type 1 corresponding to the material experiencing the least degree of reheating, and type 6 the chondritic material which experienced the most intense reheating.

Table II-2 lists the criteria which VAN SCHMUS and WOOD (1967) employed to distinguish the various types. They point out that some meteorites exhibit conflicting properties, making a choice between two adjacent types very difficult. For this reason, heavy vertical bars are drawn in the diagram for those properties they feel should be used for the ultimate decision in ambiguous cases (e.g., the distinction between types 3 and 4 can best be made on the basis of the degree of inhomogeneity of the ferromagnesian silicates). VAN SCHMUS and WOOD (1967) normalized their numerical types to the original carbonaceous-chondrite types of WIIK (1956); i.e., they accepted the WIIK division, but with minor changes to be discussed below. On this basis, they found that types 1 and 2 are missing among the enstatite and ordinary chondrites, and they classified the members of these groups into types 3 to 6. The ANDERS-KEIL division of the enstatite chondrites is maintained; types I, intermediate, and II of KEIL (1968) correspond to VAN SCHMUS-WOOD types 4, 5 and 6, respectively, although some meteorites are shiftet between types as a result of KEIL's more accurate observations.[2] VAN SCHMUS and WOOD introduced a shorthand notation for group and type in which the group abbreviation (E, H, L, LL, C) is followed without space by an arabic numeral giving the petrologic type. For example, Abee is an E4 chondrite, St. Severin an LL6 chondrite.

The most serious weakness of the classification of VAN SCHMUS and WOOD (1967) and of its forerunners is that, by placing all carbonaceous chondrites in the same group, it implies that they are closely related. In fact, there has never been much evidence for a close relationship between the three WIIK types, and recent studies by VAN SCHMUS (1969b), MASON (1971a) and MCCARTHY and AHRENS (1972) clearly indicate that the original members of type III must be separated into two distinct groups. These I will call the CV and CO groups, where the second letters are the first letters of the meteorites which VAN SCHMUS (1969b) chose as type specimens, Vigarano and Ornans. A number of criteria for distinguishing between these two groups are listed in Table II-3. The petrographic properties are taken from the work of VAN SCHMUS (1969b), the Al/Si and Ca/Si ratios from that of AHRENS (1970), the Fe/Mg ratio from MASON (1971a), and the rare gas distinctions from data given by MA-

[2] VAN SCHMUS and WOOD (1967) tentatively assigned Adhi-Kot to the E3 type; however, BAEDECKER and WASSON (1974) show that E4 is a better assignment for this meteorite.

Table II-2. Criteria for distinguishing different petrologic types of chondrites. (After VAN SCHMUS and WOOD, 1967)

	Petrologic type 1	2	3	4	5	6
I. Homogeneity of olivine and pyroxene compositions	—	Greater than 5% mean deviations		Less than 5% mean deviations to uniform	Uniform	
II. Structural state of low-Ca pyroxene	—	Predominantly monoclinic		Abundant monoclinic crystals	Orthorhombic	
III. Degree of development of secondary feldspar	—	Absent		Predominantly as microcrystalline aggregates		Clear, interstitial grains
IV. Igneous glass	—	Clear and isotropic primary glass; variable abundance		Turbid if present	Absent	
V. Metallic minerals (maximum Ni content)	—	Taenite absent or very minor (<20%)	Kamacite and taenite present (>20%)			
VI. Average Ni content of sulfide minerals	—	>0.5%	<0.5%			
VII. Overall texture	No chondrules	Very sharply defined chondrules		Well-defined chondrules	Chondrules readily delineated	Poorly defined chondrules
VIII. Texture of matrix	All fine-grained, opaque	Much opaque matrix	Opaque matrix	Transparent microcrystalline matrix	Recrystallized matrix	
IX. Bulk carbon content	3–5%	0.8–2.6%	0.2–1%	<0.2%		
X. Bulk water content	18–22%	2 –16%	0.3–3%	<1.5%		

ZOR et al. (1970). Other distinguishing features are also known, e.g., LAUL et al. (1970) report that the Vigarano group contains higher Bi/Si ratios than the Ornans group. There are exceptions or ambiguous cases for all the criteria listed with the possible exceptions of the Al/Si and Ca/

Table II-3. Criteria for distinguishing between the CV (Vigarano-type) and CO (Ornans-type) chondrites

	Chondrule size (mm)	Chondrule/ matrix ratio	Atomic ratios Al/Si	Ca/Si	Fe/Mg	Ar/Xe[b]	^{36}Ar[a] (10^{-8} cm^3 g)
Vigarano group	0.5—2	ca. 0.5	≧0.12	≧0.08	≦0.76	~100	≦30
Ornans group	0.1—0.4	>2	≦0.10	≦0.07	≧0.76	~200	≧60

[a] Not valid for the (rare) type-4 members of these groups.

Si ratios. For this reason, several properties may have to be measured before a new carbonaceous chondrite can be assigned to a group. The classification of these meteorites has been independently reviewed by VAN SCHMUS and HAYES (1974), who reach very similar conclusions.

Two CV chondrites (KABA and MOKOIA) have Ni-rich sulfides, which on the basis of the criteria listed in Table II-2, places them in petrologic type 2 (VAN SCHMUS and WOOD, 1967). To distinguish these chondrites from the other type 2 carbonaceous chondrites (which have higher Fe/Mg ratios and much higher C and H_2O contents) I will call the latter meteorites the MIGHEI group, and designate them with the symbol CM. Mighei was chosen as a type specimen because it is widely distributed and because data of WOOD (1967a) and others indicate that it is typical of the group.

In order to combine meteorite categories into a single group, one should have reasonably strong evidence that the categories are genetically related. Genetically related in the case of carbonaceous chondrites might mean that the one sort of material resulted from the heating of the other, or that they represent different degrees of a fractionation process which has taken place in the condensing and agglomerating solar nebula. There is no strong evidence linking the Vigarano, Ornans or Mighei groups with each other, nor any of these with the type 1 carbonaceous chondrites. For that reason the type 1 chondrites must also be distinguished as a separate group. Although Orgueil is the largest and most widely studied, and therefore the most logical type specimen for this group, I will instead choose the second most abundant representative, Ivuna, since this provides a letter which is different from that already chosen for the Ornans group. The complete symbol, CI, gains in mnemonic value through this choice, since it can remind the reader that all type 1 chondrites belong to this group. The designation C1 is also unique, and this usage without the letter would seem to be in order. The name carbonaceous chondrite should henceforth be used in the same way as ordinary chondrite, i.e., to gather together distinct groups which are potentially related.

Table II-4. Distribution of chondrite falls among the petrologic types

Group	Petrologic type 1	2	3	4	5	6
E	—	—	—	3	2	6
H	—	—	6	23	53	32
L	—	—	9	11	28	117
LL	—	—	6	1	7	20
CV	—	4	4	—	—	—
CO	—	—	5	1	—	—
CM	—	14	—	—	—	—
CI	5	—	—	—	—	—

The distribution of chondrites among the different petrologic types is shown in Table II-4. The classification of known enstatite and carbonaceous chondrites is essentially complete; that of ordinary chondrites is only about 40% complete. Most of the assignments are taken from the original paper by VAN SCHMUS and WOOD (1967), with the exception of the group designations for the carbonaceous chondrites. The names of the members of each group and type can be found in Appendix II B along with references dealing with their classification.

2. Differentiated Silicate-Rich Meteorites

The groups of differentiated silicate-rich meteorites differ much more from one another in their properties than do the chondrite groups, and it is therefore more difficult to discuss them in general terms. Some, like the eucrites, appear to have resulted from igneous differentiation. Others, like the mesosiderites, are breccia mixtures of materials which previously existed in different locations. Still others, like the aubrites, while highly differentiated, display some properties which indicate that they were not formed by igneous processes. Despite these difficulties, the state of knowledge regarding the petrographic and chemical properties of these meteorites has reached a reasonably high level, and most of them can now be unambiguously classified into groups.

Two key parameters for distinguishing these groups of meteorites are the Ca/Mg ratio and the Fe/(Fe+Mg) ratio of the silicate fraction; these are illustrated in Fig. II-4. Data are shown for all silicate-rich meteorites, and for the silicate portions of some metal-rich meteorites for which good data are available. For comparison purposes, data are also shown for two terrestrial basalts and four lunar crystalline rocks. The Fe/(Fe+Mg) ratio data are based on bulk-chemical analyses for all objects with Ca/Mg ratios greater than 10%, as well as for the carbonaceous

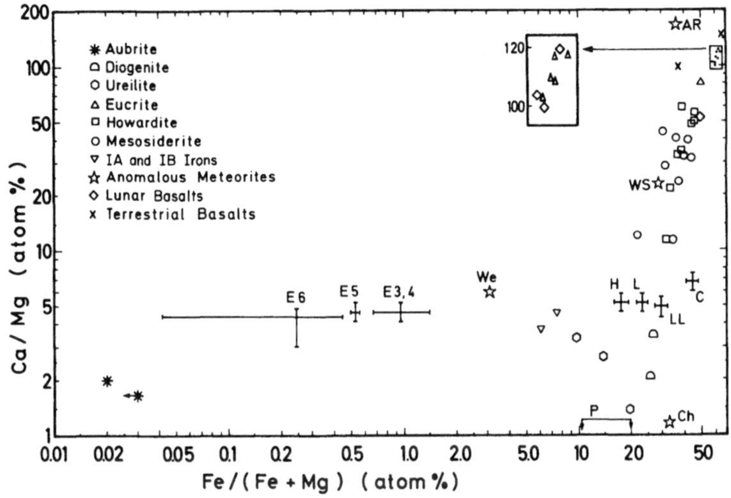

Fig. II-4. Plot of Ca/Mg vs. Fe/(Fe+Mg) ratio of silicate-rich meteorites and three lunar basaltic rocks. Chondritic groups are shown as ranges, and differentiated meteorites are plotted individually. In most meteorites the Fe/(Fe+Mg) ratio is a measure of the degree of oxidation of the mineral assemblages. The variation in Ca/Mg ratio is probably mainly a result of igneous differentiation. It rises from very low values in ultramafic, high-melting silicates (such as those in the pallasites) to very high values in low-melting basaltic silicates as found in the eucrites. The anomalous meteorites are abbreviated: Angra dos Reis (stone), AR; Chassigny, Ch; Weatherford, We; Weekeroo Station (now included in a new iron meteorite group II E), WS

chondrites, and on the composition of the low-Ca pyroxene or olivine (whichever is the more abundant ferromagnesian mineral) for the remaining objects. The mineral composition data have the advantage of being more precise, particularly for weathered samples or those with very low contents of oxidized Fe. They could not be used for meteorites in which the compositions of the ferromagnesian minerals are highly variable (unequilibrated or brecciated) or in which an appreciable fraction of the oxidized Fe or Mg is present in other phases. Numerical Ca/Mg ratios for differentiated silicate-rich meteorites will be found in Appendix II C, along with ferromagnesian mineral compositions.

The logical bases for the choice of axes in Fig. II-4 are as follows: During magmatic fractionation processes, the Ca/Mg ratio tends to increase with decreasing mean solidification temperature of the residual magma. Thus very high or very low Ca/Mg ratios are suggestive of formation by magmatic fractionation. From the classificational viewpoint, the advantage in plotting this ratio is that its range of variation within groups is considerably smaller than the factor of ca. 200 found in

Table II-5. Properties of silicate-rich meteorites

	Symbol of group (and type)	Ferromagnesian silicates			
		Name	Abund.[a] (wt. %)	Fe/(Fe+Mg) (mole %)	Inhomogen. (% MD)
Enstatite chon.	E3, E4	Clino-enstatite	ca. 45[b]	0.7—1.4[h]	med.[h]
Enstatite chon.	E5	Ortho-enstatite	ca. 46[b]	0.5—0.6[h]	low[h]
Enstatite chon.	E6	Ortho-enstatite	ca. 55[b]	0.04—0.5[h]	low[h]
H-group chon.	H	Olivine	25—40	16—19[i]	<2[i]
L-group chon.	L	Olivine	35—60	21—25[i]	<2[i]
LL-group chon.	LL	Olivine	50—60	27—32[i]	<2[i]
Vigarano-group chon.	CV	Olivine	ca. 70	6—14[j]	50—150[j]
Ornans-group chon.	CO	Olivine	ca. 70	9—23[j]	21— 70[j]
Mighei-group chon.	CM	Layer-lattice sil.	ca. 70	—[k]	high[k]
Ivuna-group chon.	CI	Layer-lattice sil.	ca. 63	—[k]	high[k]
Aubrites	Aub	Ortho-enstatite	ca. 97[b]	≦0.03	<2[l]
IAB iron met.	IA, IB	Olivine	ca. 40[c]	1—7	<2[l]
Ureilites	Ure	Olivine	ca. 85	10—25	med.[m]
Pallasites	Pal	Olivine	ca. 99[d]	11—20	<2
IIE iron met.	IIE	Ortho-pyroxene	ca. 10	15—22	<2
Diogenites	Dio	Ortho-pyroxene	ca. 95[e]	25—27	<2[l]
Meso-siderites	Mes	Ortho-pyroxene	40—80[f]	23—37	ca. 10[n]
Howardites	How	Ortho-pyroxene	40—80[g]	25—40	ca. 15[o]
Eucrites	Euc	Pigeonite	40—80[g]	50—67	<4

Legends see page 28.

Table II-5 (continued)

Metallic Fe[p] (wt. %)	$\delta^{18}O$[u] (‰)	Remarks[w] (concentrations by weight; ratios as atomic fractions)	General references
ca. 25	ca. 6.0	2.6—3.5% Si in kam; coexisting SiO_2 and Mg_2SiO_4.	KEIL (1968), MASON (1966), WASSON and WAI (1970)
ca. 24	—	2.9—3.6% Si in kam	
ca. 19	ca. 6.0	1.1—1.7% Si in kam; chond rare or absent.	
15—19[q]	4.8—5.4	gas ret ages gen ≧3.7 Gyr	MASON (1963a, 1967b), KEIL and FREDERIKSSON (1964), DODD et al. (1967)
4— 9	5.0—5.9	gas ret ages gen ≦2.5 yr	
0.3— 3	—	gas ret ages gen ≧3.7 Gyr; gen genomict brec	
0— 8	3.0—5.5	$H_2O \leqq 4\%$; see Table II-3 for other properties	MASON (1971a), VAN SCHMUS (1969b)
0— 5	ca. −0.8	$H_2O \leqq 2\%$; see Table II-3 for other properties	
0— 1[r]	8.4—11.3	6—16% H_2O; highly unequilibrated	WOOD (1967a)
ca. 0	ca. 12.2	18—22% H_2O; no chond; highly unequilibrated	BOSTROM and FREDRIKSSON (1966)
ca. 1[s]	5.9—6.0	<0.004—1.0% Si in kam; cos ray ages gen 40 Myr	WASSON and WAI (1970)
60—92	—	Silicates have chon bulk comp; neg corr of Ge and Ni in metal	BUNCH et al. (1970)
0.3— 6	ca. 8.6[v]	1.5—4% C; diamonds; 1.5—4% Ni in kam; two types	VDOVYKIN (1970)
28—88	3.4—4.0	Largest oliv crys 2 8 cm; oliv has very low minor elem cont	BUSECK and GOLDSTEIN (1969)
ca. 90	—	Feldspar crys to 11 cm; chromian augite abundant	BUNCH et al. (1970), WASSERBURG et al. (1968)
<1	3.7—4.3	Pyr crys as long as 5 cm; one unbrec; others monomict brec	MASON (1963b)
30—55	3.9—4.4[v]	0.1 < Ca/Mg < 0.5; Ca/Al < 0.66; polymict brec	POWELL (1971)
<0.3[t]	ca. 4.0[v]	0.1 < Ca/Mg < 0.6; Ca/Al > 0.68; gen genomict brec	DUKE and SILVER (1967), MASON (1967c)
<0.1	4.2—4.5	0.7 < Ca/Mg < 1.2; gen monomict brec	

Legends see page 28.

the silicate-containing meteorites as a whole. To a fair approximation, the Fe/(Fe + Mg) ratios in the silicate minerals of metal-containing meteorites are a measure of their degrees of oxidation. Alternatively, during the fractionation of a magma, the Fe/(Fe + Mg) ratio tends to be lower in the solid than in the liquid phase. Again, for purposes of classification, the practical usefulness of this ratio results from the fact that the range within groups is much smaller than the range in the meteorites as a whole.

Members of the different groups are found to occupy relatively small fields in Fig. II-4. This is particularly true of the chondrites, for which the

---

[a] Based on KEIL (1969a) unless otherwise noted.
[b] Based on bulk-chemical data of VON MICHAELIS et al. (1969), assuming that all Mg is present as enstatite.
[c] Fraction of non-opaque phases; based on CIPW norm given by BUNCH et al. (1970)
[d] Fraction of non-opaque phases.
[e] Estimated from bulk-chemical data listed in MASON (1963a).
[f] Fraction of non-opaque phases; rough estimates based on petrographic descriptions given by POWELL (1971).
[g] Rough estimates based on petrographic descriptions given by DUKE and SILVER (1967).
[h] Data of KEIL (1968). His data are not given in a form which allows an accurate estimate of the degree of inhomogeneity.
[i] Listed values refer to the equilibrated members of these groups. Unequilibrated members have % mean deviations up to 45%. Olivine in unequilibrated L and LL chondrites has mean Fe/(Fe+Mg) ratios substantially lower than the listed ranges.
[j] Data of VAN SCHMUS (1969b); CV4 chondrite Coolidge and CO4 chondrite Karoonda show <5% MD of olivine.
[k] See Chapter IV for more details.
[l] Estimated value; unequilibrated silicates have not been reported
[m] MCCALL and CLEVERLY (1968) report that the Fe/(Fe+Mg) ratio in olivine from the North Haig ureilite ranges from 0 to 30 mole %; WLOTZKA (1972) and MARVIN and WOOD (1972) report variations of 11 and 21% between rim and center of olivine domains in Haverö.
[n] Based on combined data on fine and coarse pyroxene grains in Vaca Muerta (POWELL, 1971).
[o] Based on data on low-Ca pyroxene in Kapoeta (FREDRIKSSON and KEIL, 1963).
[p] Based on bulk-chemical data in reference(s) at right, or those listed in Appendix II B.
[q] Lower values in H3 chondrites (Bremervörde, Sharps, Tieschitz).
[r] Renazzo (11% metallic Fe) is an exception.
[s] Shallowater (9% metallic Fe) is an exception.
[t] Bununu (1% metallic Fe) is an exception.
[u] Data from REUTER et al. (1965) and TAYLOR et al. (1965) except as noted; bulk analyses, except as noted; ca. indicates that only one meteorite was analyzed.
[v] Pyroxene analyses.
[w] The abbreviations in the Remarks section correspond to the italicized portions of the following words; *brec*cia, *chon*drite, *chond*rules, *com*position, *con*tent, *cor*relation, *cos*mic, *cry*stal, *elem*ent, *gen*erally (meaning — "in about 60—90% of the cases"), *kam*acite, *neg*ative, *oliv*ine, *pyr*oxene, *ret*ention, *unbrec*ciated.

best and most complete data are available. I have not plotted individual points for the chondrites, but rather the locations of the groups. The error bars correspond to the total observed ranges in Fe/(Fe + Mg) ratios and include roughly 95% of the observed Ca/Mg ratios.

That most of the genetic groups tend to fall into distinct fields in Fig. II-4 confirms that their separate classification is correct. The only cases of overlap on the diagram are: 1. a region where the howardites and mesosiderites plot together, along with lunar rock 12004; and 2. one where the eucrites plot together with lunar rocks 10017, 10020, and 12051, and not far from the terrestrial basalt BCR-1. As summarized in Table II-5, the silicate-rich genetic groups can easily be separated on the basis of other properties. For example, the eucrites, howardites, and mesosiderites have bulk $^{18}O/^{16}O$ ratios which are distinctly lower than those in the terrestrial and lunar rocks. The distinction of mesosiderites from howardites is trivial: they differ greatly in their contents of reduced metal. With one exception, they can also be distinguished on the basis of their Ca/Al ratios (McCarthy and Ahrens, 1971; Mason and Jarosewich, 1973).

Some properties especially useful for the classification of silicate-rich meteorites are listed in Table II-5. Data are given for 17 groups, four of which will also be discussed with the metal-rich meteorites. The groups are listed in the following order: chondrites in order of increasing Fe/(Fe + Mg) ratio in the silicate fraction, followed by differentiated meteorites ordered in the same manner. The enstatite chondrites are listed by type, since the listed properties vary systematically according to type. For properties other than silicate inhomogeneity, the other chondrite groups show much smaller variations between types in the listed parameters.

Special attention is accorded the description of the most abundant (in rare cases, the second most abundant) ferromagnesian silicate, since studies of these are currently the most common source of accurate classificatory information. One or more general petrologic references is listed for each group. These are generally the sources of the ferromagnesian composition data, and also have provided much of the information summarized in the remarks column. The fraction of metallic Fe in the whole rock is taken from the analyses listed in Appendix II B.

C. Classification of Metal-Rich Meteorites

Iron meteorites were first classified into clearly defined genetic groups by Wasson (1967) and Wasson and Kimberlin (1967). Prior to these publications and the succeeding ones in the series, the classification most commonly used was that listed in Table II-1, in which the irons were grouped according to the widths of their kamacite lamellae. This latter method suffered from the fact that nature proved too complex to allow a valid classification of the irons based on a single parameter. For the same reason, classifications based only on Ni concentrations have also been unsuccessful. Bandwidth or Ni concentrations are, of course, valuable parameters for the classification of irons (provided the data are accurate), but combination with two or three other parameters is generally necessary to allow definitive assignment.

The classification of iron meteorites is described in a series of seven papers by my co-workers and myself (WASSON, 1967; WASSON and KIMBERLIN, 1967; WASSON, 1969, 1970; WASSON and SCHAUDY, 1971; SCHAUDY et al., 1972; SCOTT et al., 1973). To date 15 groups have been defined, among which are four related pairs. The classificatory parameters employed are structure and concentrations of the elements Ni, Ga, Ge and Ir. The choice of compositional parameters was made as follows: About 99% of the metallic portions of iron meteorites consists of Fe and Ni, with Fe normally much more abundant than Ni. Knowledge of the Ni concentration allows the calculation of the Fe content to an accuracy comparable to that possible by a normal chemical analysis. Further, a knowledge of the Ni content is necessary for interpreting the structure of irons in terms of the Fe-Ni phase diagram. Gallium and Ge were chosen on the basis of earlier work of BROWN and co-workers (GOLDBERG et al., 1951; LOVERING et al., 1957), who analyzed large numbers of irons and showed that the distributions were non-random, with pronounced clustering at certain concentrations. Subsequent work has produced more accurate data on a much larger selection of meteorites, and has confirmed that Ge concentration is the best single classification parameter known for iron meteorites. The total range in Ge concentration among the iron meteorites as a whole is a factor of 4×10^5, while that within most of the genetic groups is less than a factor of 1.4. Iridium was chosen because it showed large fractionations among the iron meteorites as a whole, because it appeared to fractionate differently than Ga, Ge, and Ni within the groups, and because it can be determined with great sensitivity by neutron-activation analysis. It was later found to show very large fractionations within groups, and is therefore an inherently less valuable classification parameter than Ge. It is very useful for intragroup classification, however. As it must be for any valid classification system, the choice of these parameters is arbitrary. Studies of other elements can be shown to yield the same classification of irons into groups, and the same arrangement into sequences within these groups (SCOTT, 1972).

The manner in which elemental concentration data can be used in the classification of iron meteorites is illustrated in Fig. II-5. On the left

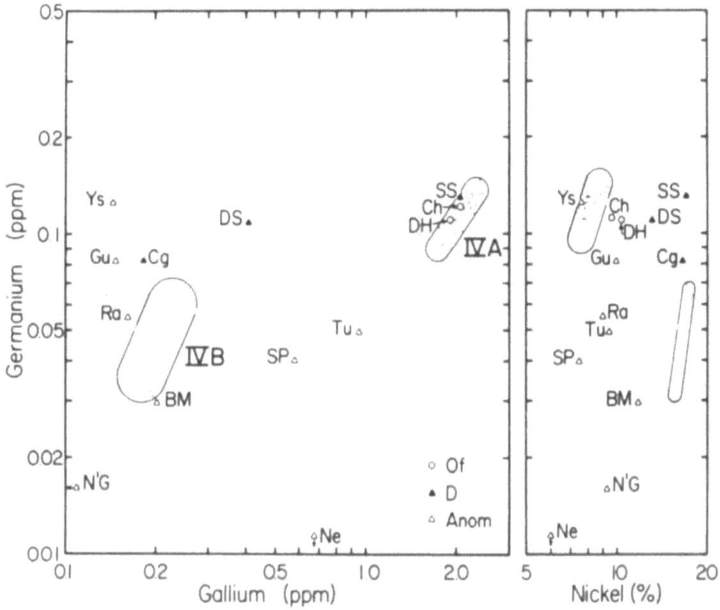

Fig. II-5. Ge-Ga and Ge-Ni plots of iron meteorites with Ge concentrations less than 0.5 ppm. Two groups, IVA, with 40 members and IVB, with 11 members, are found in this concentration range; both show positive Ge-Ni correlations. The element Ge is well suited for classifying iron meteorites because it shows small ranges within groups, but a very large range of 4×10^5 among all meteorites. The names of the symbolically represented anomalous meteorites can be determined by reference to Appendix II D

side of this figure, Ge is plotted against Ga, on the right side, Ge versus Ni. This plot, which is based on one from SCHAUDY et al. (1972), shows the Ge concentration region from 0.01 to 0.5 ppm. This compositional range includes two well-defined groups, designated IVA and IVB. Group IVA is a cluster of 35 points near 0.1 ppm Ge, 2 ppm Ga and 8% Ni. Group IVB has only 10 members, and covers slightly larger fields on the log-log plots near 0.05 ppm Ge, 0.3 ppm Ga and 17% Ni. Germanium is positively correlated with Ga and Ni in both groups. The group-IVA meteorites are octahedrites with kamacite bandwidths of about 0.3 mm; the IVB irons are ataxites, with kamacite spindles about 0.01 mm in width. The other meteorites in this Ge concentration range are classified as anomalous.

Figure II-6 is a plot of Ir versus Ni for the same Ge concentration range covered in Fig. II-5. Here we can see how the addition of one more element helps simplify the decision as to whether a meteorite lying near a well-defined group should be included in that group or not. Deep Springs, an ataxite which lies reasonably near an extrapolation of IVB in

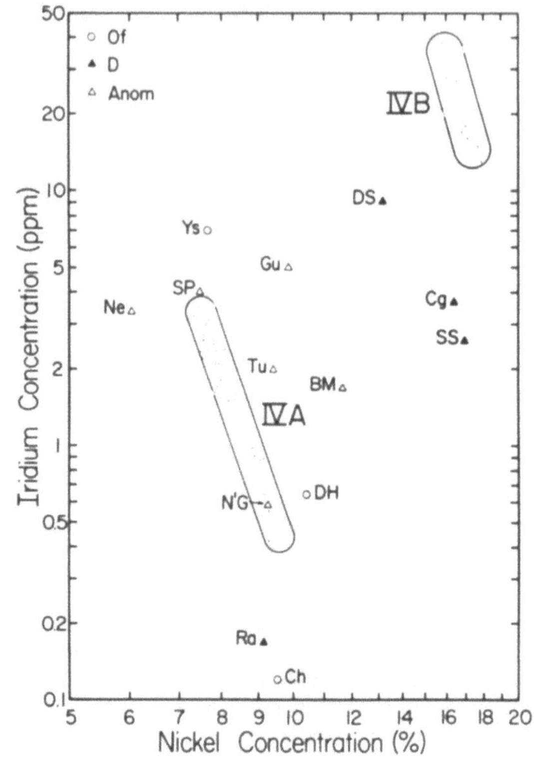

Fig. II-6. Plot of Ir vs. Ni for iron meteorites with Ge concentrations less than 0.5 ppm. The Ir concentration ranges of genetic groups of iron meteorites are large (up to a factor of 3000). As a result, Ir is useful for determining the position of an iron within a group, but since Ir ranges of the different groups overlap, it is of limited value for assigning an iron to a particular genetic group

the Ge-Ga plot of WASSON (1967), is found to fall well away from the IVB trend in Ge-Ni and Ir-Ni plots. For this reason SCHAUDY et al. (1972) have classified it as anomalous. They also note that its structure is distinctly different from that of IVB members. In contrast, Smithland, a meteorite with anomalous structure, falls well within the IVA fields on the Ge-Ga, Ge-Ni and Ir-Ni plots, and appears to be a IVA iron which lost its octahedral structure as a result of reheating by man.

In Fig. II-7 are plotted Ga, Ge and Ni data for about 500 metal-rich meteorites which have been investigated in our laboratory. As in Fig. II-5, Ge is plotted versus Ga and versus Ni. In this figure, however, individual points are plotted only for the anomalous irons, and not for those meteorites belonging to groups. The fields occupied by these groups are outlined. Note that Ge is positively correlated with Ga in all groups, whereas the Ge-Ni correlation is negative in groups IA, IB, IIB, IIIB,

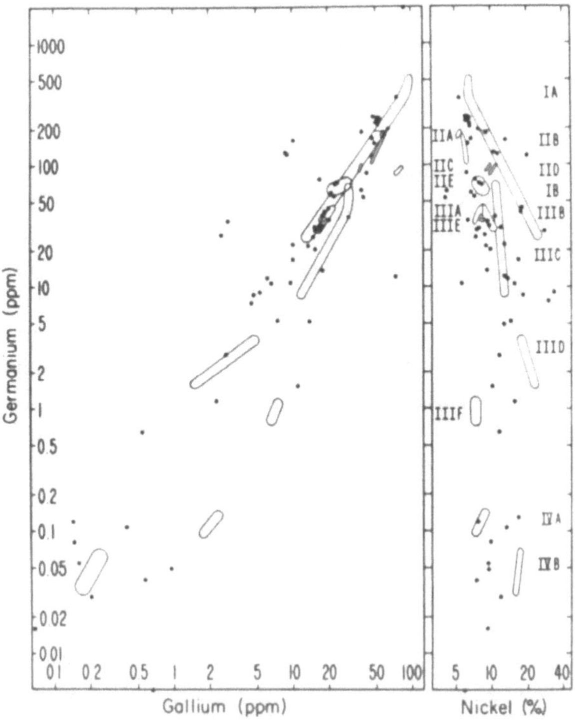

Fig. II-7. Ge-Ga and Ge-Ni plots showing the location fields of the 15 iron-meteorite groups (including four related pairs of groups). The points are anomalous irons, which account for about 16% of all irons. Concentrations of Ni, Ga, Ge, and Ir in iron meteorites are listed in Appendix II D

and III D, absent in II E, III C, III E, and III F, and positive in the remaining seven groups. The resolution of the groups is clearest on the Ge-Ni plot. Four pairs of groups (I A-I B, II A-II B, III A-III B, and III C-III D) are genetically related to one another. In each case, the mean properties of each group differ considerably from those of the group with which it is paired, and it appears best for general studies of iron meteorites to consider the groups separately. Where it is desired to refer to the combined group, the two letter suffixes can be combined (e.g., I AB). All investigations indicate that properties are in fact continuous between the paired groups (although the small group III CD bears further investigation).

Following the pioneering work of LOVERING et al. (1957), some researchers have continued to call these genetic groups of iron meteorites "Ga-Ge groups". As is obvious from Fig. II-7, a Ge-Ni plot yields much better discrimination than the Ge-Ga plot. Further, as pointed out earlier, even Ge is not absolutely necessary for the development of the classification. I have generally used the designation "chemical groups" in my papers, to emphasize the fact that a number of elements must be used to provide a unique classification for every meteorite. Equally good would be to drop the "chemical", and simply to designate them as groups.

The original classification of iron meteorites with high Ge concentrations (WASSON, 1970) has been revised and simplified. Group I was previously defined to be a heavily populated field between 190 and 520 ppm Ge on the Ge-Ga and Ge-Ni plots. In addition, anomalous "categories" were defined: I-An 1, I-An 2 and I-An 3. It seems more rational to eliminate the term "category", and redesignate these meteorites as follows: group I is now group IA: members of category I-An 2 with Ge concentrations <190 ppm are now group IB, those with Ge >190 ppm are added to group IA; I-An 1 meteorites are now designated IA-An; and I-An 3 irons are now listed as anomalous.

The group listed as IIE has not been given that designation in published papers, but as a result of the discovery of additional genetically related irons, it is now considered to meet the criteria for group designation. In SCOTT et al. (1973) these meteorites are called Weekeroo-type irons.

As noted above, earlier structural classes of iron meteorites were defined somewhat arbitrarily, since the investigators were unable to distinguish groups of genetically related irons. Recently, GOLDSTEIN (1969) and BUCHWALD (1974; quoted in WASSON, 1970) have proposed slightly different bandwidth divisions which minimize the number of structural classes within a given chemical group. BUCHWALD's classes are listed in Table II-6. They are preferred because they follow the rational procedure of having limits which increase in the form of a geometric series: by nearly constant factors of 2.5. Both the GOLDSTEIN and BUCHWALD divisions preserve the traditional bandwidth limits (HEY, 1966) of the Of and Off classes and make only minor changes in the limits of the other classes. BUCHWALD recommends fairly liberal use of the anomalous structural category, particularly for meteorites having structures which give evidence of substantial thermal alterations, or are otherwise exceptional.

The list also includes the plessitic octahedrite class which was introduced by BUCHWALD and MUNCK (1965) to cover irons which do not have continuous kamacite bands, but do have spindles of kamacite in an octahedral orientation. Although the addition of this class clearly improves our ability to resolve the irons into relatively homogeneous structural classes, there is as yet no clearly definable criterion for distinguishing between these Opl octahedrites and those ataxites which show a few rare kamacite spindles. This points out the fact that the main value of

Table II-6. Structural classification according to BUCHWALD (1974)

Class	Symbol	Kamacite bandwidth (mm)	Remarks
Hexahedrites	H	—	No octahedral orientation even in large sections
Coarsest octahedrites	Ogg	>3.3	Taenite may or may not be present
Coarse octahedrites	Og	1.3—3.3	—
Medium octahedrites	Om	0.5—1.3	—
Fine octahedrites	Of	0.2—0.5	—
Finest octahedrites	Off	<0.2	Distinct bands of kamacite
Plessitic octahedrites	Opl	<0.2	Kamacite sparks and spindles
Ataxites	D	—	Well-developed, slowly annealed plessite, kamacite spindles very rare
Anomalous	Anom	—	Includes all irons which demand individual descriptions

structural classes is to provide a convenient shorthand designation for communicating a rough description of structure, but that comparisons of similar meteorites must always be expressed in more quantitative terms.

A summary of the properties of groups of metal-rich meteorites is given in Table II-7. In addition to ranges of elemental composition and bandwidth, the table shows the fraction of investigated meteorites belonging to each group. The four groups I A, II A, III A, and IV A account for 59% of iron meteorites. The other eleven groups account for 25% of the irons, and the anomalous irons the remaining 16%. In calculating these abundances, I have included with the groups those irons which show slightly anomalous compositions, but which on the basis of structural or other evidence appear to be closely related to the more typical members of these groups (e.g., the irons designated I A-An). Metal-rich meteorites are listed according to class in Appendix II D, together with compositional and bandwidth data, and remarks regarding paired falls and genetic relationships among irons in the anomalous category.

D. Summary

The 30 groups characterized in Tables II-5 and II-7 are listed in Table II-8. In four cases in which there is strong evidence that two iron-meteorite groups are but separate portions of an extended sequence, the groups are combined (see previous section). Entries are also given in

Table II-7. Properties of metal-rich meteorite groups

Group	Freq. (%)	Bandwidth (mm)	Struc.	Ni (%)	Ga (ppm)	Ge (ppm)	Ir (ppm)	Ge-Ni corr.
IA	17.1	1.0—3.1	Om-Ogg	6.4—8.7	55—100	190—520	0.6—5.5	neg.
IB	1.7	0.01—1.0	D-Om	8.7—25	11—55	25—190	0.3—2.0	neg.
IIA	8.3	>50	H	5.3—5.7	57—62	170—185	2—60	pos.?
IIB	2.6	5—15	Ogg	5.7—6.4	46—59	107—183	0.01—0.5	neg.
IIC	1.5	0.06—0.07	Opl	9.3—11.5	37—39	88—114	4—11	pos.
IID	2.6	0.4—0.9	Of-Om	9.8—11.3	70—83	82—98	3.5—18	pos.
IIE	2.3	0.1—2	Anom[a]	7.5—9.7	21—28	60—75	1—8	abs.
MES	—	ca. 1	Anom	6.1—10.1	8.9—16	37—56	2.2—6.2	abs.
PAL	—	ca. 0.9	Om[c]	7.9—12.9	14—27	29—71	0.01—2	neg.?
IIIA	24.9	0.9—1.3	Om	7.1—9.3	17—23	32—47	0.17—19	pos.
IIIB	7.0	0.6—1.3	Om	8.4—10.5	16—21	27—46	0.01—0.17	neg.
IIIC	1.5	0.2—0.4	Off-Of	10—13	11—27	8—70	0.07—0.55	abs.
IIID	1.1	0.01—0.05	D-Off	16—23	1.5—5.2	1.4—4.0	0.02—0.07	neg.
IIIE	1.7	1.3—1.6	Og	8.2—9.0	17—19	34—37	0.05—6	abs.
IIIF	1.1	0.5—1.5	Om-Og[b]	6.8—7.8	6.3—7.2	0.7—1.1	1.3—7.9	abs.
IVA	8.3	0.25—0.45	Of	7.4—9.4	1.6—2.4	0.09—0.14	0.4—4	pos.
IVB	2.3	0.006—0.03	D	16—26	0.17—0.27	0.03—0.07	13—38	pos.

[a] Also Om and Og.
[b] Also Anom.
[c] Or Pal (= Pallasitic).

Table II-8 for two broad anomalous "categories", the silicate-rich and the metal-rich anomalous meteorites.

Listed in Table II-8 are the number of falls and finds in each class, based on evidence summarized in Appendix II. Where quantitative evidence indicates that several meteorites are paired, only one was counted in arriving at these totals.

The fraction of observed falls belonging to each class is also listed in Table II-8. These were arrived at as follows: A comparison of Appendix II data with those taken from the Hey (1966) catalog and from recent issues of the Meteoritical Bulletin indicated that Appendix II data include about 90% of all falls of ordinary chondrites and about 93% of the falls of other silicate-rich meteorite classes. The total number of iron meteorite falls appears to be 44.[3] The numbers used to calculate fall

[3] For some reason, it often happens that iron meteorites are incorrectly attributed to witnessed falls. In consultation with V.F. Buchwald, I have attempted to eliminate each incorrect attribution from Appendix II D, in which 32 irons are listed as falls. Although some bear further investigation, the additional 12 irons also appear to be falls *and* are still preserved in meteorite collections: Cabin Creek, Majorca, Mariaville, Muzaffarpur, Nyaung, Palinshih, Patos de Minas (hexahedrite), Prambanan, Puerta de Arauco, Quesa, Tandil and Winburg. Not falls are Gundaring, Helt Township and Victoria West (Buchwald, 1974). There is no good evidence that Elbogen is an observed fall.

Table II-8. Populations of known groups of (5 or more genetically related) meteorites as well as of the anomalous categories

Group abbreviation	Falls	Finds	Fall frequency (%)
Enstatite chondrites	11	5	1.5
H-group chondrites	229	230	32.1
L-group chondrites	278	192	39.0
LL-group chondrites	51	16	7.2
CV-group chondrites	8	3	1.1
CO-group chondrites	6	0	0.81
CM-group chondrites	14	0	1.9
CI-group chondrites	5	0	0.68
Aubrites	8	1	1.1
Ureilites	3	3	0.41
Diogenites	8	0	1.1
Mesosiderites	6	14	0.81
Howardites	18	1	2.4
Eucrites	20	4	2.7
IAB iron meteorites	6	80	1.0[a]
IIAB iron meteorites	5	46	0.62
IIC iron meteorites	0	7	0.09
IID iron meteorites	2	10	0.15
IIE iron meteorites	0	10	0.12
Pallasites	2	33	0.27
IIIAB iron meteorites	5	141	1.8
IIICD iron meteorites	1	11	0.15
IIIE iron meteorites	0	8	0.10
IIIF iron meteorites	0	5	0.06
IVA iron meteorites	3	37	0.49
IVB iron meteorites	0	11	0.13
Anomalous silicate-rich meteorites	9	8	1.2
Anomalous metal-rich meteorites	4	73	0.94

[a] Fall frequencies of iron meteorites calculated on basis of 44 observed falls allocated to chemical groups according to frequencies summarized in Table II-7.

frequency of silicate-rich meteorites, mesosiderites and pallasites were those listed in Table II-8, corrected for the appropriate factors. For iron meteorites, the 44 observed falls were divided among the different classes on the basis of the frequency observed for all classified falls and finds as summarized in Table II-7.

The order in which the groups are listed in Table II-8 is consistent with the following simple rules: Chondrites are listed first, roughly in order of increasing Fe/(Fe+Mg) ratio in the silicate fraction. These are followed by the differentiated silicate-rich meteorites, listed in order of increasing Ca/Mg ratio. Finally, the differentiated metal-rich groups are listed in order of decreasing mean Ge concentration. Mesosiderites are listed with the silicate-rich meteorites, pallasites with the metal-rich meteorites (in Appendix II they are listed in both places).

This listing still fails in some cases to list together closely related groups (the aubrites and the enstatite chondrites), and associates some groups which probably formed by igneous differentiation with others which did not. It should be used with these caveats in mind.

III. The Study of Meteorites: Sources, Bibliographies, and History

A number of techniques have been applied to the measurement of the properties of meteorites. Some techniques are more accurate than others for the measurement of a given property. These differences in resolving power are very important in the evaluation of data on meteorites. Space and time limitations have not permitted me to discuss each technique in detail. I have, however, in the following chapters, commented on the relative accuracies of certain techniques, and occasionally, on the special competence of certain research teams. I hope that these value judgements are objective.

There are a number of recent publications which review a wide range of meteorite properties. Among the more valuable of these reviews are those of ANDERS (1964, 1968, 1971a), DODD (1969), KAULA (1968), KEIL (1969a), LEVIN (1965, 1969), MASON (1962a, 1967a), MCCALL (1973), VAN SCHMUS (1969a), VAN SCHMUS and HAYES (1974), WÄNKE (1966), WASSON (1972a) and WOOD (1963b, 1968). These are excellent sources of information, and generally provide extensive reference lists. Most authors also attempt to assess the reliability of the data which they quote. Oddly enough, those which provide the most critical data evaluations often draw conclusions from the data which are much less certain than the text implies. Let the reader beware!

Some of the older literature is also valuable for certain types of information. Excellent descriptions of fall phenomena and meteorite morphology are given by FARRINGTON (1915a), HEIDE (1957), KRINOV (1960), and NININGER (1952). The three volumes of COHEN's (1894, 1903, 1905) unfinished *Meteoritenkunde* provide important summaries of 19th century structural and mineralogical observations; his 1905 work on iron meteorites is especially useful, although incomplete, and soon to be superseded by the detailed study of BUCHWALD (1974). The monograph of MEUNIER (1884), while only slightly older than COHEN's work, is more subjective and of much less use as a reference. FARRINGTON's (1915b) work summarizes useful information on individual North American meteorites. Reviews of the state of meteorite research at the end of the first third and first half of the 20th century are given by MERRILL (1930) and KRINOV (1960), respectively.

Table III-1. Location and most recent catalog of the eight most important meteorite collections of the world

Collection	Reference
Europe	
Academy of Sciences, USSR, Moscow	KVASHA (1962)
British Museum, (Natural History), London	HEY (1966)
Muséum National d'Histoire Naturelle, Paris	LACROIX (1928)
Naturhistorisches Museum, Vienna	BERWERTH (1903)
North America	
American Museum of Natural History, New York	MASON (1964)
Arizona State University, Tempe	KARR et al. (1970)
Field Museum of Natural History, Chicago	HORBACK and OLSEN (1965)
Smithsonian Institution, Washington	MASON (1974)

The one most important reference work for persons carrying out experimental research on meteorites is the British Museum catalog of HEY (1966). This standard reference work summarizes fall and recovery information and provides an excellent overview of the distribution of meteorite specimens among the collections of the world. Its information on other collections is not always current, however, and it is advisable for a researcher to consult the most recent catalog of each collection from which he plans to request samples. Table III-1 lists the latest catalogs for the four largest European and four largest North American meteorite collections. It is fortunate that there are meteorite researchers on the staffs of each of these collections, since this has the double benefit that the meteorites are made accessible for deserving research projects, but they are also hoarded to that degree necessary to preserve adequate samples for posterity.

The smaller collections are also important sources of research specimens. Lists of the catalogs of meteorite collections have been compiled by SPENCER (1949), MASON (1962b) and UNESCO (1968). Information on recent meteorite falls is distributed immediately by the Center for Short-Lived Phenomena of the Smithsonian Astrophysical Observatory, Cambridge, Massachusetts, USA. More detailed reports on recent falls and finds appear in *The Meteoritical Bulletin* (through No. 49 published by the Commission on Meteorites of the International Union of Geological Science, and edited by E. L. KRINOV; starting with No. 50, published in *Meteoritics* by the Meteoritical Society, and edited by R. S. CLARKE, Jr.).

A bibliography of meteorite literature for the period 1491—1950 is given by BROWN et al. (1953). The bibliographies of SABLENA and YAVNEL (1965, 1968, 1971) cover the period 1961—1968; earlier papers in this series covering the period 1950—1960 were published in *Meteoritika*, the most recent being that of MASSALSKAYA (1963). The compilations of MAGNOLIA (1962, 1963) parallel the Russian bibliographies for the period 1950—1962. The references in DODD (1971a) and WASSON (1971a) are largely confined to publications from the USA, and those of BOGARD (1971) and KIELBASINSKI and WANAT (1968) are confined to the specialized subjects of rare-gas studies and isotopic- and elemental-composition studies, respectively.

Meteorites have generally been accepted as extraterrestrial objects since the pioneering studies of CHLADNI (1794, 1819) and BIOT (1803). Although many observations have been made and many hypotheses proposed since that time, there is no general work tracing the development of meteorite research. In fact, the only area which has received careful historical attention appears to be the study of the Widmanstätten pattern of iron meteorites (PANETH, 1960; SMITH, 1962; MEHL, 1965). One man's meteoritic odyssey is recounted in NININGER (1972).

IV. Bulk Composition

Major- and minor-element concentration data are important for the classification of meteorites (particularly when the original petrologic structure has been altered), and for the detection of fractionations within genetic groups. The content of these elements together with temperature and pressure determines the mineralogical composition and O_2 fugacity of the system. Bulk-composition data are thus important reference materials for a wide range of meteorite studies. Although trace-element data (discussed in Chapter VII) are also important for classification and fractionation studies, they are less important for petrologic investigations.

The set of elements determined in bulk analyses of silicate-rich meteorites includes H (as H_2O) and all elements from Na through Ni, with the exception of Cl, Sc, and V. The amount of metallic Fe is also determined. In Appendices II B and II C are listed references to bulk analyses of silicate-rich meteorites by the following analysts: H.B. Wiik (91); M.I. Dyakonova (47, of which 27 are joint with Kharitonova); V.Y. Kharitonova (21); E. Jarosewich (59); A.J. Easton (13); H. Haramura (9); A.D. Maynes (9); O. Müller (6); H. König (4); and J. Nelen (3). The numbers in parentheses give the number of analyses cited. These analysts were chosen on the basis of having reported nine or more bulk analyses, or because they were long-term members of well-established meteorite research groups. Also listed in Appendices II B and II C are 57 X-ray fluorescence analyses of H. von Michaelis and other members of the Cape Town group, and 9 mesosiderite analyses by B.N. Powell. Although these do not include data on Na, S, Cr, Co, Ni or metallic Fe, the large number of analyses and their high precision justify their inclusion.

In order to investigate the precision and the possibility of systematic biases in the data of the leading analysts, histograms of their results on three major (Mg, Al, Si) and three minor (P, K. Ti) elements in L-group chondrites are compared in Fig. IV-1. From top to bottom are shown the results of von Michaelis, Wiik, Dyakonova and Kharitonova, and Jarosewich, respectively. Available data through 1971 are included, with no attempt made to screen out analyses of weathered specimens. It appears that each group has analyzed approximately equal proportions of weathered material. The inclusion of such data appears to affect mainly the precision of the Mg and Si results. A spot check shows that the lowest Mg and Si values in the data of Jarosewich and Kharito-

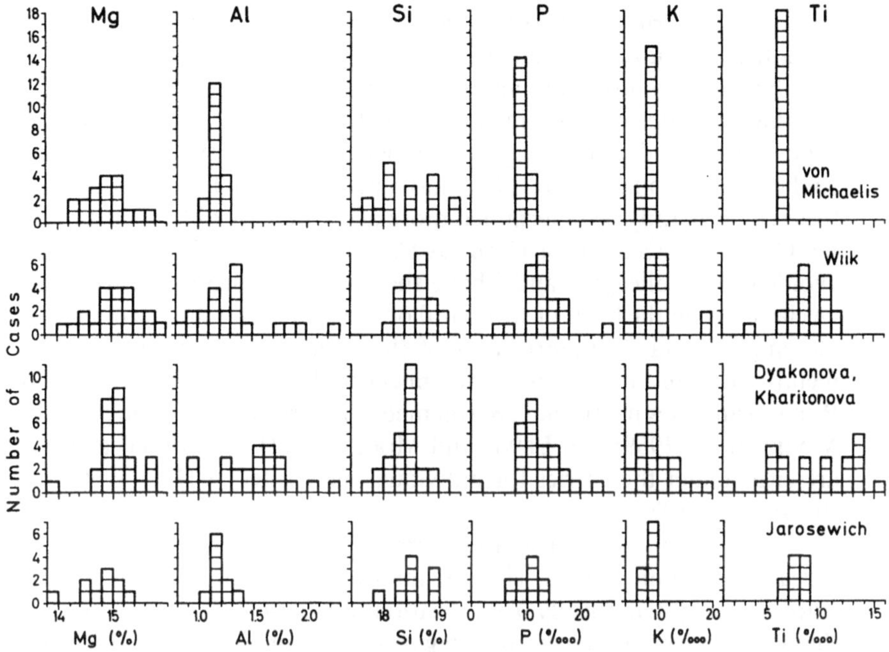

Fig. IV-1. A comparison of analytical data produced by three analysts and one pair of analysts. Published results through 1971 on three major and three minor elements in L-group chondrites are shown. Although all data are good, those by VON MICHAELIS and JAROSEWICH are clearly superior

NOVA are analyses of the weathered Kyle and New Almelo chondrites, respectively.

The data summarized in Fig. IV-1 show that, although all four data sets are of good quality, those of VON MICHAELIS and JAROSEWICH are somewhat better, especially for minor elements. The precision of VON MICHAELIS on the minor elements is remarkable, whereas his Mg and Si data show more scatter than those of JAROSEWICH. The data of VON MICHAELIS were determined by X-ray fluorescence, whereas the other data were obtained by relatively standard wet-chemical techniques. It is clear that, in the right hands, X-ray fluorescence can yield data of comparable quality to the more time consuming wet-chemical techniques, although data on some important elements are not obtained.

Average bulk compositions for silicate-rich meteorite groups are given by YAVNEL (1968) and KEIL (1969a). In addition to the references cited in Appendix II of this work, summaries of individual analyses of certain silicate-rich meteorites are given in the following papers: chondrites—UREY and CRAIG (1953); ordinary chondrites—MASON (1965a);

enstatite chondrites—MASON (1966). Further citations are given under the entries for individual meteorites in HEY (1966).

The elements normally analyzed in the iron meteorites and in the metallic separates of other metal-rich meteorites are Fe, Ni, Co, P, and sometimes S and C (see, e.g., HENDERSON and PERRY, 1958). Recent studies (e.g., MOORE et al., 1969) have not reported Fe contents since these can generally be determined more accurately by difference (i.e., Fe, Ni, Co, and P account for about 99.8% of most irons, if inclusions are avoided). My colleagues and I have shown that the iron meteorites can be classified on the basis of structural observations and the use of Ni, Ga, Ge, and Ir data. Concentrations of these elements in more than 500 metal-rich meteorites are given in Appendix II D. Tabulations of Ni, Co, P, and S analyses in 100 and 38 iron meteorites, respectively, are given by MOORE et al. (1969) and LEWIS and MOORE (1971). Additional analyses are cited under the entries for individual meteorites in HEY (1966) and BUCHWALD (1974).

Two different types of sampling errors are of importance in the analytical study of meteorites. The first involves the possibility that composition varies systematically from location to location. The second type is the more conventional question regarding the minimal size which will yield data characteristic of the whole meteorite.

Although neither of these sources of error has been adequately investigated, the information available indicates that systematic variations within most meteorites are negligible. For example, the Na and Mn data determined by OSBORN and SCHMITT (1970) on chondrite samples chosen from different locations and, sometimes, different specimens, showed dispersions which were no different from those expected from samples of the same size from adjacent locations. Also, although Ni concentrations vary from 6.4 to 9.1% among Canyon Diablo specimens (MOORE et al., 1967), the Ge and Ga contents determined on a large number of Canyon Diablo specimens are indistinguishable, within a precision of $\pm 4\%$ (WASSON, 1968). That Ge and Ga show less dispersion than Ni in iron meteorites is attributable to the fact that an appreciable portion of the Ni is in minor phases, such as taenite and schreibersite, which show substantial variations on a scale of cm in Canyon Diablo. The distribution data of GOLDSTEIN (1967) show that a much smaller fraction of the Ge is present in minor phases, and the same is probably true for Ga.

The minimum sample size necessary for adequate sampling clearly depends on the substance involved and must be carefully defined at the beginning of each new investigation. KEIL'S (1962) estimate that minimum sample sizes of chondrites should be in the range of 10—70 g appears to be unnecessarily high for most elements. JAROSEWICH (1966) obtained excellent data on samples between 9 and 14 g, and TANDON and WASSON (1968) reported a relative standard deviation of only 3.5% for Ga determinations on 0.5—0.6-g samples of L-group chondrites. The Ga and Ge analyses of the Canyon Diablo (group I A) iron discussed in the previous paragraph were made on 1-g samples. It appears that an 0.5—0.6-g sample which visually appears to be representative of the whole specimen will, for most elements, result in a sampling error no larger than the errors resulting from other sources.

Many different studies have shown that weathering can cause substantial changes in the composition of meteorites, particularly for minor and trace elements in silicate-rich meteorites, which are often quite porous. It is perhaps well to warn persons new to meteorite research that many specimens of meteorites listed as falls in the HEY (1966) catalog are weathered. For example, the widely-studied Richardton samples in the Arizona State University collection were recovered by NININGER (1952, p. 159) some 15 years after the date of fall. It would be useful if future editions of the British Museum catalog would include information regarding the current disposition of these weathered specimens of "falls".

V. Mineralogy and Phase Composition

A. General

Names and formulas of 24 common meteoritic minerals are listed in Table V-1. Their crystal structures and meteoritic occurrences are summarized by KEIL (1969a), who also summarizes information about the less common minerals. MASON (1967d; 1972) gives brief general reviews of the mineralogy of meteorites, and VAN SCHMUS (1969a) a somewhat more detailed review of chondrite mineralogy. RAMDOHR (1963a, b, 1964, 1973) gives detailed accounts of his optical observations of opaque minerals in stony meteorites.

Table V-1. The more common meteorite minerals and their formulas

Mineral	Formula
Augite	$(Ca, Mg, Fe^{2+}, Al)_2(Si, Al)_2O_6$
Chlorapatite	$Ca_5(PO_4)_3Cl$
Christobalite	SiO_2
Chromite	$FeCr_2O_4$
Clinopyroxene	$(Mg, Fe, Ca)SiO_3$
Cohenite	Fe_3C
Daubreelite	$FeCr_2S_4$
Farringtonite	$Mg_3(PO_4)_2$
Graphite	C
Ilmenite	$FeTiO_3$
Kamacite	$Fe_{0.93-0.96}Ni_{0.07-0.04}$
Magnetite	Fe_3O_4
Oldhamite	CaS
Olivine	$(Mg, Fe)_2SiO_4$
Orthopyroxene	$(Mg, Fe)SiO_3$
Pentlandite	$(Fe, Ni)_9S_8$
Plagioclase	$NaAlSi_3O_8$—$CaAl_2Si_2O_8$
Quartz	SiO_2
Rutile	TiO_2
Schreibersite	$(Fe, Ni)_3P$
Taenite	$Fe_{<0.8}Ni_{>0.2}$
Tridymite	SiO_2
Troilite	FeS
Whitlockite	$Ca_3(PO_4)_2$

Table V-2. Names and formulas of the component minerals of olivine, pyroxene and plagioclase solid-solution series, and of some pyroxenes of intermediate compositions; abbreviations are listed for end members of olivine, plagioclase and (clino- and ortho-) pyroxene solid-solution series

Mineral	Abbr.	Formula
Albite	Ab	$NaAlSi_3O_8$
Anorthite	An	$CaAl_2Si_2O_8$
Bronzite		$Mg_{0.70-0.88}Fe_{0.30-0.12}SiO_3$ [b]
Diopside		$Ca_{0.5}Mg_{0.5}SiO_3$
Enstatite [a]	En	$MgSiO_3$
Fayalite	Fa	Fe_2SiO_4
Ferrosilite [c]	Fs	$FeSiO_3$
Forsterite	Fo	Mg_2SiO_4
Hedenbergite		$Ca_{0.5}Fe_{0.5}SiO_3$
Hypersthene		$Mg_{0.5-0.7}Fe_{0.5-0.3}SiO_3$ [b]
Pigeonite		$Ca_x(Mg,Fe)_{1-x}SiO_3$ ($x \sim 0.1$)
Wollastonite	Wo	$CaSiO_3$

[a] Petrologists often consider enstatite to include all pyroxenes in the composition range $Mg_{0.88-1.00}Fe_{0.12-0.00}SiO_3$.

[b] Note that these definitions embody current petrological usage, and differ from those used by Prior in his meteorite classification – see Chapter 2, Section B.

[c] Pure ferrosilite is not stable.

The three most common silicate minerals in meteorites are olivine, pyroxene and plagioclase feldspar. Table V-2 contains the names of the end members of the solid-solution series of these minerals, as well as the petrologic names applied to low-Ca pyroxenes falling in various composition ranges. Olivines consist of solid-solutions of fayalite and forsterite, with only very minor amounts of cations other than Fe and Mg. Feldspar in meteorites generally consists of plagioclase, although in some meteorites other feldspars are found.

Pyroxene mineralogy can be very confusing to persons without mineralogical backgrounds. The three most common pyroxene phases are: a) a high-Ca, monoclinic form generally called augite, but sometimes described as diopside, the name of the Fe-free end member; b) a low-Ca, monoclinic form known as pigeonite; and c) an orthorhombic form having a very low Ca content. The compositional fields these minerals occupy in various meteorite classes are illustrated in Fig. V-1, a truncated triangular diagram of the system enstatite (En)-ferrosilite (Fs)-wollastonite (Wo). In most meteorites two of these pyroxene forms coexist. Tielines between coexisting pairs are given in Fig. V-1. The equilibrium temperatures to which these measured mineral compositions correspond are not well known, but probably lie in the range $900 \pm 200°$ C

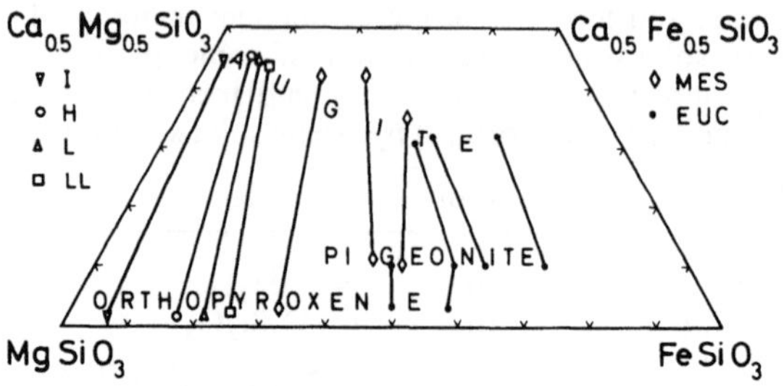

Fig. V-1. Partial phase diagram for the system $MgSiO_3$-$FeSiO_3$-$CaSiO_3$. Measured compositions of coexisting mineral pairs in meteorites from six different groups are shown. Data sources are: BUNCH et al. (1970) for IA iron Campo del Cielo; VAN SCHMUS and KOFFMAN (1967) for mean compositions of several type-6 H, L, and LL chondrites; POWELL (1971) for three unspecified mesosiderites; and DUKE and SILVER (1967) for the eucrites Moore County, Juvinas, Nuevo Laredo and Pasamonte (listed in order of increasing $FeSiO_3$)

in all cases. It is not possible to illustrate in this diagram the fact that the augitic pyroxenes often contain appreciable concentrations of Al. A fourth type of pyroxene is a monoclinic form which occupies essentially the same field as that shown for orthopyroxene; this is often found in meteorites which have experienced relatively little recrystallization. It may be that it is a metastable, high-temperature phase, which inverts to orthopyroxene under conditions of sustained annealing in the orthopyroxene stability field (VAN SCHMUS and WOOD, 1967).

The common names (hypersthene and bronzite) applied to low-Ca pyroxenes having compositions within certain ranges should be avoided in the meteorite literature, since the current petrological usage listed in Table V-1 differs from that used by PRIOR (1920) in the nomenclature of meteorite groups. Further, enstatite should only be used for the nearly Fe-Free pyroxene (Fs < 1 mole %) found in the enstatite chondrites and achondrites, and not applied to the more Fe-rich varieties which are found in the group IAB irons, even though these also fall within the enstatite range as defined by petrologists. The best way to avoid confusion is to use the more general terms, such as orthopyroxene, and give compositional information numerically.

With the exception of the eucrite data, data plotted in Fig. V-1 were obtained by use of the electron microprobe. The first major meteorite research paper based on this technique was the report of KEIL and FREDRIKSSON (1964) on the composition of the ferromagnesian silicates of the ordinary chondrites. Petrological studies are more valuable if they involve the use of the microprobe to obtain accurate phase composition data. Many reviews of this technique are available. Two by meteorite researchers are those of FREDRIKSSON and REID (1967) and KEIL (1967).

B. Pressure-Indicating Mineral Systems

The pressure inside a spherical body of uniform density is given by:

$$P = 1.4 \times 10^{-6} \varrho^2 (a^2 - r^2)$$

where the pressure is in kb, the density (ϱ) is in g cm^{-3}, and the radius of the body (a) and distance of the sample from the center (r) are in km. Fig. V-2 is a plot of this relationship for a density of 3.6 g cm^{-3} and sample locations at $r = 0$ and $r = 0.7\,a$.

The properties a mineral must have in order to record a high-pressure formation can be stated simply: it must be impossible for the mineral to form below a certain minimum pressure, and it must be capable of metastable existence during the period between pressure release on the breakup of the parent body and the investigation of the meteorite following its capture by the earth. The best known mineral fulfilling these requirements for both terrestrial and meteoritic samples is diamond, which does not exist stably below a pressure of 14 kb, but which transforms to stable graphite at a rate so slow that diamond is able to survive on the earth's surface during geologic time.

The properties which minerals must have to record the absence of pressures higher than a certain maximum value are more complex. In principle, minerals such as the silica polymorphs tridymite, which is not stable above a pressure of 3 kb, or cristobalite, which is not stable above

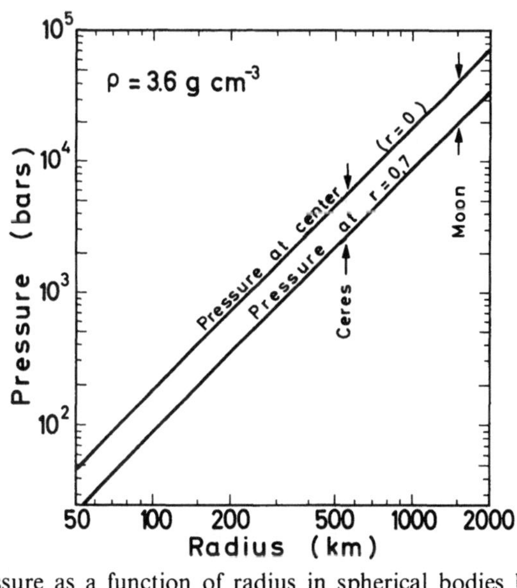

Fig. V-2. Pressure as a function of radius in spherical bodies having a uniform density of 3.6 g cm^{-3}

a pressure of 5 kb, should be able to provide such information (see SiO_2 phase diagram in ANDERS, 1964). Unfortunately, it must also be shown that such minerals cannot form metastably at high pressures, or by the decomposition of high-pressure minerals. Thus, although it is likely that meteorites containing large, "primary" tridymite crystals have never been at pressures greater than 3 kb, this interpretation must be used with some caution. As shown in Fig. V-2, a pressure of 3 kb corresponds to that at the center of a parent body with a "chondritic" density of 3.6 g cm^{-3} and a radius of about 400 km, slightly smaller than Ceres, the largest asteroid (ca. 550 km, Table XII-1).

Another means of estimating maximum pressures is provided by compositional data on coexisting minerals which share one or more components. For example, the boundary between the taenite field and the two-phase kamacite-taenite field of the Fe-Ni phase diagram is pressure-sensitive. High pressure adds stability to the close-packed taenite phase relative to the kamacite phase (RINGWOOD and KAUFMAN, 1962). For a given equilibration temperature, the higher the pressure, the lower the Ni content of the taenite at the kamacite-taenite interface. On the basis of kamacite and taenite composition, GOLDSTEIN and OGILVIE (1965) and GOLDSTEIN (1965) concluded that iron meteorites of a wide variety of types had formed at pressures lower than 12 kb which, according to Fig. V-2, indicates that they originated in parent bodies with radii less than 800 km.

All currently available evidence favors the viewpoint that meteorites formed at relatively low ambient pressures, much less than those found in the interior of a moon-sized body, and consistent with an origin in parent bodies no larger than the largest asteroids. A few years ago it was common to find persons willing to defend the ideas that the presence of diamonds (CARTER and KENNEDY, 1964, 1966), cohenite (RINGWOOD, 1965) or cliftonite, a form of graphite with cubic morphology sometimes held to be a pseudomorph after diamond (CARTER and KENNEDY, 1964; UREY, 1966), indicated that the meteorites had formed under high static pressures. In the meantime, BRETT and HIGGINS (1969) have synthesized cliftonite at low pressures. BRETT (1967) and LIPSCHUTZ and ANDERS (1964) argue that cohenite is not a high-pressure mineral and BRETT proposes that the existence of metastable cohenite together with graphite in iron meteorites can only be understood if low pressures prevailed during the cooling of these objects. The diamonds in the crater-forming group-IA iron Canyon Diablo (they are not found in other IA irons) seem attributable to shock upon impact with the earth (ANDERS and LIPSCHUTZ, 1966), whereas those in the ureilites probably resulted from collisions in space (LIPSCHUTZ, 1964; VDOVYKIN, 1970). The other rare occurrences of high-pressure minerals in meteorites such as the spinel, ringwoodite (BINNS et al., 1969), and the garnet, majorite (SMITH and MASON, 1970), appear attributable to shock during space collisions (see also Chapter VI).

Now that a low-pressure origin of the meteorites is widely accepted, it would be interesting to find mineral systems capable of recording maximum pressures lower than 3 kb, the pressure at the center of a 400-km asteroid. If most meteorites are of asteroidal origin, some, perhaps many, of them probably originated in the smaller asteroids, with central pressures less than 100 b.

C. Temperature-Indicating Mineral Systems

Meteoritic thermometry can be approached in several different ways. These tend to fall into two categories: a) those which record the maximum temperature which the material could have reached since achieving its present petrographic structure; and b) those which record an equilibrium temperature, which is usually taken to be the lowest temperature at which coexisting minerals were able to maintain equilibrium. Examples of minerals which record the absence of high temperatures are the "chlorite" and unpyrolized organics of carbonaceous chondrites (DUFRESNE and ANDERS, 1962; BOSTROM and FREDRIKSSON, 1966). Such substances could not have survived temperatures greater than about 500 °K. The parent body of these meteorites was probably never hotter than this.

Certain other mineralogical thermometers are unable to provide information about the earlier existence of high temperatures, but do give evidence regarding maximum temperatures during the period that a given geochronologic clock has been recording. For example, FLEISCHER et al. (1967) and SHIRCK et al. (1969) found charged-particle tracks in meteorites which they attribute to the decay of extinct ^{244}Pu. If this interpretation is correct, temperatures have not been high enough to anneal out the tracks during the ca. 4.4 Gyr since the meteorite minerals started to record them. FLEISCHER et al. (1967) estimate that temperatures have not exceeded about 550 °K for any extensive periods during this time. Somewhat similar arguments have been used by GOLES et al. (1960) to show that meteorites with K-Ar ages approaching 4.6 Gyr must have been stored at very low temperatures (less than about 400 °K, according to FECHTIG et al., 1963). Thermoluminescence (LIENER and GEISS, 1968; LALOU et al., 1970; CHRISTODOULIDES et al., 1971) can be used to show that meteorites have not been stored at temperatures higher than about 400 °K for appreciable periods since the thermoluminescence clock started running, presumably as a result of the exposure of the meteorite to cosmic rays.

The measurement of equilibration temperatures is probably the technique with the greatest potential. Generally the distribution of one or two elements between coexisting minerals is determined, with the most useful mineral pairs being those showing a strong temperature-dependence of the distribution coefficient. Since most meteorites have been stored in moderately large (asteroidal-sized) parent bodies, the measured temperatures are not sharp, "quench" temperatures, but rather a range of temperatures during which diffusion became too slow for the system to maintain equilibrium (UREY, 1958). Further, apparent temperatures based on mean concentrations in large grains will, in a cooling body, be higher than those based on small grains, or exsolution lamellae.

CRAIG (1964), MUELLER (1964), RINGWOOD (1966), and SUESS (1964) attempted to use the distribution of Fe and Mg between orthopyroxene and olivine for thermometry. Unfortunately, the system does not behave ideally, and recent experimental studies show that the distribution coefficient is relatively insensitive to temperature in the range 1150—1550 °K (MEDARIS, 1969). LOVERING (1957) and NICHIPORUK (1968) used the distribution of Ni between metal and olivine to estimate high (1700—2000 °K) equilibration temperatures for the pallasites. It seems likely, however, that O_2 fugacity has played a greater role than temperature in determining the distribution coefficients in this system.

The best mineral thermometer currently available seems to be the orthopyroxene-augite system (VAN SCHMUS and KOFFMAN, 1967). The system is reasonably ideal, and the distribution coefficient Kd ($Kd=(Mg/Fe)_{Opx}/(Mg/Fe)_{Aug}$) decreases appreciably with decreasing temperature. The data shown in Fig. V-2 for type-6 ordinary chondrites yield Kd values of about 0.61. These have been used by VAN SCHMUS and KOFFMAN (1967) to estimate equilibration temperatures of about 1090 °K, based on the temperature dependence of Kd given by KRETZ (1963). KRETZ' calibration is based on rough temperature estimates for natural systems and appears to me to yield systematically high temperatures. The system should be calibrated using the techniques of experimental petrology. BLANDER (1972) points out that the value of Kd in this system is very dependent on the bulk Ca content and that interpretation of all data from a relationship between Kd and temperature alone is not warranted. This system has been reexamined by BUNCH and OLSEN (1974) who find small differences in Kd values among the three ordinary chondrite groups. These are attributed to differences in Ca content, and the data interpreted in terms of similar equilibration temperatures of about 1100 °K in each group.

Equilibration temperatures can also be estimated from oxygen isotope data. ONUMA et al. (1972a) use such data to estimate temperatures of 1220 ± 100 °K for seven type 5 and 6 ordinary chondrites, 1570 °K for

L6 Shaw, and 960 °K for L4 Bjurböle. They employ two different mineral pairs in their method, — plagioclase-pyroxene and pyroxene-olivine. The latter yields temperature estimates which average about 100° higher than the former. Thus, it appears that this method also needs further refinement.

D. The System Fe-FeO-MgO-SiO_2-O_2 and Prior's Rules

The elements Fe, Mg, Si, and O comprise nearly 90% of most chondrites, and only slightly smaller proportions of the more differentiated meteoritic silicates. In contrast to terrestrial rocks, most meteorites contain metallic Fe. Thus, the thermodynamic system of greatest general applicability to the meteorites is the system Fe-FeO-MgO-SiO_2-O_2. Discussions of the thermodynamic data and the interpretation of meteoritic properties in terms of such data have been given by CRAIG (1964), MUELLER (1964, 1965), SUESS (1964), RINGWOOD (1966), LARIMER (1968a) and WILLIAMS (1971).

LARIMER (1968a) notes that the pO_2 (the partial pressure of O_2) of a metal- and olivine-bearing meteorite system can, at a given temperature, T, be determined from the mole fraction of fayalite in the olivine, X_{Fa}, by the following equation:

$$\log pO_2 = 2 \log X_{Fa} - 29{,}080/T + 7.2 .$$

The mole fraction of ferrosilite in orthopyroxene, X_{Fs}, can be substituted for X_{Fa} without greatly affecting the accuracy of the results. The Ni concentration of the metal also plays a relatively minor role for Ni concentrations below 20%, as found in most meteorites (WAI et al., 1968). Thus, for an arbitrary comparison temperature of 1200 °K, data in Table II-5 give the equilibrium log pO_2 values ranging from less than -24 in the aubrites to about -18 in the howardites or LL chondrites. These pO_2 pressures are substantially lower than those found in terrestrial magmatic rocks, although there is some overlap between the higher meteoritic values and those observed in the lunar crystalline rocks.

In 1916 PRIOR stated that "the less the amount of Ni-Fe in chondritic stones, the richer it is in Ni, and the richer in Fe are the magnesium silicates." These two relationships are known as Prior's rules. In 1920 PRIOR widened applicability of these rules to include all meteoritic stones, and apparently, also the "stony-irons".

There is no doubt that these rules are qualitatively correct for intergroup chondrite comparisons. The explanation for this is straightforward: To a first approximation, the Ni, Fe, Mg, Si, and S abundances are the same within all chondrite groups. Providing metal is present, Ni is highly concentrated in the metal; Mg is highly concentrated in the ferromagnesian silicates; and that portion of the Fe which is not bound to S is distributed between the metal and the ferromagnesian silicates. In reduced chondrites most of the Fe is present as metal, and the Ni content of the metal is correspondingly low. With increasing degree of oxidation, more of the Fe is oxidized; the Fe content of the ferromagnesian silicates increases, and since the amount of metal decreases while the amount of Ni remains constant, the Ni concentration of the metal increases as well.

PRIOR (1920) thought that his rules indicated that all meteorites had formed from a single system which was closed to all elements except O, although he did not attempt to prove this quantitatively. Modern, accurate analytical data have shown that this viewpoint is incorrect (CRAIG, 1964; MUELLER and OLSEN, 1967). For example, the fractionation in the Fe/Si ratio discovered by UREY and CRAIG (1953) demonstrated that the ordinary chondrites could not be understood in terms of such a simple model. There is increasing evidence that the quantitative model does not hold within individual groups, as shown for the enstatite chondrites by ANDERS (1964) and MASON (1966) and for the ordinary chondrite groups by TANDON and WASSON (1968) and MÜLLER et al. (1971). The qualitative trends given by Prior's rules do hold within the ordinary chondrite groups, but not within the enstatite chondrites, group IAB irons-with-silicate-inclusions, or the differentiated silicate-rich meteorite groups. The chief importance of Prior's rules today is historical, and to some degree, mnemonic.

E. The Fe-Ni System and Cooling Rates

Figure V-3 shows a polished and etched slab of the group IIIA iron meteorite, Casas Grandes. The observed structure is called a Widmanstätten or octahedral pattern, the latter because the four sets of bands generally observable in a random section through such a meteorite are really sets of lamellae oriented with respect to each other as are the faces of a regular octahedron. The precise angular relationship between the octahedral planes as a function of the sectioning plane is given by BUCHWALD (1969a). The Widmanstätten pattern in this section continues uninterrupted over the whole section (longest dimension ca. 48 cm), as is typical of octahedrites. Additional photos showing iron meteorite structures are found in Chapter XIV.

The origin of the Widmanstätten structure can be understood in terms of the Fe-Ni phase diagram shown in Fig. V-4. The dimensions over which the pattern is continuous, without change in the orientation

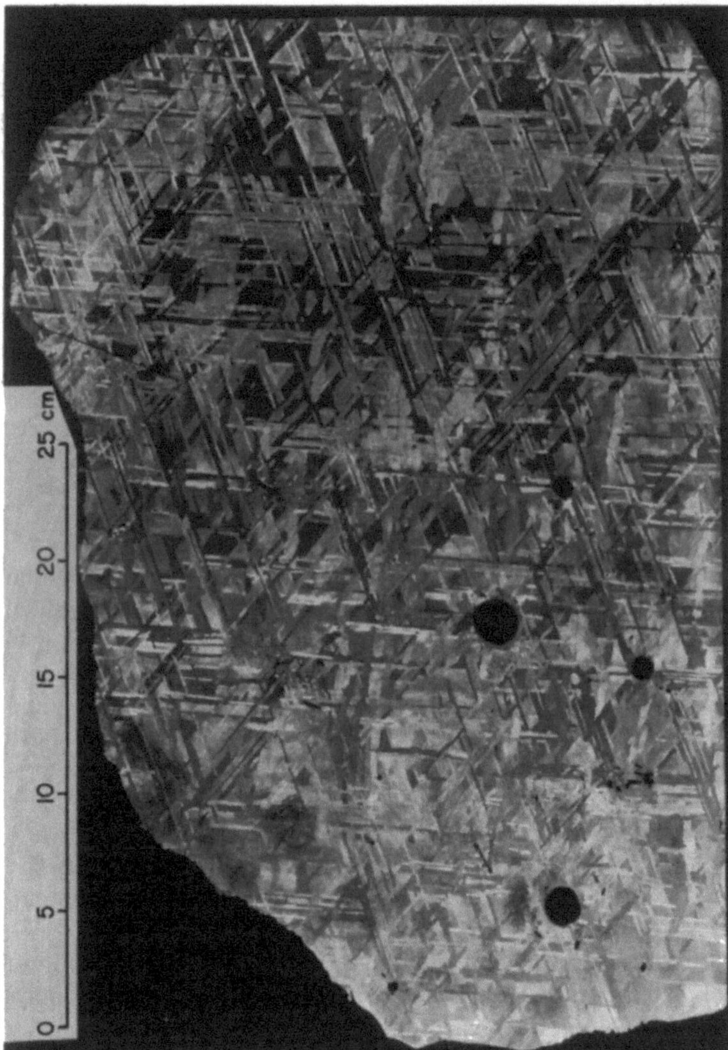

Fig. V-3. Polished and etched section of the Casas Grandes (group III A) iron meteorite. The structure consists mainly of octahedrally oriented kamacite lamellae and is known as a Widmanstätten pattern. Three sets of lamellae intersect at angles of 60°; the fourth set is parallel to the plane of the section, and forms wide bands oriented nearly parallel to the base of the section. Note that parallel kamacite bands can differ widely in reflectivity. All inclusions are troilite. (Smithsonian Institution photo)

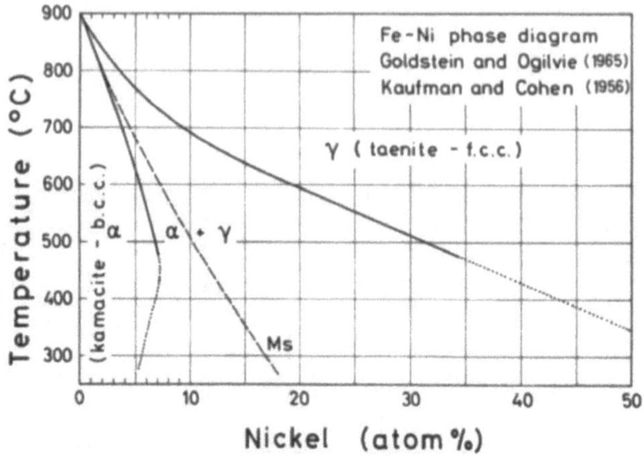

Fig. V-4. Low-temperature, low-Ni portion of Fe-Ni phase diagram. Taenite (γ-iron) which forms at high temperatures decomposes with cooling into a mixture of kamacite (α-iron) and taenite. In meteorites with less than about 7% Ni all taenite eventually disappears if cooling is slow enough to achieve equilibrium down to 500° C. The dotted portions of the phase boundaries are not experimentally determined, but based on microprobe data on coexisting meteorite phases and assumed temperatures. The dashed line labelled Ms marks the upper boundary of the metastable martensite (distorted b.c.c. structure) field

of the lamellae, correspond to the original dimensions of individual taenite (γ phase) crystals, which formed either by solid-state growth of smaller crystals within the γ field, or by direct precipitation from a slowly crystallizing melt.

As these larger γ crystals cooled, the boundary between the γ and $\alpha+\gamma$ fields was crossed, and the α phase started to precipitate. A minimum of atomic rearrangement is necessary if the 110 planes of the precipitating body-centered-cubic α phase are aligned along the 111 planes of the precursor face-centered-cubic γ phase (YOUNG, 1926; DERGE and KOMMEL, 1937). There are four sets of 111 planes in the precursor γ phase, corresponding to the four planes occupied by α lamellae in octahedrites.

Iron meteorites have Ni contents ranging from 4 to 42%, if the minor amounts of material of the 60%-Ni irons Oktibbeha County and Lafayette are neglected. Metal with Ni contents less than 7% will, with further cooling, leave the $\alpha+\gamma$ field and enter the α field. If sufficient time is available, all taenite will eventually disappear from the system. If the size of kamacite crystals becomes so large that bands cannot be resolved, even in large (> 50 cm) sections, the meteorite is designated structurally as a hexahedrite. If, on the other hand, the octahedral structure is still discernible in large sections, the correct structural designation in the

Buchwald system (Table II-6) is octahedrite, even in those rare cases in which taenite is absent. The data given in Appendix II D show that the practical border between hexahedrites and octahedrites occurs at a Ni concentration of 5.7%.

The width of the kamacite bands in octahedrites is determined mainly by three factors: the Ni concentration, the cooling rate and the nucleation temperature. That bandwidth should decrease with increasing Ni content is clear from the phase diagram. For a given equilibration temperature within the two-phase field, the lower the Ni content, the greater the amount of kamacite in the equilibrium assemblage. The effect of cooling depends on the fact that diffusion limits the ability of a given assemblage to approach equilibrium. Obviously, material which spends a longer period of time within a given temperature range will approach equilibrium more closely and will have thicker kamacite lamellae than material of the same Ni content which cooled more rapidly. The nucleation temperature also affects the bandwidth. Since diffusion coefficients decrease rapidly with decreasing temperature, most of the kamacite observed in octahedrites forms within a small temperature interval below the nucleation temperature. For a given Ni concentration and cooling rate, material which nucleates at higher temperatures will achieve coarser structures than that which nucleates at lower temperatures.

The upper portion of Fig. V-5 is a photomicrograph of the 12% Ni IIIC iron meteorite Anoka, and the lower portion is a microprobe trace of the Ni concentration between points marked P and P′ on the upper section. The broad whitish bands, most of which are vertical in this photo, are kamacite with a Ni concentration which is nearly constant at about 6.5%. A slight decrease in kamacite Ni concentration near the taenite interface is the basis for the slope of the dashed portion of the $\alpha/\alpha+\gamma$ phase boundary in Fig. V-4. The narrow whitish bands on the borders of the kamacite are taenite, with Ni contents varying from about 37 to 20%. The dark areas are finely divided mixtures of kamacite and taenite, and are called plessite. That individual kamacite and taenite grains are present in the plessite is shown by the irregular microprobe trace, although the grains were not completely resolved. A very fine Widmanstätten structure can be observed in the large plessite field near the center of the photo. If the γ phase cools to a temperature below the dashed line labelled Ms in Fig. V-4, it immediately undergoes a diffusionless transformation to the distorted-body-centered-cubic phase, martensite (α_2). It is probable that most of the plessite in Anoka (Fig. V-5) has formed by a decomposition sequence including martensite as an intermediate form $\gamma \rightarrow \alpha_2 \rightarrow (\alpha + \gamma)$.

That the slope of the Ni concentration is much steeper in taenite than in kamacite results from two factors: a lower diffusion coefficient in

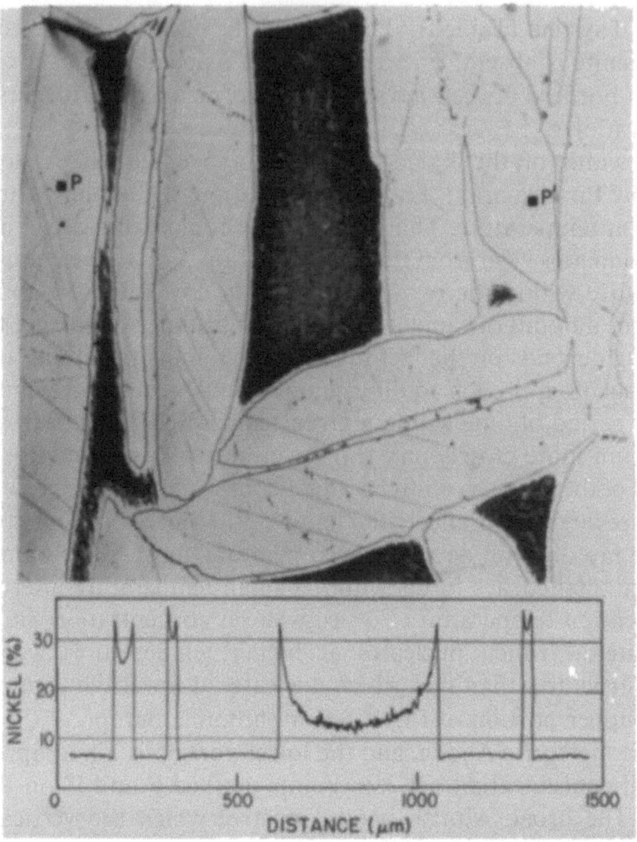

Fig. V-5. Above: Photomicrograph of Anoka IIIC iron. Below: Microprobe trace of Ni concentration between points marked P and P′ on the photomicrograph of Anoka. The Ni concentration gradient is greater in the high-Ni taenite phase, which indicates a lower diffusion coefficient in this phase. The Ni concentrations in the kamacite and taenite at their interface should correspond to isotherm compositions on the Fe-Ni phase diagram. The dark areas are plessite. (From WOOD, 1968)

taenite than in kamacite (by a factor of 100 according to GOLDSTEIN and OGILVIE, 1965), and a much steeper gradient in the $\gamma/\alpha+\gamma$ phase boundary than in the $\alpha/\alpha+\gamma$ phase boundary. Microprobe Ni traces from kamacite across taenite-plessite-taenite and back into kamacite have shapes resembling the letter M and have been termed "M-profiles" by WOOD (1964). Figure V-6 shows the way such an M-profile grows during the cooling of a 10%-Ni alloy (WOOD, 1968). Time runs from the bottom to the top of the figure. Note that about 70% of the growth of the kamacite occurs within 70° of the nucleation temperature of 690° C.

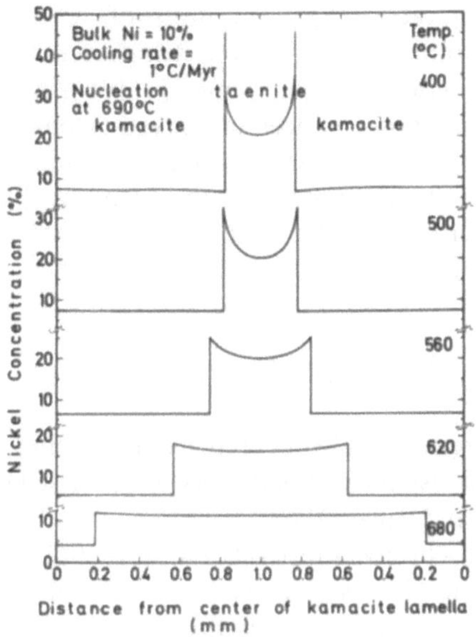

Fig. V-6. Calculated shapes of Ni concentration profiles across kamacite and taenite phases in a cooling 90% Fe-10% Ni alloy. Time runs from bottom to top in this diagram, which is adapted from WOOD (1968)

Detailed microprobe determinations of the Ni distribution in meteoritic metal can, in principle, be used to estimate the rates at which the meteorite parent bodies cooled through the temperature range of about 700—500° C. WOOD (1964) and GOLDSTEIN and OGILVIE (1965) derived similar computer programs for estimating cooling rates, and they as well as GOLDSTEIN (1965), GOLDSTEIN and SHORT (1967a) and SHORT and ANDERSON (1965) tabulated cooling rates for a number of iron meteorites. Similar estimates of cooling rates are given by WOOD (1967b) for chondrites, and by POWELL (1969) for mesosiderites.

REED (1965a) estimated relative cooling rates on the basis of the Ni concentrations measured in the kamacite and taenite near the interface between the two phases. SHORT and GOLDSTEIN (1967) developed an equation for estimating absolute values from such data, and BUSECK and GOLDSTEIN (1969) applied the method to a large number of pallasites.

SHORT and GOLDSTEIN (1967) developed a second rapid method for cooling rate estimation based only on the Ni concentration and mean kamacite bandwidth. WASSON (1971b) showed that the graphical data of these authors can be fitted with the equation

$$\log CR = -2.040 \log BW - 8.940 \log [\mathrm{Ni}] + 8.700$$

where the cooling rate (*CR*) is in units of °C/Myr, the bandwidth (*BW*) is in units of mm, and the Ni concentration in wt %. The equation is valid for the Ni concentration range 7—14%, which includes most octahedrites. The method has been employed extensively by GOLDSTEIN and SHORT (1967b) and GOLDSTEIN (1969) to estimate cooling rates for iron meteorites. Partly because the method ignores the effect of variable nucleation temperatures on the bandwidth, it is not very precise (WASSON, 1971b; 1972b).

F. Other Mineral Systems

MASON (1968) reviewed the occurence of pyroxenes in meteorites, and the mineralogy of feldspars in chondrites was discussed by MASON (1965b) and VAN SCHMUS and RIBBE (1968).

The phosphide, schreibersite (REED, 1965b, 1969; GOLDSTEIN and OGILVIE, 1963; DOAN and GOLDSTEIN, 1969) and several phosphates (FUCHS, 1969; VAN SCHMUS and RIBBE, 1969) are common accessory minerals in meteorites. Phosphides and phosphates often coexist in the same meteorite, providing a mineral assemblage which is useful for estimating pO_2 (OLSEN and FREDRIKSSON, 1966; OLSEN and FUCHS, 1967).

Sulfides are very common accessory minerals in meteorites. A comprehensive review of meteoritic sulfides does not appear to exist, but important studies of their occurrence in iron meteorites have been published by EL GORESY (1965, 1967), GOLDSTEIN and OGILVIE (1963), BUNCH et al. (1970) and ORSINI and CENTO (1967). A reinvestigation of the system Fe-Cr-S and its application to meteoritic minerals is given by EL GORESY and KULLERUD (1969). A series of rare but important sulfide minerals is found in the highly reduced enstatite chondrites and achondrites (KEIL, 1968), 1969b; LARIMER, 1968b). These meteorites also contain Si-bearing kamacite and the silicide perryite (REED, 1968; KEIL, 1968; WASSON and WAI, 1970).

Surveys of the different occurrences of chromite and ilmenite in several meteorite classes are given by BUNCH et al. (1967), RAMDOHR (1967), SNETSINGER and KEIL (1969) and BUNCH and KEIL (1971).

Special importance attaches to the occurrence of some rare minerals because of their ability to concentrate heavy radioactive elements (Th, U, and Pu). Examples are zircon (FLEISCHER et al., 1965) and the phosphate whitlockite (WASSERBURG et al., 1969a).

Other important mineralogical studies are cited in the general review works of MASON (1967b, 1972) and KEIL (1969a).

VI. Petrology

A. Petrographic Descriptions of Individual Meteorites

A careful petrographic description of an individual meteorite can serve for many years as the foundation upon which research in widely divergent fields is built. The series of publications by MASON and WIIK (1966; and others listed in WIIK, 1969) are to be cited in particular as examples of careful and complete studies of silicate-rich meteorites. Each of these papers includes a bulk chemical analysis, a density determination, a careful petrographic description of the meteorite including olivine and/or pyroxene composition, and on the basis of these observations, an accurate classification. More recent papers (e.g., MASON, 1973) also include estimates of the VAN SCHMUS-WOOD (1967) petrologic type (see Sections II B and VI C).

Examples of complete and informative descriptions of individual iron meteorites include the work of BUCHWALD (1967) or the ca. 400 descriptions in BUCHWALD (1974). These include determinations of Ni, Ge and other elements, detailed metallographic descriptions based on large polished sections, an accurate classification, and citation of those meteorites which are most similar in composition, with a careful consideration of the possibility that some of these might be paired with the investigated iron.

B. Metallographic and Metallurgical Studies of Metal-Rich Meteorites

In this section I will discuss only studies of general application to the understanding of iron meteorites. A review of iron meteorite studies prior to about 1966 is given by AXON (1968). Detailed descriptions of most iron meteorites available for scientific research are given in the book by BUCHWALD (1974), together with literature citations. The book by PERRY (1944) was epoch-making in its time and is still of value for its extensive photomicrographs.

A number of iron meteorites show the effects of cosmic heat treatment followed by cooling too rapidly to produce typical hexahedral or octahedral structures. In most cases these effects postdate the period during which the irons were stored in the deep interior of a parent body, and probably reflect shock heating during collisions with other interplanetary debris or heating experienced during passage close to the sun. Studies of the effects introduced during the laboratory heating of iron-meteorite specimens are reported by BRENTNALL and AXON (1962), MARINGER and MANNING (1962), WOOD (1964) and STAUB and MCCALL (1970), and the results used to interpret observed structures in iron meteorites. All too often irons have been heated by man, and it is necessary to separate these effects from the truly cosmic ones (BUCHWALD, 1965, 1974). Discussion of shock effects and additional discussion of cosmic reheating are given in Section VI F.

There seems to be general agreement in the current literature that the octahedral orientation of kamacite lamellae results from the mechanism described in Section V E. The formation of plessite in all its morphological forms is more complex, however. Martensite is currently believed to be an intermediate precursor of certain forms of plessite (MASSALSKI et al., 1966), and in some cases, the plessite as presently observed may represent martensite decomposed during a subsequent reheating event (WOOD, 1964; STAUB and MCCALL, 1970). SCOTT (1973) discusses the formation of intermediate, dark-etching zones between taenite and plessite.

A problem that has received insufficient attention on the part of metallurgists is the presence of features which would indicate whether meteoritic metal originally crystallized from a melt. For example, AXON and FAULKNER (1967) note that the original dendritic grains formed during such crystallization (of a non-convecting magma) should be outlined by massive nonmetallic segregations, and that such occurrences are not known. Another property which would be of interest in the mean size of the precursor taenite grains in the various groups of iron meteorites, since these should provide information about the rate of crystallization, or the duration of high-temperature annealing in the γ field, or both. Statistical evidence regarding these taenite grain sizes would be helpful in choosing between igneous and non-igneous origins for the different groups of iron meteorites.

The petrology of the iron meteorite inclusions is also a somewhat underdeveloped field. Among the recent discoveries of note are the massive chondritic silicate inclusions in IAB irons (see BUNCH et al., 1970) and the presence of oriented ellipsoidal troilite inclusions in the anomalous iron Santa Rosa (BUCHWALD and WASSON, 1968) and in the IIIA iron Cape York (BUCHWALD, 1971a). In the latter, the direction of the gravitational field is uniquely determined, since "light" phosphates are concentrated in the "upper" end of the inclusions, and "heavy" metal inclusions are found in the "lower" end.

C. Chondrules, Chondritic Petrology, and Metamorphism

There are three basic questions which have received extensive discussion in petrologic studies of chondrites: 1. How did the chondrules originate? 2. What is the significance of the differing degrees of recrystallization observed within genetic groups of chondrites? and 3. How did the chondritic breccias originate?

Petrologic evidence bearing on the first question is discussed briefly in Chapter XVIII, together with evidence from other types of studies. The key facts about chondrules are: 1. they consist mainly of high-temperature minerals or glass; 2. they are generally spheroidal; 3. their bulk compositions vary, but except for a lower metal content, their average composition is essentially the same as that of the non-chondrule, "matrix" portion of the meteorite; and 4. they show a very limited range of radii within each meteorite group. Brief reference here might be made to the papers by WOOD (1963a), REID and FREDRIKSSON (1967) and DODD (1971b), which present much of the available petrologic evidence together with some well-developed views regarding their interpretation.

A thin section of the chondrule-rich L3 chondrite Bishunpur is illustrated in Fig. VI-1. The chondrules show a fascinating variety of textures. The more common types of chondrule textures have been named; some are illustrated in Fig. VI-2. According to VAN SCHMUS (1969a), the most common types are:

Barred Olivine Chondrules (Fig. VI-2a). The inner rim is olivine. The interior contains parallel skeletal crystals of olivine which are often crystallographically a single crystal. The interstitial material consists of soda-alumina-silica glass or microcrystalline material. Outer rims of these chondrules generally contain droplets of mixed metal-sulfide.

Radiating Pyroxene Chondrules (Fig. VI-2b). These are nearly monomineralic, consisting of radiating laths of low-Ca pyroxene with occasional glassy or microcrystalline material between the laths. Individual laths range from 1—10 μm in width.

Porphyritic Chondrules (Fig. VI-2c). These are probably the most common chondrules, but their textures exhibit greater variation than those in the preceding two types. They consist of subhedral (partially bounded by own crystal faces) to euhedral (completely bounded by own crystal faces) crystals of olivine and pyroxene set in glassy or microcrystalline material. The relative proportions of olivine and pyroxene are infinitely variable. The pyroxene is generally monoclinic and shows twinning lamellae when viewed in polarized light. Sometimes a coating of augite is found on the pyroxene grains, and sometimes the olivine is enclosed poikilitically (each crystal independently oriented) inside pyroxene. In most type 3 chondrites the olivine and pyroxene grains are zoned, with Fe content increasing towards the exterior.

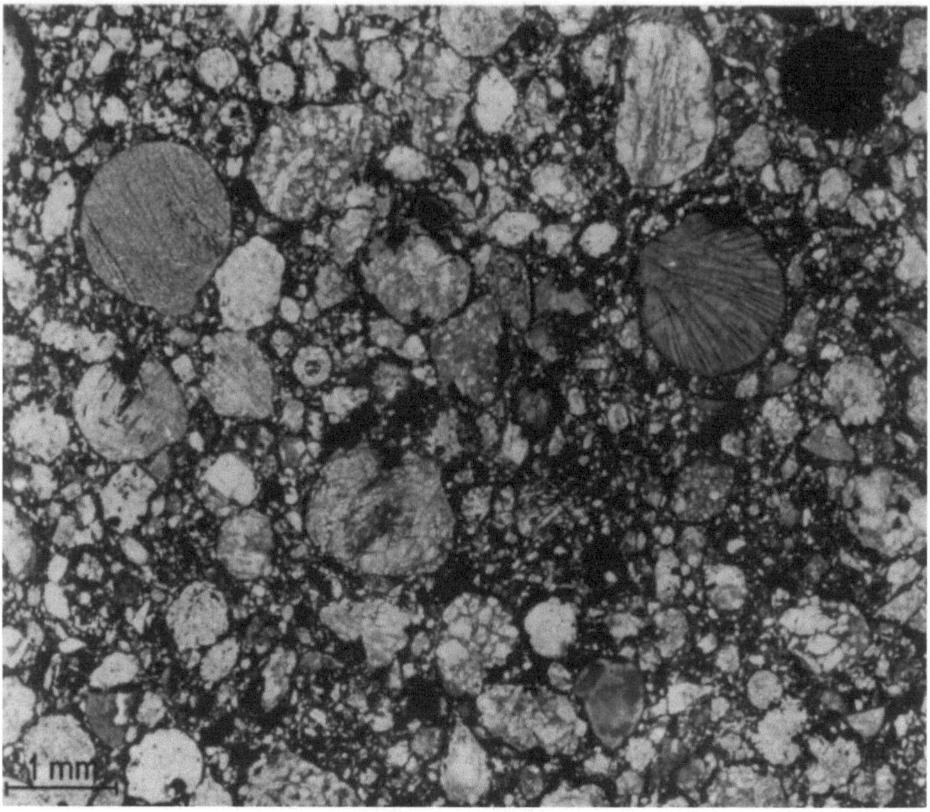

Fig. VI-1. Thin section of Bishunpur L 3 chondrite photographed in transmitted light. Note the diversity of sizes, shapes, and textures of chondrules. (Photo by W. R. VAN SCHMUS)

Glassy Chondrules (Fig. VI-2d). These consist entirely or almost entirely of glass. They are rare and are found only in chondrite types 2 and 3. The glass is usually rich in soda, alumina and silica. These chondrules often contain skeletal, feathery, radiating or euhedral crystals of olivine, pyroxene or spinel ($MgAl_2O_4$) and droplet-like metal-sulfide inclusions.

Dark-zoned Chondrules (Fig. VI-2e). Although VAN SCHMUS (1969) suggested that these were low-temperature agglomerates rather than true chondrules, a more recent study by DODD and VAN SCHMUS (1971) indicates an origin from more typical chondrule material during a reheating event – possibly as a result of shock. They are generally dark, with the darkness (and the concentration of opaque minerals such as metal and sulfide) increasing towards the surface.

Lithic Fragment "Chondrules" (Fig. VI-2f). These are chondrule-sized bodies with irregular or angular shapes, which appear to be fragments of larger crystalline masses. Their olivines and pyroxenes are always of constant composition, in support of the idea that they formed by comminution of larger masses of recrystallized chondritic material. It appears best not to call these chondrules.

The chondrules are embedded in fine-grained material generally known as the "matrix". Because of its small grain size, this material has not been well characterized. In CM chondrites the matrix appears to be very similar in bulk mineralogy to that found in primitive CI chondrites (which are chondrule-free). On the other hand, REID and FREDRIKSSON (1967) state that the bulk of the matrix in ordinary chondrites consists of comminuted chondrule material of high-temperature origin.

The highly recrystallized texture of the L6 chondrite Peace River is illustrated in Fig. VI-3. A few outlines of what must have been chondrules are still preserved. The bulk chemical composition of Peace River is essentially the same as that of Bishunpur except for highly volatile elements such as In and the rare gases, which are about 100 times more abundant in Bishunpur.

As discussed in Chapter II, DODD et al. (1967) and VAN SCHMUS and WOOD (1967) have shown that all intermediate textures between that of Bishunpur and that of Peace River can be found among the ordinary chondrites. They propose that this is a metamorphic sequence and that the textures observed in higher petrologic groups are a direct result of reheating material in the lowest petrologic groups. Thus, according to their viewpoint, all ordinary chondrite textures were formed by metamorphic reheating of material resembling type 3 material, but possibly still more primitive, since some evidence of reheating is found even in many type 3 chondrites. WOOD (1967b), for example, finds evidence of Ni redistribution in L3 Chainpur, L3 Mezö-Madaras and H3 Tieschitz, although not in L3 Bishunpur and L3 Krymka.

All chondrite petrologists seem to agree that type 6 material can be produced by metamorphic heating of type 4 material. Several petrologists, however, argue that it is not possible for metamorphism to convert type 3 material into type 4 (e.g., REID and FREDRIKSSON, 1967). They note that the mean fayalite content of olivine in L3 chondrites (e.g., 16.6% in Bishunpur) is much lower than that in L4 chondrites (26.2% in Bjurböle), and that the mean deviation of the olivine composition changes from about 40% in L3 to <2% in Bjurböle. Since Fe and Mg are two of the most abundant elements, a large amount of mass transport is involved. REID and FREDRIKSSON (1967) argue that this would have necessarily destroyed "the delicate internal chondrite textures, and the chondritic texture itself". Bjurböle is very chondrule-rich and is also a very friable assemblage.

SUESS and WÄNKE (1967) argue that the planetary rare gases (cf. Chapter IX) which MERRIHUE (1966) showed to be trapped in the interi-

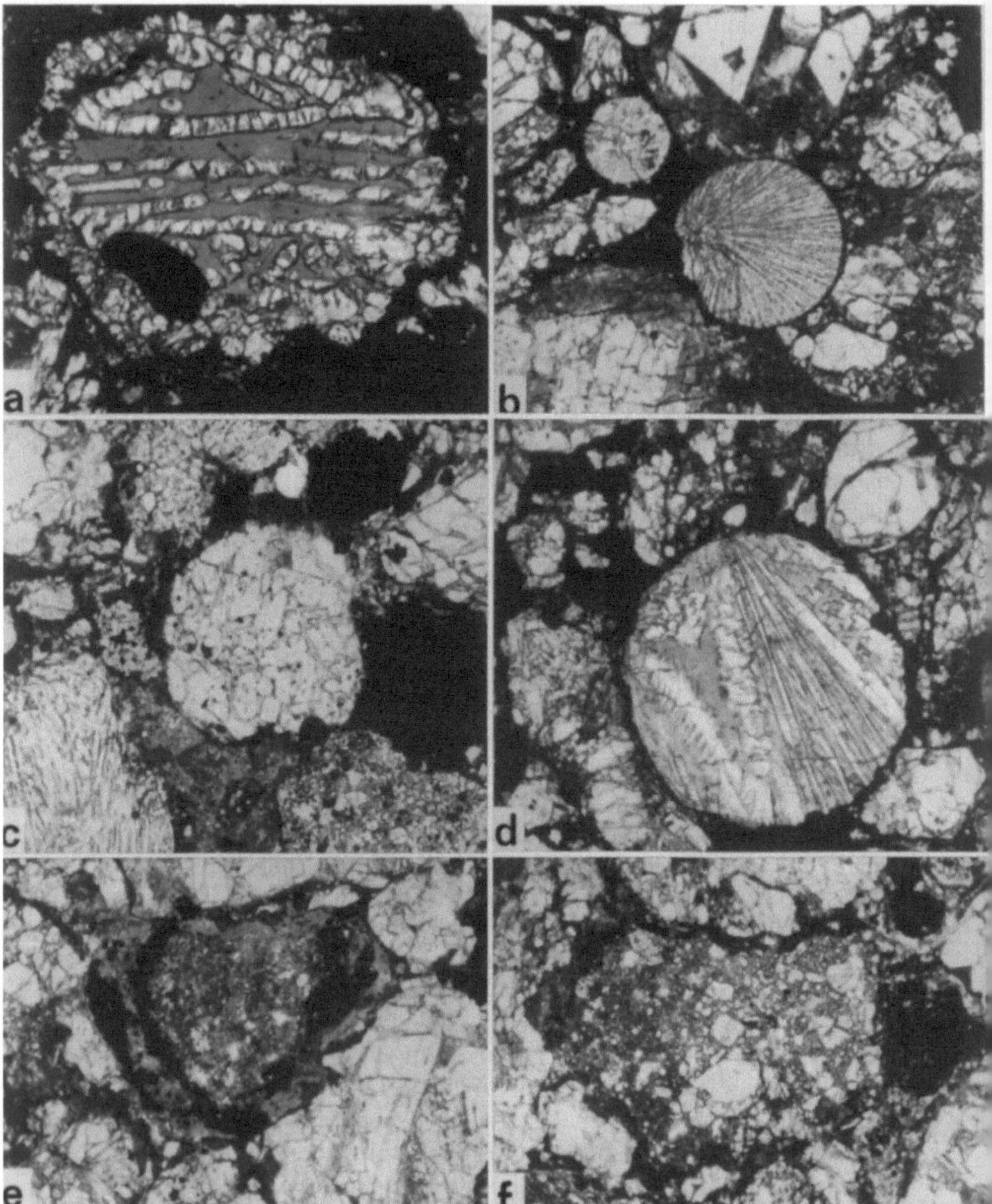

Fig. VI-2. Thin sections of chondrules and a lithic fragment photographed in transmitted light. Field of view approximately 0.6 mm in all cases. (a) Barred olivine chondrule; bars of olivine (white) are embedded in clear glass (gray). Mezö-Madaras L 3 chondrite. (b) Radiating pyroxene chondrule; such chondrules are nearly monomineralic. Sharps H 3 chondrite. (c) Porphyritic chondrule containing olivine (white) crystals in a clear isotropic glass (gray). Hallingeberg L 3 chondrite. (d)

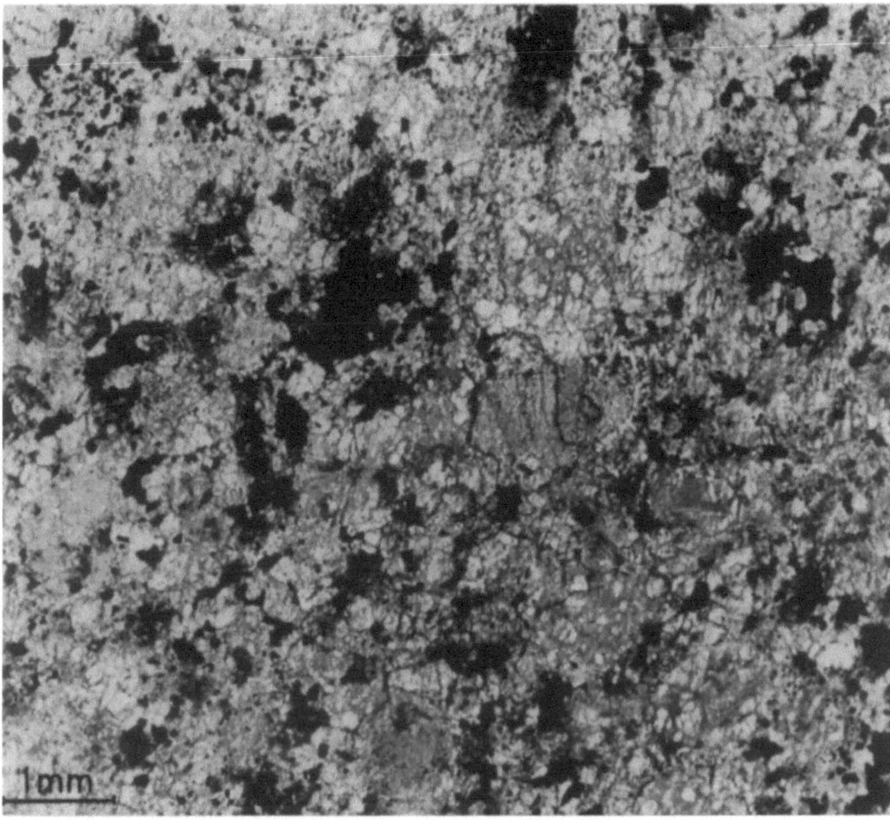

Fig. VI-3. Thin section of Peace River L6 chondrite photographed in transmitted light. Although the bulk compositions of Peace River and Bishunpur are the same except for highly volatile elements, Peace River is more coarsely grained and shows only vague outlines of chondrules. (Photo by W. R. VAN SCHMUS)

ors of chondrules would also necessarily be lost during this redistribution of Fe and Mg.

The counter arguments include the following: 1. It is clear that you cannot make type 3 material from type 4. Therefore, if you cannot do the inverse, you must hypothesize separate origins for two types of material

Glassy chondrule; some skeletal and feathery olivine crystals (white) have crystallized from the isotropic glass(gray). Sharps H3 chondrite. (e) Dark-zoned chondrule. Some evidence suggests that these objects are nebular agglomerates and were never molten (and thus not true chondrules). Other evidence indicates production from typical chondrules types by reheating and possibly also shock. The opaque surface material is mainly troilite. Hallingeberg L3 chondrite. (f) Lithic fragment; this seems to be a fragment of recrystallized and brecciated chondrite material. The roundish shape probably resulted from tumbling. Hallingeberg L 3 chondrite. (Photos by W. R. VAN SCHMUS) from a single genetic group. This runs into trouble with Occam's razor.[4] 2. Some chondrites (e.g., Cynthiana, Tennasilm) contain olivine of constant composition coexisting with pyroxene showing mean deviations of 6—7%. This is easily explained by the metamorphic model in terms of more rapid diffusion of cations in olivine (in which the SiO_4^{-4} are isolated tetrahedra) than in pyroxene (a chain silicate). It is not easy to propose a means of producing the differences in olivine and pyroxene composition by condensation, shock, or any other reasonable nebular or parent-body process. 3. Although major loss of planetary gases must have occurred during recrystallization, WOOD (1967c) points out that some gas must have been trapped in voids in well-crystallized materials and that, because of their large atomic radii, rare gases would find it very difficult to undergo diffusional escape even though the crystal framework was undergoing a solid-state equilibration.

Although the issue is by no means settled, I find that the evidence weighs in favor of all ordinary-chondrite types 3—6 being a single metamorphic sequence. More detailed arguments are found in DODD (1969) and WASSON (1972a). I do not agree with VAN SCHMUS and WOOD (1967) that the enstatite chondrites represent a single metamorphic sequence; the large variations in Fe/Si and Mg/Si ratios between the E5 and E6 types are impossible to produce by metamorphism, even though textures similar to those in the E6 chondrites could be produced by reheating E4,5 material (BAEDECKER and WASSON, 1974). VAN SCHMUS and WOOD (1967) and VAN SCHMUS and HAYES (1973) show that different petrologic types are represented among chondrites, but hold a metamorphic relationship between different groups of carbonaceous chondrite to be unlikely.

D. Calcium-Aluminum-Rich Inclusions

During the past few years a number of high-temperature phases have been discovered in CV and CO carbonaceous chondrites. These include spinel ($MgAl_2O_4$), melilite, a solid-solution mixture of gehlenite ($Ca_2Al_2SiO_7$) and akermanite ($Ca_2MgSi_2O_7$), and perovskite ($CaTiO_3$) (CHRISTOPHE MICHEL-LEVY, 1968, 1969; KURAT, 1970; KEIL et al., 1969). The importance of these first became apparent with the fall of several tons of CV3 material at Allende, Mexico in 1969. About 5—10% of Allende consists of white inclusions; some are spheroidal and some irregular,

[4] The 14th century scholar William of Occam stated that "entities should not be multiplied except out of necessity"; if two models are equally able to explain the facts, that which does so with the fewer processes, starting materials, etc. is the preferable.

Fig. VI-4. Fractured surface of Allende CV3 chondrite showing numerous light-colored inclusions in a dark gray matrix. Some inclusions are oriented with their long axes nearly parallel. Some are spheroidal, whereas some have forms suggesting deformation while plastic. The inclusions consist mainly of high-temperature phases such as $MgAlO_4$, $CaTiO_3$, and $Ca_2Al_2SiO_7$.

some appear to have been deformed plastically as a result of compression. A typical specimen of Allende is illustrated in Fig. VI-4. Because of the ready availability of Allende material, a large number of petrologic studies of these white inclusions have been made. Although consisting mainly of high-temperature minerals such as those cited above, some also contain "low-temperature" feldspathoids such as sodalite ($3NaAlSiO_4 \cdot NaCl$). There is some evidence that these are of later origin than the high-temperature minerals.

Thermodynamic calculations by LORD (1965) and LARIMER (1967) showed that the first solids to condense from a cooling solar nebula should be Ca-, Al-, and Ti-rich minerals such as spinel, melilite and perovskite. Similarly, a residual material following volatilization of 99% of the interstellar solids during solar-system formation would consist of the same minerals (since thermodynamics does not depend on the reaction path). Thus it has been hypothesized that the white, Ca-Al rich inclusions were high-temperature residues (KURAT, 1970) or condensates (LARIMER and ANDERS, 1970; MARVIN et al., 1970). The high-temperature nature has been confirmed by observations of high concentrations of Ir, a siderophilic element with a high condensation temperature, despite the fact that the inclusions are essentially free of Fe-Ni metal, the usual site of Ir (GROSSMAN, 1973). GROSSMAN (1972, 1973) has carried out detailed calculations and shown that these materials last equilibrated with the solar nebula at a (pressure-dependent) temperature of about 1500 °K. The outermost portions appear to have reacted at lower temperatures either with nebular gas or, following accretion, with the chondrite matrix to form the feldspathoids and some other low-temperature phases.

Although the preceding interpretation is generally accepted, there are differences in detail. For example, FUCHS and BLANDER (1973) argued that the material condensed as liquids rather than solids, and ARRHENIUS (1972) proposed that the feldspathoids are evidence of plasma condensation.

Various authors have invoked these Ca-Al rich materials as parental matter for the moon, since they seem to provide a nebular source for the high abundance of refractory elements found in lunar rocks. This raises another sticky problem, however—how did these materials become isolated from 95% of the remaining condensable matter of the solar system in order to form the moon?

E. Differentiated Silicate-Rich Meteorites

It is difficult to generalize about the differentiated silicate-rich meteorites, since they display a wide variety of compositions and textures. Perhaps their most prevalent common feature is their coarse crystallinity relative to the chondrites. Some (e.g., the eucrites) appear to be unquestionably of igneous origin (DUKE and SILVER, 1967). Others (e.g., ureilites, howardites) appear to consist mainly of igneous silicates with a seasoning of low-temperature material of nebular origin (WLOTZKA, 1972; JÉROME and GOLES, 1971). Some (e.g., the silicate portions of Netschaëvo and the group IAB irons) are chondritic and appear to have escaped any appreciable melting.

Most of the meteorites traditionally classified as achondrites have very low metal contents. However, as shown in Table II-5 and Appendix IIC, they are not metal-free, and the classification of meteorites as achondrites on this basis is artificial and should be discontinued. The silicate portions of the howardites, which generally have less than 0.3% metal, and of the mesosiderites, which contain about 50% metal, are similar both chemically and petrologically.

Another common practice is to distinguish between Ca-rich and Ca-poor achondrites. However, the hiatus in Ca contents (or Ca/Mg ratios—see Fig. II-4) does not appear to have great genetic significance. There are several features suggesting a genetic link between Ca-rich howardites and Ca-poor diogenites, whereas there is virtually no evi-

Fig. VI-5. Thin section of Pasamonte eucrite taken in transmitted light. The clast on the left shows an igneous texture, with needles of plagioclase crossing pigeonite crystals. Minor SiO_2 is present as tridymite (T). The microbreccia on the right is compositionally identical to the igneous clast. (Photo by M. B. DUKE)

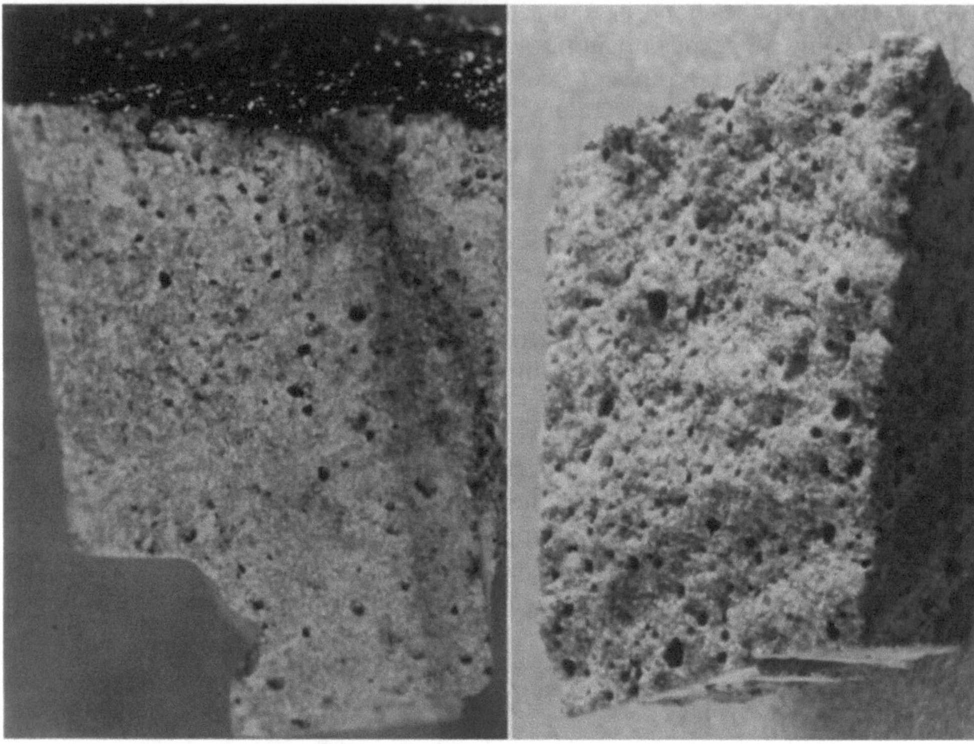

Fig. VI-6. Two faces of Ibitira eucrite showing vesicular structure. Note glossy fusion crust in left photo. The void spaces appear to have resulted from the internal evolution of vapor following extrusion onto a parent-body surface. Face on right is about 2.0 × 2.3 cm. (Photo by F. WLOTZKA and P. DEIBELE)

dence suggesting a close relationship between the two Ca-poor groups of diogenites and aubrites.

A more significant possible relationship which has not been pursued with the diligence it deserves is the possibility that several of the differentiated silicate-rich groups are complementary differentiates (RINGWOOD, 1961; MASON, 1967c; LOVERING, 1962; JÉROME and GOLES, 1971). Detailed chemical studies should make it possible to definitively resolve such relationships.

One feature shared by most differentiated silicate-rich meteorites is brecciation. This will be discussed more fully in the following section. In fact, unbrecciated members of some groups are so rare that they always receive special attention in petrographic surveys (e.g., the Tatahouine diogenite, the Moore County and Serra de Magé eucrites, the Shallowater aubrite). It is curious that brecciation should be so much more common in these groups than among the chondrites. Perhaps this is an indication that the differentiated groups originated in larger asteroids (such as Vesta, which has a eucritic reflection spectrum—see Chapter XVII), where shock effects almost invariably accompanied ejection from the surface with a velocity exceeding the escape velocity.

One of the most thorough studies of differentiated silicate-rich meteorite groups is that of DUKE and SILVER (1967). On the left side of Fig. VI-5, which is taken from their paper, is an ophitic (refers to the presence of lath-shaped plagioclase crystals embedded in pyroxene crystals) fragment which shows evidence of the crystallization sequence expected during solidification of an igneous melt. The breccia texture on the right has had nearly all record of the crystallization sequence erased, but composition studies indicate that it was formed by comminution of material such as the ophitic fragment.

The extrusive origin of some eucrites is confirmed by their vesicular structures. Figure VI-6 shows a fragment of Ibitira. The voids are formed by the expansion of gas evolved from the magma as a result of the pressure drop during extrusion onto a parent body surface, and are not found in igneous rocks which crystallize at depth.

F. Shock and Brecciation

Shock has played an important role in the history of the meteorites. Most meteorites appear to be asteroidal fragments which were removed from their parent bodies by hypervelocity impacts. In some cases several such events may have been necessary to remove the material, reduce it to the sizes typical of meteorites, and deflect it into orbits which intersect that of the earth. JAIN and LIPSCHUTZ (1971) estimate that about 65% of the iron meteorites have been subjected to shock pressures greater than 130 kb, and TAYLOR and HEYMANN (1969) find that only 12 of 103 ordinary chondrites contain olivine crystals giving completely sharp Laue X-ray diffraction patterns, indicating the absence of appreciable shock.

Investigations during the past few years show that many common petrological features of meteorites are of shock origin. In Table VI-1 are listed a number of these shock indicators, with some estimations of the shock intensity necessary to produce the effect, and literature references. Some of these properties are produced only by shock (e.g., Neumann lamellae or maskelynite), whereas others are purely thermal effects (e.g., melted troilite) which could be produced by heat from sources other than shock. Shock effects vary from location to location and from phase to phase. Thus, to completely characterize the shock history of a given meteorite it is best to study a number of shock-related properties in as large a sample as possible. A comprehensive survey of shock processes

Table VI-1. Some petrological indicators of shock in meteorites

Shock indicator	Minimum shock pressure (kb)	References
Silicate systems		
Blackening	—	Heymann (1967)
Veining	—	Heymann (1967); Binns (1967a)
Plagioclase→maskelynite	300	Binns (1967a); Stöffler (1972)
Plagioclase, pyroxene→glass	450	Fredriksson and Kraut (1967); Stöffler (1972)
Disordering of orthopyroxene	—	Pollack (1968)
Orthopyroxene→clinopyroxene	—	Wlotzka (1969)
Pyroxene→majorite	135	Smith and Mason (1970); Stöffler (1972)
Deformation of olivine	50	Carter et al. (1968); Müller and Hornemann (1969); Stöffler (1972)
Polycrystallinity of olivine	~450	Fredriksson et al. (1963); Heymann (1967); Stöffler (1972)
Olivine→ringwoodite	450	Binns et al. (1969); Stöffler (1972)
Opaque systems		
Metallic flow structures	—	Axon (1969); Bunch and Cassidy (1968)
Melted troilite	—	Vogel (1965); El Goresy (1965)
Melted metal	—	Begemann and Wlotzka (1969)
Neumann lamellae in kamacite	~10	Heymann et al. (1966); Zukas (1969)
Shock hardening of kamacite	—	Axon et al. (1968); Zukas (1969)
Acicular kamacite	130	Maringer and Manning (1962); Heymann et al. (1966)
Recrystallization of kamacite	750	Taylor and Heymann (1969); Heymann et al. (1966)
Polycrystallinity of schreibersite	—	Lipschutz (1968)
Polycrystallinity of cohenite	—	Lipschutz (1967)
Graphite→cubic diamond	130	DeCarli and Jamieson (1961); Lipschutz and Anders (1961)
Graphite→hexagonal diamond	700	Hanneman et al. (1967); Stöffler (1972)

and effects is given by Stöffler (1972): although Stöffler's study is primarily concerned with effects produced in lunar and terrestrial materials (and thus mainly silicates), a good survey of meteorite observations is also included.

Many meteorites are breccias (some were discussed in the previous section). A classic example of a breccia consisting of materials having very different properties and, probably, different origins is Cumberland Falls, which is shown in Fig. VI-7. The light portion of Cumberland Falls

Fig. VI-7. Section through Cumberland Falls meteorite, an example of a polymict breccia. The light portion is typical aubritic material; the dark portion is chondritic material of anomalous composition (oxidation state intermediate between E and H groups). (Smithsonian Institution photo)

is a typical aubrite, with <1% metal and only 0.05 mole % Fs in the pyroxene; the dark portion is a chondrite of anomalous type, with 12% metal and unequilibrated silicates with pyroxene compositions ranging between Fs2 and Fs17 (BINNS, 1969).

WAHL (1952) proposed that the breccias be divided into two categories, the *monomict* breccias, in which the xenoliths are of the same material as the surrounding, more finely divided host, and the *polymict* breccias, in which xenoliths are present which are not closely related to the interstitial host, and in which additional types of material are sometimes found. According to the preceding definition (DUKE and SILVER, 1967), most eucrites are monomict breccias whereas Cumberland Falls is a polymict breccia *par excellence.*

Many ordinary chondrites and some meteorites of other groups are breccias consisting of light-colored xenoliths in a darker colored host (WLOTZKA, 1963; FREDRIKSSON and KEIL, 1963; BINNS, 1968; VAN SCHMUS, 1967). WAHL (1952) called these polymict breccias. Recent research (e.g., BINNS, 1968; FREDRIKSSON et al., 1968; KURAT et al., 1969) has shown that the xenoliths and host invariably belong to the same chemical group, but that they often differ in petrologic grade and in the detailed composition of individual minerals. Thus, the properties of these meteorites fit the above definition of monomict breccia better than that of polymict breccia, but the former is also not really suitable, since the clasts and matrix may have come from widely differing locations (e.g., from different planetesimals, WASSON, 1972a). I propose that a new word—*genomict* breccias—be coined for breccias consisting of materials from the same chemical group but which have had different histories. Genos is the Greek word for race.

Many of the light-dark breccias found in chondrites, aubrites and howardites are of special interest because of the presence of large concentrations of solar-type rare gases in the dark, "host" portions. This point is discussed in more detail in Chapters IX and XVIII.

VII. Trace Elements

A. Introduction: Mean Solar-System Abundances

For many years the main motivation for trace-element studies of meteorites was the determination of abundances. This interest was first stimulated by GOLDSCHMIDT's (1937) monumental study of the silicate, metal and troilite phases of meteorites, which he thought were analogous to the mantle, the core, and a troilite shell above the core in the earth. A second impulse was provided by use of smoothed chondritic abundances to estimate mean solar-system abundances by SUESS and UREY (1956) and the interpretation of these abundances in terms of nucleosynthetic processes by BURBIDGE et al. (1957). Recent studies have used trace elements as diagnostic indicators of solar-nebula (e.g., LARIMER and ANDERS, 1967) or parent-body (e.g., SCHNETZLER and PHILPOTTS, 1969) fractionation processes, or as a means for defining genetically related groups of meteorites (e.g., SCHAUDY et al., 1972).

Reviews of chondritic trace-element data are given by UREY (1964) and LARIMER and ANDERS (1967). SCOTT (1972) reviewed trace-element concentrations in iron meteorites. A comprehensive review of abundances for all meteorite classes is provided by the articles in the book edited by MASON (1971 b).

In many cases significant insights into fractionation processes which have affected meteoritic material can be gained by normalizing elemental abundances in a fractionated class to those in a relatively unfractionated class. The most common choice for a normalization standard is the CI chondrites, since these appear to provide the best available estimate of mean solar system abundances (UREY, 1964; ARNOLD and SUESS, 1969; ANDERS, 1971 b). CI elemental abundances from MASON (1971 b) are listed in Table VII-1. Available data are given for all stable elements between atomic numbers 3 and 83 as well as for Th and U.

CI abundances for the highly volatile elements such as C and the rare gases are substantially lower than mean solar system values expected on the basis of astronomical observations and nucleosynthesis models (CAMERON, 1968). The CI abundances of such elements are italicized in Table VII-1 and are of limited use for the evaluation of fractionation processes.

Table VII-1. Elemental abundances in CI chondrites. These are thought to be approximations of mean solar-system abundances except for italicized values, which appear to be low as a result of incomplete retention of highly volatile elements. For some elements a second value is listed in boldface. These were obtained from Fig. VII-1 by interpolation and are thought to be better approximations of the mean solar-system abundances

Element	Abundance	Element	Abundance	Element	Abundance
^{3}Li	49.5	^{32}Ge	135[b]	^{56}Ba	4.8
^{4}Be	0.81[a]	^{33}As	6.6	^{57}La	0.43
^{5}B	140	^{34}Se	69	^{58}Ce	0.987
^{6}C	*$7.0 \cdot 10^{5}$*	^{35}Br	19[b]	^{59}Pr	0.147
^{7}N	*$4.9 \cdot 10^{4}$*		**15**	^{60}Nd	0.664
^{8}O	*$7.5 \cdot 10^{6}$*	^{36}Kr	*$1.9 \cdot 10^{-4}$*	^{62}Sm	0.212
^{9}F	2500[b]		**53**	^{63}Eu	0.0840
^{10}Ne	*0.0042*[b]	^{37}Rb	7.0	^{64}Gd	0.441
	$2.4 \cdot 10^{6}$[f]	^{38}Sr	27	^{65}Tb	0.0756
^{11}Na	6000	^{39}Y	4.8	^{66}Dy	0.298
^{12}Mg	$1.06 \cdot 10^{6}$	^{40}Zr	27[b]	^{67}Ho	0.0903
^{13}Al	$8.5 \cdot 10^{4}$	^{41}Nb	—	^{68}Er	0.225
^{14}Si	$1.0 \cdot 10^{6}$		**1.1**	^{69}Tm	0.0351
^{15}P	$1.0 \cdot 10^{4}$[b]	^{42}Mo	4.0	^{70}Yb	0.21
^{16}S	$5.0 \cdot 10^{5}$	^{44}Ru	1.9	^{71}Lu	0.0370
^{17}Cl	2500[b]	^{45}Rh	—	^{72}Hf	0.47
^{18}Ar	*0.0097*[c]		**0.30**		**0.30**
	$2.3 \cdot 10^{5}$[f]	^{46}Pd	1.3	^{73}Ta	0.019[a]
^{19}K	3500[b]	^{47}Ag	0.96	^{74}W	0.16[a]
^{20}Ca	$7.21 \cdot 10^{4}$		**0.72**	^{75}Re	0.052
^{21}Sc	31	^{48}Cd	2.4	^{76}Os	0.72
^{22}Ti	2400	^{49}In	0.20[b]	^{77}Ir	0.8[b]
^{23}V	254	^{50}Sn	3.6	^{78}Pt	1.2
^{24}Cr	$1.27 \cdot 10^{4}$	^{51}Sb	0.36	^{79}Au	0.19
^{25}Mn	9200		**2.9**	^{80}Hg	29[d]
^{26}Fe	$9.0 \cdot 10^{5}$	^{52}Te	5.8		**1.4**
^{27}Co	2300	^{53}I	1.0[b]	^{81}Tl	0.121
^{28}Ni	$4.7 \cdot 10^{4}$		**0.70**	^{82}Pb	4[e]
^{29}Cu	540	^{54}Xe	*$3.3 \cdot 10^{-4}$*	^{83}Bi	0.160
^{30}Zn	1600		**2.2**	^{90}Th	0.043
^{31}Ga	43[b]	^{55}Cs	0.38	^{92}U	0.013

[a] No CI analysis; CM abundance is tabulated.
[b] Arbitrary average of several values listed in MASON (1971b).
[c] Arbitrary average of data listed in MAZOR et al. (1970).
[d] The CI Hg values are much higher than expected from interpolation from neighboring elements. The interpolated value should be used in abundance comparisons.
[e] Includes Pb produced by decay of ^{232}Th, ^{235}U and ^{238}U.
[f] Ne and Ar abundances from CAMERON (1968).

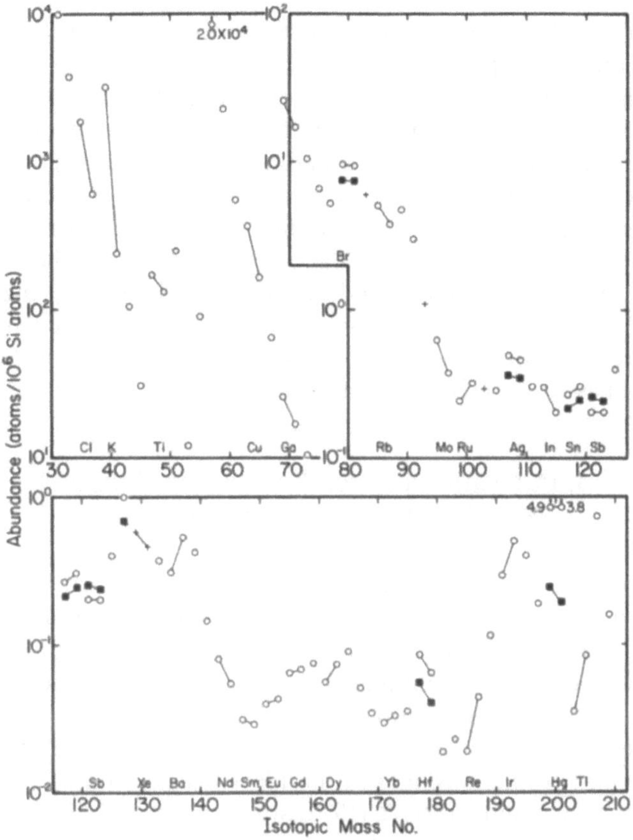

Fig. VII-1. Abundances of the odd-mass isotopes in CI chondrites (open circles; MASON, 1971b). Interpolated abundances (filled squares) are shown for those elements which have not been studied or which have CI abundances which appear to deviate from mean solar-system abundances

SUESS and UREY (1956) showed that the abundances of odd-mass-number nuclides tend to vary smoothly, and that this fact makes it possible to interpolate abundances with a fair accuracy (perhaps to ±30% if the elements on either side are well determined). Elements with two odd-mass isotopes are especially useful for evaluating the local slope of the curve. The smooth variation in abundance is illustrated by Fig. VII-1; all abundances represented by open circles are taken from MASON (1971b). This plot is the basis for interpolated abundances of the rare gases Kr and Xe, and of the unanalyzed elements Nb and Rh. Interpolated abundances are shown in boldface in Table VII-1. The variation in the curve is not well defined below mass 40, and Ar has no odd-mass isotopes; therefore Ne and Ar are taken from CAMERON (1968), and are based on other types of systematics.

A number of elements have proven very difficult to analyze reproducibly, and some volatile elements seem to be overabundant in CI chondrites for unknown reasons. Outstanding among the latter is Hg, for which the mean CI abundance is greater by a factor of 20 than the extrapolated value. Other elements in these categories are Br, Ag, Sn, Sb, I, and Hf; their interpolated abundances are listed in boldface in Table VII-1. Since both the reported and extrapolated abundances are plotted in Table VII-1, the reader can readily check the proposed changes. The interpolated abundances of these elements appear somewhat better for purposes of assessing cosmochemical fractionations.

B. Partition between Coexisting Phases; Studies of Rare-Earth Elements

A knowledge of the manner in which elements partition between phases can provide a powerful tool for the evaluation of igneous processes (e.g., GAST, 1971). Certain solar nebula processes (e.g., partial agglomeration) can also be investigated in this fashion if the applicable partition coefficients are known. For these reasons, an increasingly larger number of meteorite trace element investigations involve the analysis of separated fractions, and in the ideal case, separated phases. An example is CHOU and COHEN's (1973) study of Ga and Ge in magnetic (mainly metal) and non-magnetic (mainly silicate) fractions of L-group chondrites. They showed that Ga in the metal increased as evidence of recrystallization increased within their suite of meteorites, and concluded that Ga condensed chiefly from the nebula in oxidized form and moved into the metal during later metamorphic reheating. Another example is the study of GROSSMAN and CLARK (1972), who found the white, Ca, Al-rich inclusions of the Allende CV3 chondrite to contain high concentrations of refractory trace elements (see Section D). ALLEN and MASON (1973) and MASON and GRAHAM (1970) determined the concentrations of about 35 elements in eight major minerals separated from a series of meteorites and interpreted these in terms of the geochemical affinities of these elements.

The rare earths are the most important single group of elements for the evaluation of igneous formational histories. Their importance arises from the fact that their chemical properties are nearly identical except as they are affected by the monotonic decrease in ionic radius from 1.14 Å (La) to 0.85 Å (Lu). The normal rare-earth oxidation state is +3. The only one which is known to have an oxidation state other than +3 in

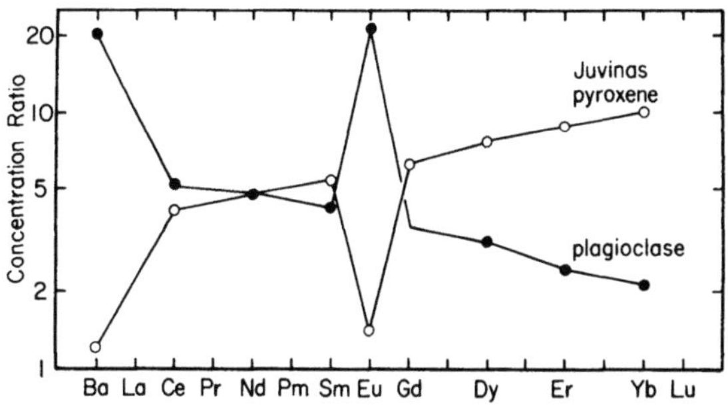

Fig. VII-2. Chondrite-normalized rare earth and Ba abundances in pyroxene and plagioclase phases of the Juvinas eucrite (SCHNETZLER and PHILPOTTS, 1969). Barium and Eu are concentrated in the plagioclase, the heavy rare earths in the pyroxene

meteoritic minerals is Eu, which is reduced to the +2 state in the reducing environments prevailing in most metal-bearing meteorites. Eu (II) tends to concentrate in plagioclase if this mineral is present. The differences in properties among the trivalent rare earths are generally not great enough to affect their gross distribution; they all tend to concentrate in the same minerals, such as phosphates or pigeonite. The differences are great enough to affect the detailed distribution between two phases with similar rare-earth affinities, however. Figure VII-2 shows the distribution of Ba and the rare earths between coexisting plagioclase and pyroxene (mainly pigeonite) in the Juvinas eucrite (SCHNETZLER and PHILPOTTS, 1969). All concentrations are normalized to chondritic concentrations; this eliminates the zig-zag pattern which otherwise results as a reflection of the fact that even-Z elements are more abundant than odd (MASUDA, 1962; CORYELL et al., 1963). We see that Ba and Ce (and presumably La) are concentrated in the plagioclase, and the other trivalent rare earths mainly in the pigeonite, with the pyroxene/plagioclase distribution coefficient increasing towards higher atomic numbers.

Figure VII-3 shows chondrite-normalized rare-earth and Ba abundances for members of several different differentiated silicate-rich meteorite groups as well as for a K-rich Apollo-11 basalt. I will discuss these briefly from bottom to top. The sample from the Brenham pallasite (MASUDA, 1968) consists mainly of olivine. Similar patterns have been observed in olivine from other meteorite samples (SCHNETZLER and BOTTINO, 1971; T.D. COOPER and R.A. SCHMITT, private communication). Distribution ratios favor other abundant silicate minerals over olivine, with the effect being most pronounced near the center of the spectrum. The data on the Shalka diogenite are from SCHMITT et al. (1963). The diogenites may be cogenetic with the eucrites and howardites (see, e.g.,

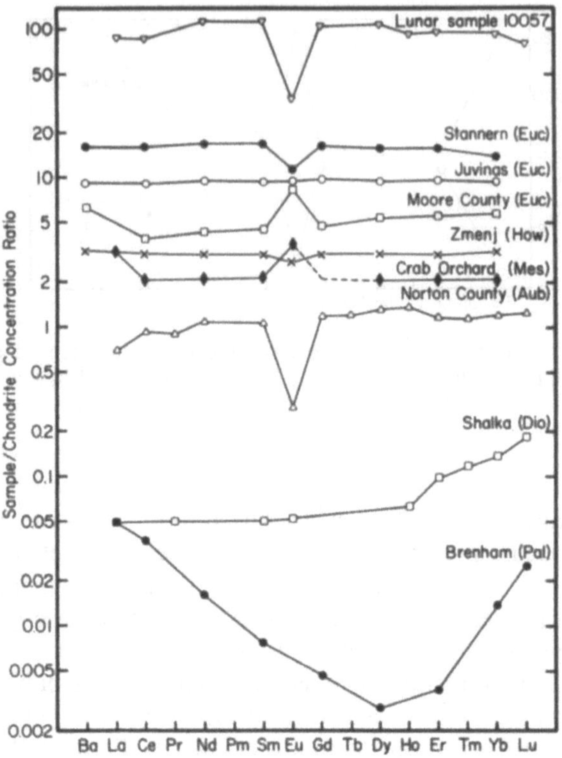

Fig. VII-3. Chondrite-normalized rare-earth abundance patterns in several differentiated silicate-rich meteorites and one lunar rock. Most of the patterns for rare earth-rich meteorites are nearly flat, except for Eu anomalies. The Norton County aubrite pattern resembles that of the pyroxene in Fig. VII-2. The Brenham pallasite sample consists mainly of olivine; meteoritic olivine tends to have very low abundances of all trace elements. See text for further details

JÉROME and GOLES, 1971; MCCARTHY et al., 1972). If we compare the Shalka data to those of the Juvinas eucrite (SCHNETZLER and PHILPOTTS, 1969), which are typical for that class, we find that 1. if the parental material had chondritic rare-earth abundances, then a mass balance indicates that there should be nine times as much Shalka-like as Juvinas-like material, and 2. the abundances of the heaviest rare earths in eucrites should decline slightly. Since the latter trend is not observed in Juvinas, and meteorite fall statistics favor eucrites over diogenites (Table II-8), it is clear that this simple two-component model does not account for the observations. Nonetheless, it illustrates the type of application of such data and does not rule out the possibility that the data could be accounted for with a more complex model.

The Norton County (aubrite) pattern (SCHMITT et al., 1963) is similar to that of the Juvinas pyroxene. Norton County consists mainly of low-Ca, high-Mg clinopyroxene (REID and COHEN, 1967). The resemblance to Juvinas pigeonite suggests that 1. the Norton County pyroxene has rare-earth affinities similar to those of "normal" Juvinas pigeonite, and 2. a plagioclase-like material has been lost by Norton County. This evidence is consistent with the proposal of WASSON and WAI (1970) that a partial agglomeration process in the solar nebula accounted for the fractionated composition of the aubrites. The Moore County eucrite is unusual in that it is unbrecciated and shows crystal layering (HESS and HENDERSON, 1949). SCHNETZLER and PHILPOTTS (1969) interpret the Moore County spectrum to indicate that Moore County is a cumulate of pigeonite and plagioclase from a eucritic magma. The accumulated plagioclase accounts for the positive Eu anomaly.

The Zmenj howardite (SCHNETZLER and PHILPOTTS, 1969) shows a minor Eu deficiency, but otherwise a remarkably flat distribution. The silicate fraction of the Crab Orchard mesosiderite (WÄNKE et al., 1972) shows a flat distribution with a positive Eu anomaly. The Juvinas (eucrite) and Stannern (eucrite) distributions (SCHNETZLER and PHILPOTTS, 1969) are flat, with a slight Eu deficiency observed in Stannern. The distribution in lunar sample 10057 (HASKIN et al., 1970) is also flat, but with a large Eu deficiency, wich apparently indicates loss of major amounts of plagioclase-rich material.

Flat distributions (Eu excepted), such as are observed in howardites, eucrites, mesosiderites and lunar basalts, are currently interpreted to result during magma formation by moderate-to-extensive partial melting of a source with a chondritic relative distribution of rare earths (e.g., GAST and MCCONNELL, 1972). Any phase subsequently lost by crystal settling, etc. either had a flat rare-earth distribution or, more likely, very low concentrations of rare earths other than Eu.

The high concentrations of rare earths in meteoritic and lunar basalt-like materials are remarkable; note that concentrations in Stannern are higher than chondritic levels by a factor of about 20, and in 10057 by a factor of about 100. There are a number of arguments indicating that such high concentrations are difficult to achieve during the formation of a basaltic magma from a source with chondritic concentration levels. For example, many workers now believe that the source material of the lunar basalts had concentrations of rare earths and other refractory elements which were at least five times greater than chondritic levels (see, e.g., GAST and MCCONNELL, 1972; RINGWOOD and ESSENE, 1970). Such materials may have formed by fractionation processes in the solar nebula or in orbit about the earth (RUSKOL, 1972).

C. Volatile Elements

A number of trace elements form stable compounds which are volatile at relatively low temperatures. Abundance patterns of these elements can often be interpreted in terms of temperature history. Thus, the low abundance of the rare gases in meteorites which appear to have formed igneously (e.g., eucrites, pallasites) is an indication that these materials have been much hotter than, say, a silicate inclusion in a group IA iron. A rough volatility scale for volatile elements can be obtained by dividing the abundances in type-6 ordinary chondrites by those in CI chondrites (WASSON, 1971c). The most volatile elements (other than the rare gases, halogens, and Hg) listed roughly in order of decreasing volatility are Tl, In, Bi, Pb, Cd, Zn, Te, Sn, S, and Ge.

It appears best to treat Hg as a special case. As noted previously, it is overabundant in CI chondrites by a factor of 20, and REED and JOVANOVIC (1967) have shown by heating experiments that a large fraction occupies very low-temperature sites; some of this is probably terrestrial contamination. The high-temperature Hg (that expelled above 450° C) in the CI chondrites is roughly that expected on the basis of interpolated elemental abundances. However, even this fraction is not systematically correlated with abundances of other volatile elements or with petrological temperature indicators. At present there seems to be no way to use Hg data to unravel the history of meteorite groups.

The halogens Cl, Br, and I are also volatile elements which are of limited usefulness because of the tendency of their data to scatter. Part of this scatter may result from water transport, since they tend to form soluble compounds. In contrast to Hg, the halogens do tend to be less concentrated in those chondrites showing greater evidence of metamorphism.

As discussed in Sections IIB and VIC, VAN SCHMUS and WOOD (1967) divided the chondrites into petrologic types on the basis of microscopic and chemical evidence (e.g., degree of recrystallization increases with increasing type number). A number of volatile elements are inversely correlated with petrologic type. The most extensive studies reported are on the ordinary chondrites, in which the concentrations of B (QUIJANO-RICO and WÄNKE, 1969), C (MOORE and LEWIS, 1967; OTTING and ZÄHRINGER, 1967), In (SCHMITT and SMITH, 1968; TANDON and WASSON, 1968; KEAYS et al., 1971; LAUL et al., 1973; Tl, Bi (CASE et al., 1972; LAUL et al., 1973), ^{204}Pb (see LARIMER, 1973) and the rare gases (ZÄHRINGER, 1966a; MARTI, 1967a) all are negatively correlated with petrologic type.

Figure VII-4 shows a plot of ^{132}Xe versus In in L-group chondrites. Indium data are from TANDON and WASSON (1968) and KEAYS et al.

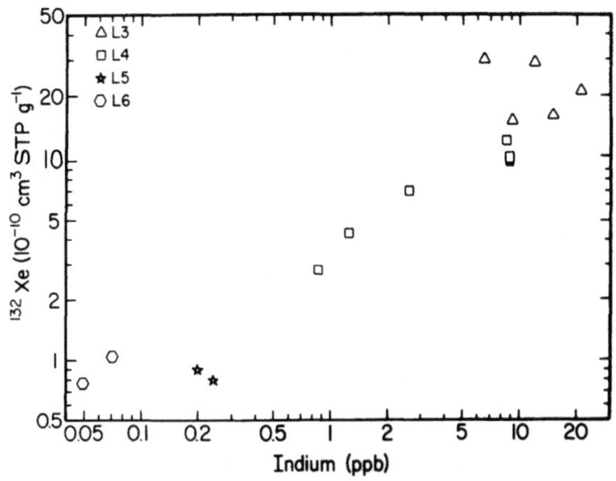

Fig. VII-4. Plot of ^{132}Xe versus In for L-group chondrites of the four different petrologic types. Despite the fact that the geochemical properties of these elements are greatly different, they show very similar degrees of fractionation. The highest abundances are in type 3, which shows the least petrographic evidence of metamorphism

(1971); ^{132}Xe data are from references cited in these papers. The petrologic type is given by the number of corners on the symbol. The correlation of ^{132}Xe and In with each other and their anticorrelation with petrologic type is strong. There is, however, essentially no difference in ^{132}Xe concentration between petrologic types L 5 and L 6, even when the more extensive data of ZÄHRINGER (1968) are compared.

There seems to be no doubt that the volatile element fractionations between different petrologic types of the ordinary chondrites are related to differences in the thermal history of the material. As discussed in Section XVIII E, the major debate is whether the fractionation reflects differing degrees of condensation of nebular material (LARIMER and ANDERS, 1967; BLANDER and ABDEL-GAWAD, 1969; LAUL et al., 1973; LARIMER, 1973) or whether it resulted from later metamorphism in the parent body (DODD, 1969) or in planetesimals (WASSON, 1972a).

In the light-dark meteorites there is some tendency for volatile elements to be more concentrated in the dark portions when these are also enriched in solar-type rare gases (see Chapter VIII). For example, REED (1963) observed Bi concentrations in (H-group) Pantar dark material which were about three times higher than those in Pantar light, and RIEDER and WÄNKE (1969) found In concentrations 9—15 times higher in the dark portions of five light-dark H-group chondrites. On the other hand, BEGEMANN and HEINZINGER (1969) found that the concentration of C was only about 10—20% higher in the dark portions of two H-group light-dark chondrites. LAUL et al. (1972) found enrichments by factors of 2 to 4 in In, Bi, Cd, Br, Cs, and Ag in the dark portions of the light-dark howardite Kapoeta and in Bi, Zn, and Ag in the dark portions ot the light-dark aubrite Pesyanoe. It is generally believed that these differences in volatile element content reflect the presence of an additional volatile-rich component in the dark portions of the light-dark meteorites. This component may be solar wind or interplanetary dust or both. Currently available data show too much scatter to allow the definition of the composition of this material.

D. Refractory Elements

Attention was drawn by LARIMER and ANDERS (1970) to the meteoritic evidence for a nebular fractionation of refractory trace elements. As discussed in Section VI D, the Ca-Al-rich inclusions in Allende have the mineralogy expected of high-temperature nebular condensates or residues, and such an origin appears definite as a result of the pioneering trace-element studies of GROSSMAN (1973). GROSSMAN showed (and others have confirmed) that elements which have calculated nebular condensation temperatures higher than those of the most refractory major minerals Fe-Ni and olivine are 10—25 times more concentrated in Ca-Al-rich inclusions than in CI chondrites. Particularly interesting was the enrichment of Ir, a strongly siderophilic element, in the nearly metal-free white inclusions. The results of multi-element investigations by WÄNKE et al. (1973) are listed in Table VII-2. Again, note that a number of strongly siderophilic elements (e.g., Re, Os, Ir, Pt) have high inclusion/CI ratios even though Fe, Co, and Ni have very low ratios.

CHOU et al. (1973) studied the distribution of Ga, Ge, Ni and Ir between magnetic and non-magnetic portions of H-group chondrites and showed that the metal/silicate concentration ratios of Ge and Ni ranged from 200 to 1100, while those for the refractory element Ir never exceeded 60. They interpreted these data to mean than an appreciable portion of the Ir condensed from the nebula prior to the condensation of Fe and Ni, and that some of this Ir was trapped in silicate lattices, and therefore did not enter the metal during metamorphic recrystallization.

In the enstatite chondrites the refractory siderophiles Re, Os (MORGAN and LOVERING, 1967) and Ir (EHMANN et al., 1970; BAEDECKER and WASSON, 1974) are about equally abundant in all types despite the fact that the abundances of metallic Fe, Co, and Ni are about 1.5 times greater in E4,5 than in E6 chondrites. EHMANN and BAEDECKER (1968) suggested that this was related to the high condensation temperatures of these elements. BAEDECKER and WASSON

Table VII-2. Concentration ratios, Allende Ca-Al-rich inclusion/CI chondrite (WÄNKE et al., 1973). Elements which appear to be refractory and "nonrefractory" based on these data are listed separately in order of increasing atomic number

Refractory elements	Ratio	Non-refractory elements	Ratio
Al	21	Mg	0.68
Ca	17	Si	1.4
Sc	24	Cr	0.16
Ti	21	Mn	0.040
Sr	15	Fe	0.046
Y	21	Co	0.08
Zr	23	Ni	0.08
Nb	22	Cu	0.06
Ru	12	Zn	0.27
Ba	20	Ga	0.075
La	26	Ge	0.66
Eu	23	Pd	0.11
Yb	22	Au	1.8
Hf	23		
Ta	15		
W	20		
Re	21		
Os	19		
Ir	17		
Pt	14		
U	13		

(1974) present new enstatite chondrite data and a review of older data, and suggest that at least half of these siderophiles was present in the silicate-rich component at the time of the metal-silicate fractionation which is recorded in the enstatite chondrites.

E. Siderophilic Elements

The siderophilic elements are those which are normally concentrated in the metallic phases of meteorites. The metallic phase generally consists of either the kamacite or taenite Fe-Ni alloys. Thus, the siderophilic elements are those which are more easily reduced to the metal than Fe. Roughly in order of decreasing siderophilic behavior, these include Pt, Os, Ir, Pd, Ru, Rh, Ni, Re, Co, and Ge. Some additional elements (e.g., Fe, Ga) have important siderophilic tendencies, even though in some meteorites the major fraction exists in other oxidation states.

The most extensive siderophilic element studies to date have been those of Ni, Ga, Ge, and Ir to characterize the iron meteorite groups (see Section II C and Appendix II D). A surprising result of these studies was the discovery that concentrations of some of these elements vary over large ranges within the iron meteorites as a whole (Ge from 0.005 ppm in Nedagolla to 2000 ppm in Butler), and in some cases, within single groups of irons (Ir from 0.01 ppm in the high-Ni extreme of group II AB to 60 ppm in the low-Ni extreme). As shown by SCOTT's (1972) review, with rare exceptions, all elements are intercorrelated within the iron-meteorite groups. In addition to the data tabulated in Appendix II D, comprehensive trace element studies have been reported by GOLDBERG et al. (1951), LOVERING et al. (1957), COBB (1967), SMALES et al. (1967), FOUCHÉ and SMALES (1966), and CROCKET (1972). SCOTT (1972; SCOTT and BILD, 1974) argues that the fractionation trends in all iron-meteorite groups except IAB and IIICD are consistent with fractionation by fractional crystallization of metallic magmas.

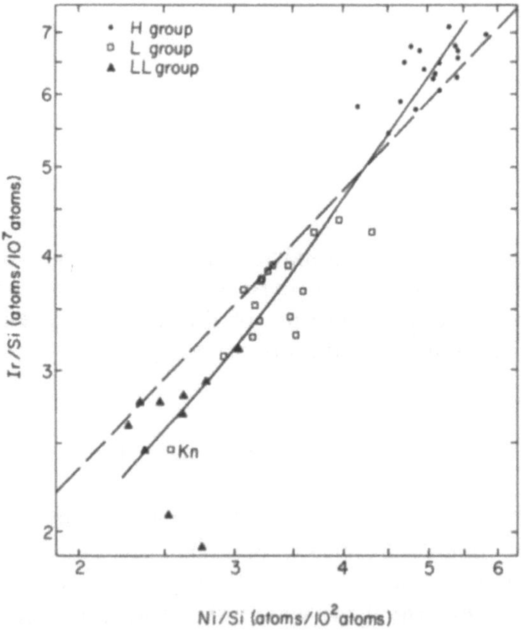

Fig. VII-5. Plot of Ir abundance versus Ni abundance in the ordinary chondrites. Although various lines of evidence indicate that the three ordinary chondrite groups are closely related, the Ir/Ni ratio is distinctly higher in the H group than in the other two groups. The dashed line corresponds to a constant Ir/Ni atom ratio of 1.18×10^{-5}

Past studies of siderophilic elements in chondritic meteorites were often concerned with the testing of PRIOR'S (1916, 1920) rules: "the less the amount of nickel-iron in meteoritic stones, the richer it is in nickel, and the richer in iron are the magnesium silicates" (see discussion in Section V D). PRIOR suggested that the meteorites originated from the progressive oxidation of a common parental magma. Calculations, however, show that even finely-divided metal settles out of a liquid magma in a short time (less than a year, even in km-sized parent bodies—ANDERS, 1964), and no one has found a good way to oxidize a magma *in situ*. As a result, interest in PRIOR'S rules has waned.

The most important siderophilic-element problems today are the definition of trends within chemical groups and the resolution of intergroup differences in the relative abundances of these metals. Studies of siderophilic elements in chondritic groups have been made by GREENLAND and LOVERING (1965), FOUCHÉ and SMALES (1966), EHMANN et al. (1970) and LARIMER and ANDERS (1970). MÜLLER et al. (1971) made a careful study of the variations of Ni and Ir within the three groups of ordinary chondrites. Their abundance data are plotted in Fig. VII-5. The Ir/Ni ratio is distinctly higher in the H-group than in the other two groups. The detailed implications of this result are discussed by WASSON (1972a). The basic conclusion is that there appears to have been not only a metal-silicate fractionation among the ordinary chondrite groups, but also a fractionation of one siderophilic element from another. As discussed further in Section XVIII H, the refractory nature of Ir has played an important role in this fractionation.

F. Miscellaneous Elements

It was earlier thought that chalcophilic elements (those which are enriched in sulfide phases, especially FeS) were fractionated because of their ability to form strong bonds with S (FISH et al., 1960). Further data showed that volatility is more important than bonding ability to any particular element (LARIMER, 1967). It now appears that the term "chalcophilic" could be dropped without harm; it has no mnemonic value, and FeS is simply one of many meteoritic phases which must be studied if the partition of trace elements is to be properly understood.

The triad of rare light elements Li, Be and B is of interest because they are burned at low temperatures in the interiors of stars (BURBIDGE et al., 1957). This plus the suggestion that they were created by immense solar-particle storms in the solar nebula (FOWLER et al., 1962; BERNAS et al., 1967) stimulated a number of studies of these elements. A surprising result is that chondritic Be concentrations reported in meteorites (SILL and WILLIS, 1962) are at least a factor of 3 lower than those observed in the earth's crust or predicted by nucleosynthesis models. Current evidence indicates that the last burst of nucleosynthesis recorded in the meteorites occured in the galaxy, not in the solar nebula (REEVES, 1972b), and the investigation of these elements will probably be less intensive during the next few years.

VIII. Stable Isotopes: Elements Other than Rare Gases

Probably the single most important conclusion resulting from the study of the isotopic composition of relatively nonvolatile meteorite elements is that the meteorites are samples of normal solar system material. This conclusion is based on the fact that, as of now, all nonvolatile elements have the same isotopic compositions in meteorites as in terrestrial and lunar samples. In fact, the recent history of isotopic investigations of meteorites is a story of anomalies discovered and anomalies lost. For a period of about five years starting in 1962 various elements in certain meteorites were reported to have isotopic compositions which differed from those found in terrestrial samples. Later, more painstaking research has invariably shown the reverse. Detailed reviews are given by REYNOLDS (1967) and VOSHAGE (1968a, b). Since the publication of these papers, previously reported anomalies have been disproved for the elements Li (GRADSZTAJN and GUEZ, 1970), Mg (SCHRAMM et al., 1970), and Ba (EUGSTER et al., 1969). Elements having identical isotopic compositions in meteoritic and terrestrial samples include Cr (MURTHY and SANDOVAL, 1965), Ga (DE LAETER, 1972), Mo (WETHERILL, 1964) and Sn (DE LAETER and JEFFERY, 1967).

Isotopic fractionations of the volatile light elements and rare gases are commonly observed, but are explained in terms of chemical equilibria or nuclear processes occurring *in situ*. The remaining exotic isotopic anomalies are in Hg (REED and JOVANOVIC, 1969) and O (CLAYTON et al., 1973). The latter is particularly intriguing. Whereas the O isotopic compositions of hydrated silicates in CM, CV, and CO chondrites fall near the trend observed for chemical fractionation processes in terrestrial samples, the high-temperature minerals form a trend interpreted as a mixing line between a component with earth-like composition and one with extremely low contents of ^{17}O and ^{18}O. CLAYTON et al. (1973) suggest that this latter component might be interstellar dust. It will be of great interest to see if the same minerals also yield Mg isotopic anomalies, as might be expected on the basis of nucleosynthesis systematics.

Many of the isotopic investigations of the last decade were stimulated by the theoretical paper of FOWLER et al. (1962; later modified by BURNETT et al., 1965) in which appreciable isotopic effects were predicted to have resulted from irradiation by an intense flux of energetic solar charged particles very early in the history of the solar system. BERNAS et al. (1967) proposed a variant of this model. The absence of detectable isotopic anomalies in Li (KRANKOWSKY and MÜLLER, 1967; DEWS, 1966; GRADSZTAJN and GUEZ, 1970), V (PELLY et al., 1970), K (BURNETT et al., 1966) and certain rare-earth elements (MURTHY and SCHMITT, 1963) indicate either that such an intense irradiation did not occur, or that solar system material was thoroughly mixed following the irradiation.

Other isotopic investigations were undertaken in order to search for anomalies produced by the decay of primordial radionuclides. The absence of anomalies in Mg (SCHRAMM et al., 1970), Ag (CHAKRABURTTY et al., 1964; DEWS and NEWBURY, 1966), and Tl (ANDERS and STEVENS, 1960; OSTIC et al., 1969; HUEY and KOHMAN, 1972) indicates that only very small amounts of "extinct" ^{26}Al, ^{107}Pd, and ^{205}Pb were present in the investigated meteorites at the time they cooled to a temperature too low to allow isotopic exchange between minerals, and that substantial improvements in mass-spectrometric techniques must be made before such anomalies can be detected. Careful studies of Xe confirm the former presence of extinct ^{129}I, ^{244}Pu, and possibly a superheavy nuclide in certain meteorites. More details about such investigations as well as on isotopic changes resulting from the decay of long-lived primordial radionuclides are given in Chapter X.

Cosmic-ray spallation-type interactions can produce substantial isotopic changes in elements in very low abundance in meteoritic materials (such as Ca in iron meteorites—STAUFFER and HONDA, 1962; HINTENBERGER et al., 1965a) or can produce neutrons which interact to reduce the isotopic abundance of isotopes having very high neutron-capture cross sections (such as Gd—EUGSTER et al., 1970). A more thorough discussion of cosmic ray effects is given in Chapter XI.

Interest in the isotopic composition of H in meteorites centers on two points: 1. the D/H ratio in primitive solar system material, and 2. information regarding the chemical processes involved in the formation of H-containing compounds. The only isotopic study of meteoritic H appears to be that of BOATO (1954), who applied his data only to point 1. He found that most carbonaceous chondrites contained H with a D/H atomic ratio similar to that observed on earth (about 160 ppm). For this reason it was difficult for him to resolve meteoritic H_2O from that caused by terrestrial contamination. However, the CI 1 chondrites Ivuna and Orgueil, the CV 2 chondrite Mokoia, and the CM 2 chondrite Murray all contain H_2O with D/H ratios distinctly higher (by as much as 36%) than any values observed in natural waters at the surface of the earth; an appreciable fraction of their H_2O must therefore be extraterrestrial.

BOATO noted that the observed enrichments could easily occur in nebular processes and concluded that the D/H ratio in primordial solar system material was similar to that currently observed on the surface of the earth.

The classification of carbonaceous chondrites discussed in Section II B is supported by the data of BOATO (1954). The CI chondrites have D/H ratios distinctly higher than those of all other carbonaceous chondrites. The CV2 chondrite Mokoia has a D/H ratio which is much higher and a $^{13}C/^{12}C$ ratio which is much lower than the ratios observed in the CM chondrites. Interestingly enough, the CM2 chondrite Murray also has a D/H ratio considerably higher than those observed in other CM2 chondrites.

Study of the isotopic composition of meteoritic C has brought to light the remarkable fact that the $^{13}C/^{12}C$ ratio in the carbonate component of carbonaceous chondrites is about 7% higher than that found in reduced forms of C (CLAYTON, 1963; KROUSE and MODZELESKI, 1970; SMITH and KAPLAN, 1970). Expressed in terms of $\delta^{13}C$[5], the carbonate in CI chondrites has a composition of $+60^0/_{00}$, much higher than the highest values observed in natural C-containing samples on earth [about $+5^0/_{00}$ (CLAYTON, 1963)]. Terrestrial biological systems containing both reduced and carbonate C also show enrichments of ^{13}C in the carbonate, but the maximum known enrichments are about a factor of 3 lower than those in the CI chondrites and a factor of 2 lower than those observed in the CM chondrites. If the ^{13}C enrichment is produced by equilibrium reactions, these must have taken place at very low temperatures, of the order of 300 °K (UREY, 1947). A similar enrichment can also be produced by kinetic effects in low-temperature reactions which do not go to equilibrium. For example, LANCET and ANDERS (1970) showed that the Fischer-Tropsch process can produce such effects at a temperature of 400 °K. Unfortunately, it is not possible to prove that the different C-containing compounds in these meteorites are cogenetic. They may have had different origins, and their present juxtaposition may have been caused by mechanical mixing processes rather than by chemical processes (KROUSE and MODZELESKI, 1970).

Differences between the different carbonaceous chondrite groups can be seen in the data of SMITH and KAPLAN (1970). The CM chondrites have consistently lower $\delta^{13}C$ values in the carbonate component than do the CI chondrites; the CV chondrite Mokoia does not contain detectable carbonate. Their data also suggest that Murray is anomalous relative to

---

5 $\delta^{13}C$ (in units of $^0/_{00}$) is defined as follows:

$$\delta^{13}C = 1000((^{13}C/^{12}C)_{sample}/(^{13}C/^{12}C)_{standard} - 1).$$

The standard in all these studies is the PDB carbonate (CRAIG, 1957).

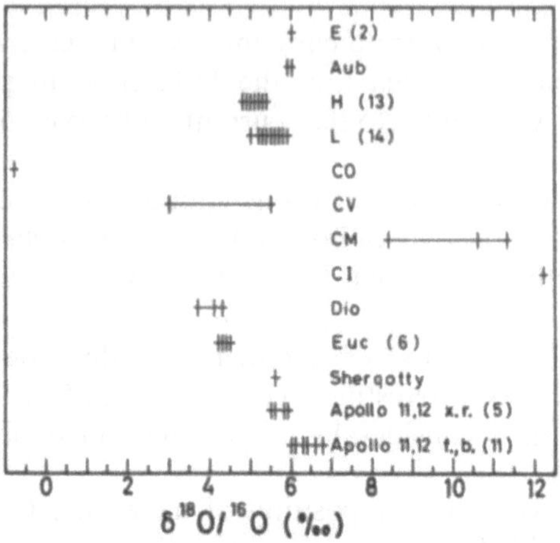

Fig. VIII-1. Comparison of $\delta^{18}O$ compositions of meteorites and lunar samples. The number in parentheses is the total number of individual values for those cases in which these exceed the number shown on the graph. The abbreviations x.r. and f., b. refer to lunar crystalline rocks, and lunar fines and breccias, respectively

the other CM chondrites. KROUSE and MODZELESKI (1970) report carbonate in Felix, and a difference of 20‰ in the bulk $\delta^{13}C$ values of Felix and the other CO chondrite, Lancé. These facts indicate a need for more detailed studies of the C isotopic composition of Ornans-group chondrites.

Data on the O isotopic composition of meteorites are especially useful for the classification of meteorites. Bulk compositional data (expressed in $\delta^{18}O$, which is defined in analogous fashion to $\delta^{13}C$, except that the standard is SMOW, standard mean ocean water) given by TAYLOR et al. (1965) and REUTER et al. (1965) are shown in Fig. VIII-1, together with data on lunar samples reported by EPSTEIN and TAYLOR (1971).

The bulk $\delta^{18}O$ values for each meteorite group fall within rather small ranges, with the exception of those for the carbonaceous chondrites, where the ranges are greater (and the data scarce). Among the items of special note are: 1. The enstatite chondrites and the aubrites have identical O-isotope compositions. 2. Although there is considerable overlap between the H and L chondrite ranges, there seems to be a trend of increasing $\delta^{18}O$ as one proceeds from H to L. ONUMA et al. (1972a) reported $\delta^{18}O$ data on separated minerals which indicate slightly higher bulk values in LL chondrites relative to L. 3. Not only are the ranges within the carbonaceous chondrite groups large, there are also some remarkably large separations between groups—e.g., between the CO and CV chondrites. One would like to see more extensive bulk $\delta^{18}O$ data on the carbonaceous chondrites (ONUMA et al. (1972b) reported data on separated phases, but no bulk data). 4. The diogenites, eucrites, howardites and mesosiderites have $\delta^{18}O$ ranges which overlap (for the latter two groups see Table II-5), suggesting that the different groups may be related. 5. The $\delta^{18}O$ samples in lunar crystalline rocks from the first two landing sites are very similar, suggesting that bulk $\delta^{18}O$ for lunar samples of basaltic composition is similar everywhere on the moon. This makes it very unlikely that the Ca-rich achondrites are of lunar origin. 6. The Shergotty meteorite, which is classified by some authorities as a eucrite, has clearly had an independent origin. The $\delta^{18}O$ data indicate that it could be from the moon. 7. The lunar fines and breccias have higher $\delta^{18}O$ values than the basaltic rocks from the same landing sites. Apparently the exotic components (e.g., anorthosite, norite) in these samples have very high $\delta^{18}O$ values which account for this difference, or solar-wind H has preferentially stripped away O as H_2O from these samples.

The distribution of the oxygen isotopes between two coexisting phases can be used to estimate an equilibration temperature for an igneous or metamorphic rock (see also Section V C). For example, CLAYTON et al. (1971) used the plagioclase-ilmenite pair to estimate equilibration temperatures of about 1440 °K for Apollo 12 basalts. ONUMA et al. (1972a) applied this technique to type-5 and type-6 ordinary chondrites, for which they report equilibration temperatures of about 950—1200 °K. These may be metamorphic temperatures.

ONUMA et al. (1972b) assumed a mean $^{18}O/^{16}O$ ratio for the solar nebula, and on this basis calculated nebular condensation temperatures of 800—900 °K for chondrules from carbonaceous chondrites and temperatures of 728—743 °K for chondrules from ordinary chondrites (see Chapter XVIII). This interpretation has recently been withdrawn on the basis of evidence for incomplete isotopic equilibration (ONUMA et al., 1974).

TAYLOR et al. (1965) pointed out that terrestrial basaltic magmas have $\delta^{18}O$ values 1 to 2‰ higher than ultramafic rocks thought to represent good approximations of the upper-mantle parent materials. It seems reasonably likely that the direction of this fractionation will be the same under all planetary settings, since it is known that minerals with lower melting points tend to have higher $\delta^{18}O$ values than the more refractory coexisting phases. Since the "basaltic" eucrites, howardites, and mesosiderites have lower $\delta^{18}O$ values than the enstatite and ordinary chondrites, it seems definite that the latter were not the parent material for the former.

The bulk $^{34}S/^{32}S$ ratios in meteorites of many different classes are constant to within about 1‰ (KAPLAN and HULSTON, 1966; MONSTER et al., 1965; HULSTON and THODE, 1965). However, within the CI chondrites a measurable fractionation of S isotopes between phases is observed. The two major S-bearing components, free S and water-soluble sulfates, have $\delta^{34}S$ values (relative to Canyon Diablo troilite) of about +1 and −2‰, respectively. If these phases are cognetic, this relatively large distribution coefficient indicates that they were formed at low temperatures, of the order of 300 °K (LEWIS and KROUSE, 1969; MONSTER et al., 1965).

IX. Stable Isotopes of the Rare-Gas Elements and Related Particle-Track Studies

The rare gases are a chemically coherent group of very inert elements. Their inertness makes it possible to separate them from other elements by vacuum techniques, and makes them ideal for study in gas-source mass spectrometers. These techniques have been applied to many meteorites, and a large amount of high quality data has resulted. The data have bearing on various meteoritic properties. This section contains a discussion of information provided by the primordial rare gases: i.e., those which have not resulted from nuclear processes occurring within the meteoritic material. Radiogenic isotopes are discussed in Chapter X, and cosmogenic isotopes in Chapter XI. Reviews of rare-gas studies of meteorites are given by PEPIN and SIGNER (1965) and REYNOLDS (1967).

Closely related to the study of certain types of rare gases are track studies of certain meteoritic minerals. When energetic charged particles pass through solids, they leave behind a trail of radiation-damaged material. Etching increases the size of these tracks and allows them to be observed by optical microscopy. A review of studies of tracks produced in meteorites by solar and galactic cosmic rays is given by LAL (1969).

Figure IX-1 shows a comparison of the rare-gas elemental abundance patterns for several meteorites which are typical of their classes. Abundance patterns are also shown for the atmospheres of the earth and the sun[6]. One abundant isotope is taken as representative of each element. The cosmogenic contribution has been estimated and removed. The Lancé and Pesyanoe ^{4}He values may still include appreciable radiogenic contributions, but these are negligible for the other meteorites. All abundances are normalized to ^{36}Ar, which is set equal to 10^{0}. The meteoritic data are taken from the following sources: Bremervörde (HEYMANN and MAZOR, 1968); Fayetteville (various sources summarized in PEPIN and SIGNER, 1965); Murray and Lancé

---

[6] No abundance is given for terrestrial He, since most primordial He has escaped during geologic time. That currently in the atmosphere is produced by radioactive decay. The solar abundance pattern is based on $^{4}He/^{20}Ne$ and $^{20}Ne/^{36}Ar$ ratios of 520 and 37 in the solar wind measured by GEISS et al. (1970, 1971); a solar-wind $^{36}Ar/^{84}Kr$ ratio of 2500 indicated by the Pesyanoe results of MARTI (1969) and lunar soil data of HEYMANN and YANIV (1970); and a $^{84}Kr/^{132}Xe$ ratio of 20 as proposed by CAMERON (1968), obtained by interpolation between abundances of neighboring elements in CI chondrites.

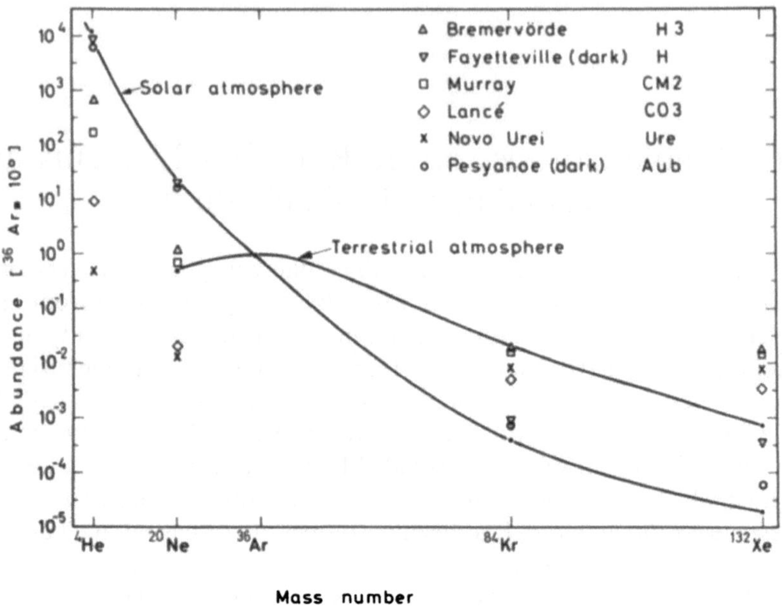

Fig. IX-1. Comparison of the abundances of major isotopes of the five rare-gas elements in meteorites of various classes and in the terrestrial and solar atmospheres

(various sources summarized by MAZOR et al., 1970), Novo Urei (MÜLLER and ZÄHRINGER, 1969); Pesyanoe (MARTI, 1969).

About 20—40% of the meteorites which are genomict breccias are found to have high rare-gas concentrations in their host portions, which are generally somewhat darker than the xenoliths enclosed by the host. The dark portions of two such light-dark meteorites, the H chondrite Fayetteville and the aubrite Pesyanoe, have rare gas/^{36}Ar ratios which, with the exception of Xe, are within a factor of 2 of those believed to be present in the atmosphere of the sun. SIGNER and SUESS (1963) called rare gases showing this distribution *solar-type* rare gases. The other meteorites show a different rare-gas distribution, in which ^{4}He/^{36}Ar and ^{20}Ne/^{36}Ar ratios are more than an order of magnitude lower than the solar ratios, the ^{84}Kr/^{36}Ar ratio more than one order of magnitude higher, and the ^{132}Xe/^{36}Ar ratio more than two orders of magnitude higher than the solar ratio. SIGNER and SUESS (1963) called rare gases with this distribution pattern *planetary-type* rare gases because of their rough resemblance to the distribution found in the terrestrial atmosphere (where only the ^{132}Xe/^{36}Ar ratio is outside the meteoritic range for this type of gas). Although the SIGNER-SUESS nomenclature is most commonly used, a confusing abundance of synonyms is found in the rare-gas literature[7].

SUESS et al. (1964), SIGNER (1964), and WÄNKE (1965) proposed that the solar-type gases were incorporated into the meteoritic material as a result of irradiation by the solar wind. Etching experiments by EBERHARDT et al. (1965a) and HINTENBERGER et al. (1965b) and microprobe release profiles of ^{4}He by ZÄHRINGER (1966b) showed that the gases are concentrated towards the surfaces of grains, lending support to this hypothesis. No comparable experiments regarding the siting of the planetary gases have been performed.

PELLAS et al. (1969) and LAL and RAJAN (1969) studied the particle-track record in individual mineral grains from the dark portions of gas-rich meteorites and found that many of these grains were also rich in tracks produced by energetic particles, with track density decreasing with distance from the surface of the grain. These observations were confirmed and extended by WILKENING et al. (1971), who showed that track-rich grains have the same composition as other grains in the meteorite and are distributed randomly within the gas-rich portions. Figure IX-2 is taken from the WILKENING et al. (1971) paper. The track density falls by nearly a factor of 10 from the edge of the crystal to its center. The particles which produced these tracks are thought to be solar-flare accelerated nuclei with an atomic number of about 26 ± 4, whereas the solar-type rare gas was probably of solar-wind origin.

All meteorites rich in solar rare gases seem to contain planetary-type gases as well. For example, the high Xe content of Fayetteville (dark) is probably attributable to a planetary component.Some planetary gas-rich meteorites also contain solar He and Ne components; Bremervörde

---

[7] The *solar* rare-gas component is called "light primordial rare-gases" by several authors (e.g., MÜLLER and ZÄHRINGER, 1969), "unfractionated rare-gases" by others (e.g., ANDERS, 1964), and meteorites containing appreciable amounts of this component are often called "gas-rich meteorites" (e.g., REYNOLDS, 1967). Because there is evidence that this component is surface-correlated, ARRHENIUS and ALFVÉN (1971) prefer the designation "implanted component". The *planetary* component is referred to as "heavy rare gases" by some authors (e.g., MÜLLER and ZÄHRINGER, 1969), "fractionated rare gases" by others (e.g., ANDERS, 1964) and meteorites containing larger amounts of this component are called "xenon-rich meteorites" by REYNOLDS (1967). ARRHENIUS and ALFVÉN (1971) call this component the "occluded" or "trapped" component, based on the assumption that it is not surface-correlated. Most other authors (e.g., MARTI, 1967b) use *trapped* as a synonym for *primordial*, to distinguish those gases which are present as a result of chemical or physical processes from those resulting from nuclear processes. I find the terms *solar* and *planetary* the best choices for the following reasons: 1. they are mnemonically useful; 2. they are historically established; and 3. they are reasonably vague—a meteorite researcher can easily remember that they are not necessarily identical to the atmosphere of the sun or of any particular planet.

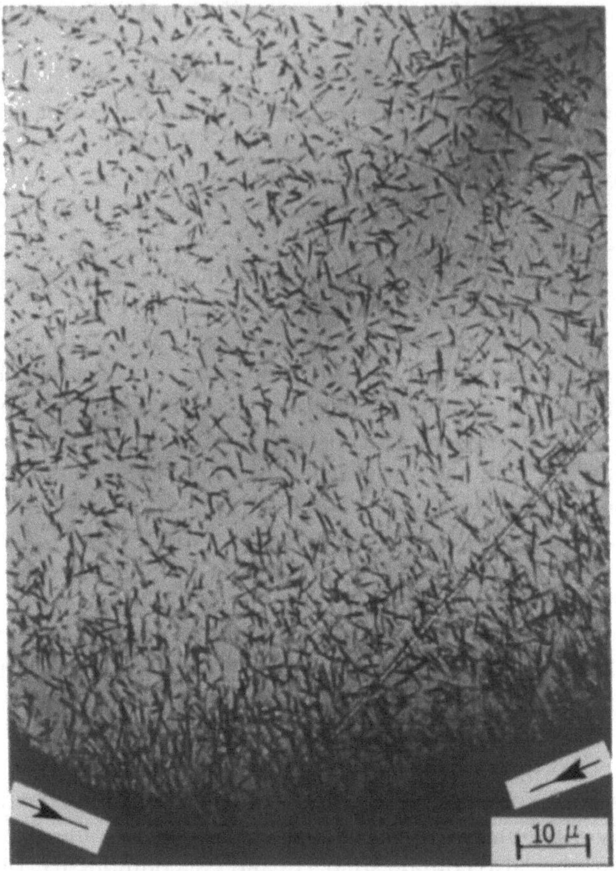

Fig. IX-2. Etched pyroxene crystal from Kapoeta howardite showing particle-track density increasing toward the surface. The location of the original surface prior to etching is shown by the arrows. (Photo by L. WILKENING)

and Murray seem to fall in this category. As will be discussed later, they have Ne isotopic compositions which are intermediate between those associated with solar and planetary rare gases. Other meteorites, such as Lancé, Novo Urei, and most ordinary chondrites, show no observable solar rare-gas component.

The origin of planetary-type rare gases is still one of the major unsolved problems of meteorite research. Since the solar-type distribution seems roughly representative for the major parts of the primitive solar system, it is likely that the planetary-type gases were produced from solar-type gases by some process or combination of processes. Perhaps the planetary gas distribution is the result of diffusive escape from mineral phases (ZÄHRINGER, 1962), other mass-fractionation mechanisms (KURODA and MANUEL, 1970), or selective adsorption effects (MAZOR et al., 1970; WASSON, 1972a); FANALE and CANNON (1972) have carried out adsorption experiments which indicate that a planetary-type elemental fractionation is produced. They find a seemingly unrealistically low temperature of 113 °K necessary to explain the amounts of trapped gas, however. LANCET and ANDERS (1973) measured the solubility of rare gases in magnetite and argue that equilibrium solubility is able to produce the planetary-type pattern. Some evidence, such as the variability in $^{84}Kr/^{36}Ar$ ratios despite a near constancy in $^{132}Xe/^{36}Kr$ ratios, suggests that planetary-type rare gas varied in composition depending on the conditions prevailing during the formation of the different groups of meteorites. As pointed out by MAZOR et al. (1970), variation within a single group is much smaller by comparison.

Planetary rare-gas concentrations generally increase with decreasing petrologic grade among the ordinary chondrites and possibly within other chondrite groups as well (ZÄHRINGER, 1966a; MARTI, 1967a; HEYMANN and MAZOR, 1968). Certain highly volatile elements (especially In) show parallel variations and are strongly correlated with the concentrations of primordial ^{36}Ar and ^{132}Xe (TANDON and WASSON, 1968; KEAYS et al., 1971). These variations provide important if poorly understood clues to the origin of the ordinary chondrites (see, e.g., ANDERS, 1971a; WASSON, 1972a).

Evidence of isotopic fractionations between different meteorites or between different phases of the same meteorite are to be found for all rare-gas elements. The largest well-defined effects are found in Ne, where solar, "planetary" (component A of PEPIN, 1967a), and cosmogenic components can be resolved, and in Xe, where meteoritic Xe is strongly fractionated with respect to terrestrial Xe.

Figure IX-3 shows the meteoritic Ne isotopic data. All bulk Ne isotopic data from meteorites fall in or near a triangular field with the composition of solar Ne, planetary Ne, and cosmogenic Ne at the apexes (PEPIN, 1967a). The solar Ne composition was taken from the solar wind composition determinations reported by BÜHLER et al. (1971), that of planetary Ne is based mainly on the data on the CM chondrites Cold Bokkeveld and Haripura, after correction for the cosmogenic contribution. The cosmogenic apex is a bit diffuse, since the exact cosmogenic isotopic ratios depend on the bulk composition of the meteorite. In Fig. IX-3 this area is chiefly defined by meteorites such as Aubres, Kandahar and Karoonda, which are essentially devoid of primordial Ne. The Breitscheid point (a residue from a selective dissolution experiment) which falls well outside the triangular field is

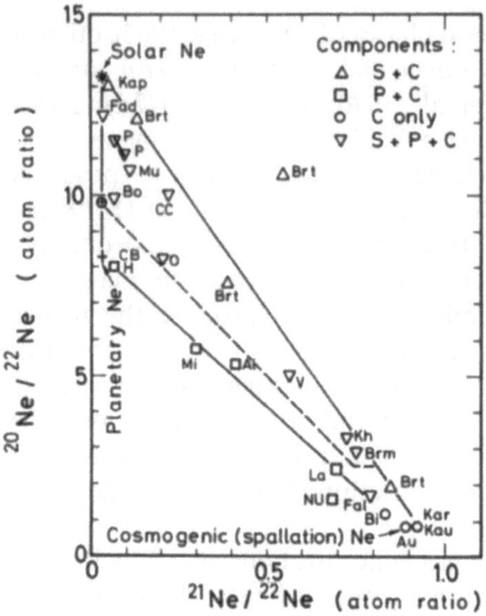

Fig. IX-3. Plot of bulk $^{20}Ne/^{22}Ne$ versus $^{21}Ne/^{22}Ne$ for silicate-rich meteorites and from different fractions of the solar gas-rich meteorite Breitscheid. Meteoritic Ne consists of mixtures of three main components—one produced by cosmic-ray interactions, and two primordial components. The meaning of the abbreviations and references to the original publications follow: Al – Alais, MAZOR et al. (1970); Au – Aubres, EBERHARDT et al. (1965b); Bi – Bishunpur, ZÄHRINGER (1968); Bo – Boriskino, MAZOR et al. (1970); Brm – Bremervörde, HEYMANN and MAZOR (1968); Brt – Breitscheid (dark), HINTENBERGER et al. (1965b); CB – Cold Bokkeveld, KIRSTEN et al. (1963); CC – Campo del Cielo, HINTENBERGER et al. (1969); Fad – Fayetteville (dark), 400° and 600° C fractions of MANUEL (1967); Fal – Fayetteville (light), sample 4 of MANUEL and KURODA (1964); H – Haripura, MAZOR et al. (1970); Kan – Kandahar, EBERHARDT et al. (1966); Kap – Kapoeta (dark), MÜLLER and ZÄHRINGER (1966); Kar – Karoonda, PEPIN (1967); Kh – Khohar, HEYMANN and MAZOR (1968); La – Lancé, STAUFFER (1961); Mi – Mighei, KIRSTEN et al. (1963); Mu – Murray, MAZOR et al. (1970); NU – Novo Urei, MÜLLER and ZÄHRINGER (1969); O – Orgueil, MAZOR et al. (1970); P – Pesyanoe (dark), GERLING and LEVSKII (1956) and MÜLLER and ZÄHRINGER (1966); V – Vigarano, MAZOR et al. (1970)

probably erroneous. Heating experiments of BLACK and PEPIN (1969) and BLACK (1972a, b) show that primordial Ne with $^{20}Ne/^{22}Ne$ ratios as low as 3.4 and as high as 14 are observed in certain temperature fractions of some meteorites. Such extremes are not found in whole rock samples. The reader interested in the implications and possible origins of these components is referred to the latter two papers by BLACK.

Some of the interesting facts revealed by Fig. IX-3 are as follows: Although some meteorites (Kapoeta, Breitscheid) with very high solar rare-gas contents lie, as expected, near the edge of the triangle connecting the solar and cosmogenic points, others (Fayetteville, Pesyanoe) contain primordial Ne with a distinctly lower $^{20}Ne/^{22}Ne$ ratio. Most authors (e.g., BLACK, 1972a) interpret these lower ratios as resulting from the mixing of another primordial Ne component with solar Ne. Some of the CM chondrites (Cold Bokkeveld, Haripura, Mighei) define the planetary-cosmogenic boundary in Fig. IX-3, whereas others (Murray, Boriskino) contain appreciable amounts of solar Ne. Primordial Ne in CI chondrites includes a small solar component, larger in Orgueil than in Alais. Despite their large planetary ^{36}Ar and ^{132}Xe contents, type 3 ordinary chondrites generally contain no detectable primordial Ne (MAZOR et al., 1970). The exceptions are Khohar and Bremervörde, in which primordial Ne consists mainly of solar Ne.

The etching gas-release studies and the track evidence indicate that the solar rare gases in most meteorites resulted from an irradiation with solar particles. Petrologic studies and a similarity to lunar breccias indicate that this irradiation took place on the surface of a planet-like body. Since it is very important to know which meteorites spent an appreciable part of their history on parent-body surfaces, and since it is not likely that a large number of meteorites will be subjected to the more definitive track studies in the near future, I propose the following arbitrary definition of a solar gas-rich meteorite: The $^{20}Ne/^{22}Ne$ ratio should be greater than 2.5, and the Ne isotopic data should plot above the dashed line extending from the point $^{20}Ne/^{22}Ne = 2.5$, $^{21}Ne/^{22}Ne = 0.75$ through the composition of terrestrial atmospheric Ne, marked by a cross in a circle on Fig. IX-3. The equation of this line is

$$^{20}Ne/^{22}Ne = -10.1\,(^{21}Ne/^{22}Ne) + 10.1\,.$$

Further, the ^{4}He content should exceed 2×10^{-5} $cm^3\ g^{-1}$, and the $^{20}Ne_p/^{36}Ar_p$ (the subscripts indicate that these two isotopes are primordial, i.e., corrected for the cosmogenic contribution) ratio should be greater than 0.3 (see Fig. IX-1). In fact, this cosmogenic contribution will be negligible for most meteorites with $^{20}Ne/^{22}Ne$ ratios greater than 2.5. In Table IX-1 are listed meteorites which are solar-gas-rich based on these criteria.

Statistical studies have shown that the primordial $^{3}He/^{4}He$ ratio undergoes significant variations which correlate with variations in the primordial $^{20}Ne/^{22}Ne$ ratio (BLACK, 1970; ANDERS et al., 1970). Curiously, the correlation is positive among CM and CI chondrites and negative among meteorites rich in solar rare gases. If we are dealing with the mixing of components, it appears that at least three types of primordial gases are involved. BLACK (1972a, b) argues that there are at least four and suggests the following origins: solar flares, recent solar wind, ancient solar wind and interstellar dust.

Table IX-1. List of meteorites containing solar-type rare gases. If more than one reference is available, the most recent or complete one is cited. Data compiled by L. SCHULTZ (see text for details)

Meteorite		References	Meteorites		References
H	Benoni	ZÄHRINGER (1968)	L	Ghubara (1954)	VINOGRADOV and ZADOROZHNYI (1964)
H	Breitscheid	SIGNER and SUESS (1963)	L	Grassland	ZÄHRINGER (1968)
H 3	Bremervörde	HEYMANN and MAZOR (1968)	L 3	Rio Negro	HINTENBERGER et al., unpub.[a]
H 5	Cangas de Onis	HINTENBERGER et al., unpub.[a]	L 4	Rupota	R. O. PEPIN, unpub.
H	Cee Vee	BLACK (1972a)	LL	Holman Island	BLACK (1972a)
H	Culbertson	ZÄHRINGER (1966c)	LL 6	St. Mesmin	HEYMANN and MAZOR (1966)
H 3	Dimmitt	EBERHARDT et al. (1966)	CV 2	Mokoia[b]	MAZOR et al. (1970)
H	Djermaia	SCHULTZ (1973)	CV 3	Vigarano	MAZOR et al. (1970)
H	Dwaleni	REYNOLDS et al. (1971)	CM 2	Boriskino	MAZOR et al. (1970)
H 4	Elm Creek	HINTENBERGER et al. (1965b)	CM 2	Essebi	MAZOR et al. (1970)
H	Fayetteville	MÜLLER and ZÄHRINGER (1966)	CM 2	Mighei	VINOGRADOV and ZADOROZHNYI (1964)
H	Gütersloh	HINTENBERGER et al., unpub.[a]	CM 2	Murchison	BOGARD et al. (1971a)
H	Hainaut	HINTENBERGER et al., unpub.[a]	CM 2	Murray	MAZOR et al. (1970)
H 5	Heredia	ZÄHRINGER (1968)	CM 2	Nawapali	MAZOR et al. (1970)
H 5	Kilbourn	HINTENBERGER et al., unpub.[a]	CM 2	Nogoya	BLACK (1972b)
H 5	Leighton	HINTENBERGER et al., unpub.[a]	CM 2	Pollen	MAZOR et al. (1970)
H	Nulles	HINTENBERGER et al., unpub.[a]	CI 1	Alais	MAZOR et al. (1970)
H 5	Pantar	SIGNER and SUESS (1963)	CI 1	Orgueil	MAZOR et al. (1970)
H	Phum Sambo	MÜLLER and ZÄHRINGER (1969)	Aub	Bustee	EBERHARDT et al. (1965b)
H 5	Pultusk	BLACK (1972a)	Aub	Khor Temiki	EBERHARDT et al. (1965a)
H 5	Tabor	SIGNER and SUESS (1963)	Aub	Pesyanoe	MÜLLER and ZÄHRINGER (1966)
H 4	Tysnes Island	EBERHARDT et al. (1966)	How	Bununu	GANAPATHY and ANDERS (1969)
H 4	Weston	SCHULTZ (1973)	How	Jodzie	MAZOR and ANDERS (1967)
L 5	Assam	SCHULTZ et al. (1971a)	How	Kapoeta	MÜLLER and ZÄHRINGER (1966)

[a] H. HINTENBERGER, L. SCHULTZ, and H. WÄNKE, unpublished data.

[b] The meteorite *pseudo* St. Caprais also contains solar gases, but appears to be mislabelled Mokoia (P. PELLAS, private communication, 1973).

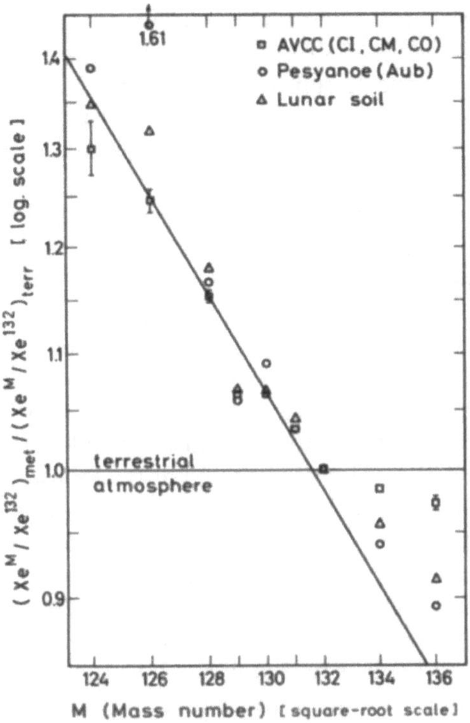

Fig. IX-4. Ratio (log scale) of meteoritic and lunar Xe isotopic abundances to those in terrestrial atmospheric Xe. The straight line drawn through the AVCC points at masses 126, 128, and 130 is thought to represent the mean solar system relative abundance of non-radiogenic Xe isotopes. Terrestrial Xe is fractionated in favor of the heavy isotopes. Terrestrial Xe contains relatively more radiogenic ^{129}Xe than most meteorites; meteoritic Xe contains more fissiogenic ^{132}Xe, ^{134}Xe, and ^{136}Xe than does terrestrial Xe

The primordial ^{36}Ar/^{38}Ar ratio is nearly constant in meteorites. BLACK (1971) presents evidence indicating a positive correlation between ^{36}Ar/^{38}Ar and ^{20}Ne/^{22}Ne ratios among the solar gas-rich meteorites. His data also suggest that there is a very slight increase in the ^{36}Ar/^{38}Ar ratio of CM and CI chondrites with increasing ^{20}Ne/^{22}Ne. These trends are also indicative of three or more primordial components.

The isotopic composition of Xe in meteorites differs from that of terrestrial Xe in a number of important respects. Figure IX-4 illustrates these differences. Plotted is the ratio of the meteoritic abundance of isotope M to the terrestrial abundance of the same isotope (PODOSEK et al., 1971), normalized to a ^{132}Xe ratio of 1. Also shown is the spectrum of surface-correlated Xe measured in lunar soil (PODOSEK et al., 1971). This

Xe is mainly of solar origin. The meteoritic data are the average carbonaceous chondrite (AVCC) data of EUGSTER et al. (1967a), which were obtained by replicate runs on one member each of the groups CI, CM, and CO, and the Pesyanoe data of MARTI (1969). The latter is important because, according to the abundance data plotted in Fig. IX-1, it seems to contain about 30% solar Xe, the highest fraction of this component known in a meteorite. Error bars are shown on some of the AVCC points. Their lengths decrease with increasing isotopic abundance. Reported errors for Pesyanoe are slightly higher than the AVCC values; those for the lunar soil are slightly lower.

The abundances of the light Xe isotopes in meteorites and lunar soil are higher than those in the terrestrial atmosphere by large, nearly constant factors. The two heaviest isotopes are higher in the meteoritic and lunar samples by factors which show considerable variation. The consensus of current opinion seems to favor the following explanation of these differences: Meteoritic and terrestrial Xe are fractionated by a mass-dependent process with respect to each other, and, in addition, meteoritic Xe contains a fission contribution to the unshielded isotopes 131, 132, 134, and 136 (KRUMMENACHER et al., 1962; EUGSTER et al., 1967b). A straight line is drawn through the fission-shielded isotopes 126, 128, and 130 of Fig. IX-4 corresponding to an enrichment factor of about 1.04 per mass unit. The difference between this line and the plotted points for the isotopes 131—136 is attributed to a fissiogenic contribution. That the meteoritic and lunar ^{129}Xe abundances fall below this line indicates that terrestrial Xe contains a larger radiogenic ^{129}Xe contribution than does the Xe in the extraterrestrial samples. Excesses observed for some of the light isotopes in the Pesyanoe and lunar soil samples are of cosmogenic origin. The reason for the low abundance of AVCC ^{124}Xe relative to the fractionation line is not understood; it may be due to an accidental accumulation of errors, including that in the choice of the slope of the fractionation line.

Additional Xe data from meteorites belonging to a number of different groups are listed in Table IX-2. Concentrations vary over a factor of 1000 in these meteorites, which include the chondrite with the lowest published Xe concentration, Saint-Séverin and Novo Urei, which together with the other ureilites, Goalpara and Haverö, has the highest Xe content known in meteorites. Note that two of the meteorites, Cold Bokkeveld and Lancé, were studied by EUGSTER et al. (1967a) in their remeasurement of the AVCC Xe spectrum, and that Woodbine is a group IA iron meteorite.

The isotopic data in Table IX-2 are normalized to ^{132}Xe, which is defined as 100. Isotopic abundances are reasonably constant among the different meteorites, with the exception of ^{129}Xe which shows a highly

Table IX-2. Bulk concentrations and normalized isotopic compositions of Xe in meteorites

Meteorite[a]	Group	Isotopic composition ($^{132}Xe = 100$)								
		^{132}Xe (10^{-12} cm^3)	^{124}Xe	^{126}Xe	^{128}Xe	^{129}Xe	^{130}Xe	^{131}Xe	^{134}Xe	^{136}Xe
Abee	E4	735[b]	0.46 ± 0.01	0.42 ± 0.01	8.64 ± 0.10	627[c]	16.2 ± 0.2	82.0 ± 0.5	38.2 ± 0.2	32.1 ± 0.2
Pine River	IA	950	0.47 ± 0.04	0.42 ± 0.02	8.31 ± 0.09	120	16.2 ± 0.1	82.2 ± 0.5	37.9 ± 0.2	31.8 ± 0.3
Pantar (dark)	H	710	0.46 ± 0.02	0.45 ± 0.03	8.45 ± 0.20	229	16.0 ± 0.1	81.5 ± 0.4	38.0 ± 0.2	32.2 ± 0.2
Tieschitz	H3	1650	0.46 ± 0.01	0.42 ± 0.01	8.08 ± 0.05	140	16.1 ± 0.1	81.5 ± 0.4	37.9 ± 0.2	31.9 ± 0.2
Krymka	L3	2450	0.45 ± 0.02	0.41 ± 0.01	8.05 ± 0.10	106	16.1 ± 0.1	81.4 ± 0.5	38.3 ± 0.3	32.1 ± 0.3
Bruderheim	L6	99	0.56 ± 0.02	0.59 ± 0.02	8.29 ± 0.08	125	15.8 ± 0.1	81.8 ± 0.5	38.6 ± 0.2	32.5 ± 0.2
Saint-Séverin	LL6	20	0.64 ± 0.02	0.72 ± 0.02	8.17 ± 0.10	174	15.8 ± 0.1	80.6 ± 0.5	39.7 ± 0.3	34.0 ± 0.3
Cold Bokkeveld	CM2	6200	0.46 ± 0.02	0.41 ± 0.01	8.20 ± 0.04	106	16.1 ± 0.2	81.8 ± 0.4	38.2 ± 0.2	32.1 ± 0.2
Lancé	CO3	4800	0.45 ± 0.02	0.41 ± 0.01	8.23 ± 0.08	114	16.0 ± 0.1	81.3 ± 1.0	38.0 ± 0.3	31.9 ± 0.3
Novo Urei	Ure	23000	0.45 ± 0.02	0.41 ± 0.01	8.08 ± 0.08	102	16.1 ± 0.1	81.5 ± 0.4	38.1 ± 0.3	32.0 ± 0.3

[a] The data are taken from the following publications: BOGARD et al. (1971b) – Abee, Pine River; EUGSTER et al. (1967a) – Cold Bokkeveld, Lancé; EUGSTER et al. (1969) – Tieschitz; MARTI (1967b) – Krymka, Novo Urei; MARTI et al. (1966) – Bruderheim; MARTI et al. (1969) – Saint-Séverin; MERRIHUE et al. (1962) – Pantar.

[b] The absolute ^{132}Xe concentrations are accurate to about $\pm 20\%$.

[c] The abundance of ^{129}Xe varies from sample to sample within meteorites. Ratios near 110 are known to about $\pm 1\%$, values >200 to about $\pm 10\%$.

variable radiogenic component. This will be discussed further in Chapter X, as will the fissiogenic component of the heavy Xe isotopes. The high ^{124}Xe and ^{126}Xe abundances in Bruderheim and Saint-Sevérin reflect a cosmogenic component which is the more noticeable because of the low primordial Xe contents of these meteorites. The high ^{128}Xe abundances observed in some meteorites result from neutron capture in I: the neutrons are cosmic-ray secondary particles, and the effect is highest in meteorites with long cosmic-ray ages and high ratios of ^{127}I to primordial Xe (MARTI et al., 1966; BOGARD et al., 1971b). Other cosmic-ray effects will be discussed in Chapter XI.

MERRIHUE et al. (1962) reported a ^{132}Xe concentration in Pantar (light) which is a factor of 12 lower than that observed in Pantar (dark) listed in Table IX-2. Except for radiogenic ^{129}Xe, the isotopic compositions of the two Xe samples are indistinguishable. The ^{132}Xe/^{36}Ar ratio of 156 indicates that the additional Xe in the dark portion is planetary, and thus that both solar and planetary gases are enriched in this portion.

The most important point to be learned from the primordial stable isotopes of Xe is that a remarkable fractionation distinguishes terrestrial from meteoritic and solar Xe. Because the latter types are so similar it appears that they are representative of mean solar system Xe, and that the fractionation is to be associated with the formation of the earth and its atmosphere. This interpretation is complicated, however, by data of LIGHTNER and MARTI (1974) which show that the isotopic composition lunar primordial Xe is nearly identical to that of terrestrial Xe.

Meteoritic Kr also differs from terrestrial Kr (EUGSTER et al., 1967a, b; MARTI, 1967b). The difference is attributed by EUGSTER et al. (1967b) to a combination of a mass-dependent fractionation and a small additional fission component in meteoritic Xe. The mass fractionation is an order of magnitude smaller than that observed in Xe, and in the opposite direction; the light isotopes are less abundant in meteoritic than in terrestrial Kr. The composition of the surface-correlated Kr in lunar soil is not well defined. According to EBERHARDT et al. (1970) and PEPIN et al. (1970), it is similar to terrestrial Kr, whereas PODOSEK et al. (1971) report it to be more similar to AVCC Kr. Needless to say, the fact that the fractionation of Kr is opposite that of Xe introduces a complicating element into any attempt to understand the origin of the isotopic differences between terrestrial, meteoritic and solar rare gases.

Microprobe mass-spectrometry experiments indicate that some primordial He is concentrated on grain surfaces (ZÄHRINGER, 1966b) and that solar rare gases undergo large variations from point to point within the gas-rich dark portions of light-dark meteorites (MEGRUE, 1969). These techniques appear to have great potential.

The rare gases are susceptible to diffusional loss during periods of relatively high temperature or during shock events. It is of obvious importance to know where such losses have occurred and to correct for their effects as much as possible. A good indicator of gas loss is provided by the cosmogenic $^{3}He/^{21}Ne$ ratio. In chondrites this varies with depth in the meteoroid from about 3.5 to about 8, with most observed values near 5.3. Appreciable gas loss since the cosmic-ray clock started running will drive this ratio much lower than 3. For example, EBERHARDT et al. (1966) reported a cosmogenic $^{3}He/^{21}Ne$ ratio of 1.0 in the Cumberland Falls aubrite. Since Rb-Sr evidence indicates that meteorites of all classes were formed very early in the history of the solar system, low K-Ar and U, Th-He ages are also indicative of gas loss.

Some authorities have attempted to avoid the effects of diffusional loss by extracting temperature fractions of the rare gases during controlled heating experiments. As will be discussed further in Chapter X, this has resulted in highly precise K-Ar ages (TURNER, 1969) and I-Xe formation intervals (HOHENBERG et al., 1967a). This technique is of doubtful value for determining the relative abundances of different rare-gas elements, however, since the release behaviors probably vary from element to element.

Diffusional studies are of some use in untangling the effects of diffusional loss. A review is given by FECHTIG and KALBITZER (1966). The most useful data are those based on separated minerals (HUNEKE et al., 1969). Diffusional evidence and/or controlled temperature release experiments can be used to infer the past temperature history of meteorites. For example, PODOSEK (1971) reports data on the LL6 chondrite Saint-Séverin showing that no more than 10% of the radiogenic ^{40}Ar has been lost from the minerals which outgas during heating to 700° C for approximately 1 hr.

X. Primordial Radionuclides and Associated Chronologies

A. Introduction

Formation ages[8] measure the time elapsed between the first resolvable isotopic-equilibration event in the sample's history and the present. Most meteorite formation ages are based on the Rb-Sr, K-Ar, and U, Th-He techniques. If later isotopic-equilibration events can be resolved, these can be called *metamorphism ages.*

Ages known as *formation intervals* can be determined by I-Xe, Pu-Xe and Pu-fission track techniques. These measure the time elapsed since the end of nucleosynthesis and the start of a formation age. Because of inadequate understanding of the rate of nucleosynthesis processes, it is not possible to measure formation intervals accurately. Relative differences in formation intervals precise to about ± 2 Myr can be measured, however, and are of great importance.

BURNETT (1971) and BOGARD (1971) provided recent bibliographic reviews of meteoritic age studies, the latter limited to those involving rare-gas isotopes. Both reviewers also surveyed studies of lunar rocks. ZÄHRINGER (1968) reviewed rare-gas formation-age data. REYNOLDS (1967), ZÄHRINGER (1968) and ANDERS (1963) reviewed age data based on all techniques. The latter review also gives good explanations of the theoretical bases for each technique.

B. Formation and Metamorphism Ages

A number of highly precise meteorite formation ages based on the Rb-Sr technique have been reported during the past several years. If a

[8] The assumption is normally made that radioactive ages date brief, high-temperature events. Unfortunately, the event which is dated is not definable with any precision. For example, in a slowly-cooled igneous rock system we do not know whether the Rb-Sr method dates 1. when the igneous melt solidified, 2. when the last major Sr-bearing phases formed, or 3. some later time when the temperature became so low that diffusional exchange ceased. In many meteorites the situation is still more complex, since there is no unequivocal evidence that the whole mass was ever molten, and it is possible that individual grains may record different events. Because of these uncertainties it appears best to use the somewhat vague term, formation age, rather than solidification age or gas-retention age, which have more specific connotations.

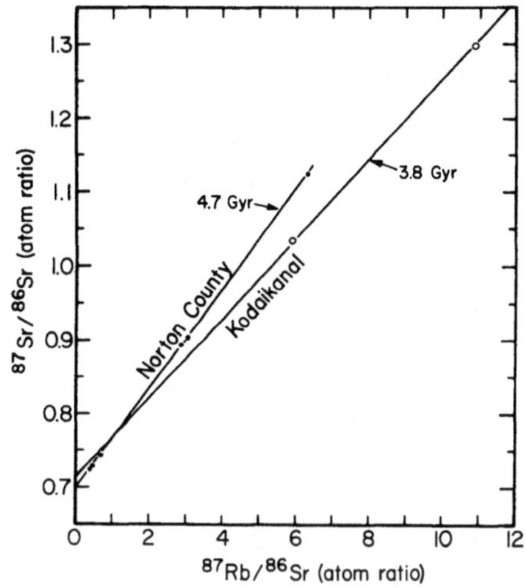

Fig. X-1. Rb-Sr isochrons for Norton County aubrite (BOGARD et al., 1967) and Kodaikanal II E-An iron (BURNETT and WASSERBURG, 1967). Each point represents data from a separate fraction. Six additional Kodaikanal fractions with $^{87}Rb/^{86}Sr$ ratios up to 153 lie off the scale of this drawing. The ages of the meteorites are any precision. For example, in a slonly-cooled igneous rock system we do not know meteorites known to have Rb-Sr formation ages below 4 Gyr

plot of $^{87}Sr/^{86}Sr$ vs. $^{87}Rb/^{86}Sr$ gives a straight line, this line is called an isochron. The equation of this line is

$$^{87}Sr/^{86}Sr = {}^{87}Rb/^{86}Sr\,(e^{\lambda t} - 1) + ({}^{87}Sr/^{86}Sr)_I$$

where $(e^{\lambda t} - 1)$ is the slope of the line, and, since the decay constant λ is known (λ is generally taken to be 1.39×10^{-11} yr^{-1}), the age t can be calculated. The y intercept $({}^{87}Sr/^{86}Sr)_I$, gives information regarding metamorphism or more severe fractionation events in the earlier history of the meteoritic material.

Figure X-1 illustrates Rb-Sr isochrons determined for the II E-An iron Kodaikanal (BURNETT and WASSERBURG, 1967) and the achondrite Norton County (BOGARD et al., 1967). The Kodaikanal isochron yields a formation age of 3.8 ± 0.1 Gyr; this and the recently determined 3.6—3.9 Gyr ages of the Kapoeta howardite (PAPANASTASSIOU et al., 1974) are the only Rb-Sr formation ages which differ significantly from the 4.6 ± 0.1 Gyr age[9] of the solar system. The formation age of Norton

---

[9] COMPSTON et al. (1965) report a Rb-Sr age of 3.5—3.9 Gyr for Bishopville, a member of the same class as Norton County. As a result of analytical difficulties associated with this determination (BOGARD et al., 1967), it should probably be considered as a tentative value until a new study has been completed. PODOSEK (1973) reports ages of 1.3 Gyr obtained by the $^{39}Ar - ^{40}Ar$ method for Nakhla and Lafayette, two anomalous achondrites of essentially identical composition. It would be very interesting to have Rb-Sr isochrons for these meteorites.

County is 4.7 ± 0.1 Gyr and indicates that the process which produced the differentiated composition of this meteorite occurred during a very short time span at the time of formation of the solar system as a whole. The very high ages of most meteorites are to be contrasted with the ages of about 4.0 Gyr obtained for the oldest terrestrial and lunar rocks.

The measured points on the Kodaikanal and Norton County isochrons were obtained from separated mineral fractions. Other recent Rb-Sr isochrons are based on whole-rock samples of a number of meteorites from a single class. GOPALAN and WETHERILL (1969, 1970, 1971) report ages for the LL, E, and L chondrites, and KAUSHAL and WETHERILL (1969, 1970) report ages for the H and carbonaceous chondrites. These whole-rock isochrons involve the assumption that all members of a class experienced high-temperature events at the same time. That the reported ages are all about 4.6 Gyr indicates that this assumption is essentially correct.

As discussed in Chapter VIII, there appears to have been a thorough isotopic mixing of the primordial matter from which the solar system formed. An important parameter for evaluating Sr isotopic evidence of metamorphism and for calculating so-called model ages is the $^{87}Sr/^{86}Sr$ ratio at the time of formation of the solar system. The $(^{87}Sr/^{86}Sr)_I$ intercept based on whole-rock analyses of eucrites (PAPANASTASSIOU and WASSERBURG, 1969) provides a widely used estimate of this value. The $^{87}Rb/^{86}Sr$ ratios in the eucrites are so low that a small difference in time between the formation of these achondrites and other old solar-system samples produces a negligible change in the eucrite $(^{87}Sr/^{86}Sr)_I$ value. The eucrite initial ratio of 0.6990 is called BABI (basaltic achondrite best initial). Inital ratios slightly lower than BABI are found in the Angra dos Reis anomalous achondrite (WASSERBURG et al., 1970) and in the Ca-Al rich inclusions in Allende (GRAY et al., 1973; WETHERILL et al., 1973).

Values of $(^{87}Sr/^{86}Sr)_I$ which are significantly higher than BABI can be interpreted in terms of metamorphism events. WASSERBURG et al. (1969b) note that the 0.7000 intercept of the isochron for the H6 chondrite Guareña (Fig. XVIII-2) can be interpreted in terms of an event which homogenized the Sr isotopes (but did not change the mean $^{87}Rb/^{86}Sr$ ratio) 74 Myr after the time when the mean $^{87}Sr/^{86}Sr$ ratio equaled BABI.

More recent metamorphism events are observable in the slopes of the isochrons. The low Kodaikanal age probably reflects a late metamorphism. As discussed further below, GOPALAN and WETHERILL (1971) presented Rb-Sr isochronic data to show that the shocked L-chondrites experienced metamorphism several hundred Myr ago.

A decade ago the most extensive formation-age data was based on the measurement of isotopic changes in Pb resulting from the decay of U and Th. These were mainly "model" ages based on an assumed composition of primordial Pb and have largely been superseded in recent years by Rb-Sr ages. The accuracy of Pb-based ages is severely limited by the low U/Pb and Th/Pb ratios found in most meteorites. Precisely because of these low ratios, however, meteorites have yielded the best information on the composition of primordial Pb; i.e., Pb as present in solar-system material at the time of formation of the solar system (PATTERSON, 1956; OVERSBY, 1970; TATSUMOTO et al., 1973). Recent data obtained by a new technique involving the suspension of a sample in silica gel are reported by HUEY and KOHMAN (1973) and TATSUMOTO et al. (1973); the latter data confirm that Angra dos Reis is one of the oldest meteorites and indicate that formation ages increase through the sequence eucrites-ordinary chondrites-CM, CV chondrites.

Another method for formation-age determination is based on the decay of ^{187}Re ($T_{\frac{1}{2}} \sim 40$ Gyr) to ^{187}Os. The long half-life and the fact that Re and Os are geochemically very coherent elements in meteorites limit the usefulness of this technique. HERR et al. (1961) determined a whole-rock Re-Os isochronic age for iron meteorites of 4.0 ± 0.8 Gyr.

The K-Ar age-determination method involves the relationship:

$$^{40}\mathrm{Ar_r} = {}^{40}\mathrm{K} f (e^{\lambda t} - 1)\,.$$

The branching ratio f (the average number of ^{40}Ar atoms per ^{40}K decay) is 0.110, and λ is normally taken to be 5.30×10^{-10} yr^{-1}. The subscript r indicates that this is the radiogenic portion, determined by subtraction of the primordial and cosmogenic (cosmic-ray produced) contributions. Most age data for meteorites are based on single extractions of Ar from whole-rock samples. In most cases the K concentration is not determined; the average value for the entire meteorite group is assumed to hold. ZÄHRINGER (1968) assumes a K content of 800 ppm to hold for all members of the E, H, and L groups of chondrites.

Ages based on U, Th-He are calculated from a similar relationship:

$$^{4}\mathrm{He_r} = 8\,{}^{238}\mathrm{U}\,(e^{\lambda_{238}t} - 1) + 7\,{}^{235}\mathrm{U}\,(e^{\lambda_{235}t} - 1) + 6\,{}^{232}\mathrm{Th}\,(e^{\lambda_{232}t} - 1)\,.$$

The integers preceding the symbols for the radionuclides represent the number of α decays produced in each decay series. The decay constants λ_{238}, λ_{235} and λ_{232} are 1.525×10^{-10} yr^{-1}, 9.848×10^{-10} yr^{-1} and 4.948×10^{-11} yr^{-1} respectively (JAFFEY et al., 1971). Most U, Th-He ages are calculated from group-average U concentrations, assuming a normal ^{235}U/^{238}U ratio of 0.00725 atoms atom^{-1} and a ^{232}Th/^{238}U ratio of 3.7 atoms atom^{-1}. Most chondrite ages are based on an assumed U concentration of 11 ppb, independent of group. FISHER'S (1972) fission-track data indicate this number to be too low in all chondrite groups, with the difference amounting to a factor of about 1.4 in the most serious

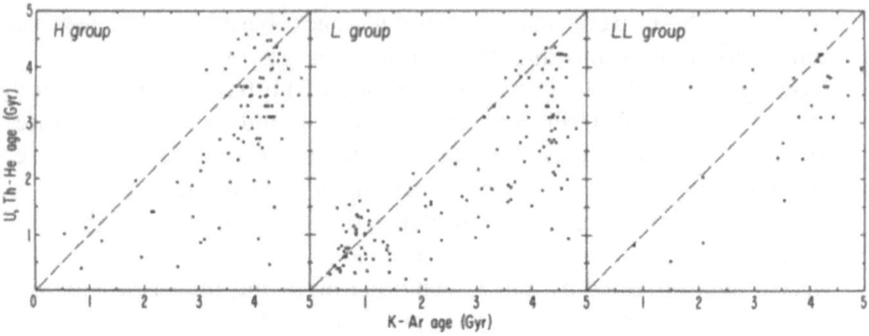

Fig. X-2. Plot of U, Th-He age vs. K-Ar age for the three ordinary chondrite groups, calculated from ^{4}He and ^{40}Ar data assuming K, U and Th concentrations of 800 ppm, 12 ppb and 45 ppb, respectively. The dashed lines show loci of concordant ages. The downward trend from concordancy at about 4.5 Gyr reflects gas loss; the effect is greater for U, Th-He ages, both because of the higher diffusibillity of ^{4}He relative to ^{40}Ar and the shorter half-life of ^{40}K relative to ^{238}U and ^{232}Th; this trend is observed in each ordinary chondrite group. An additional cluster of points near concordancy at about 0.4—1.3 Gyr is attributed to strong metamorphism during this period, probably near 0.5 Gyr

cases, the L and LL groups. Further studies are needed to establish this important parameter.

Because of the inert character of the rare gases, they are only efficiently trapped in the interiors of mineral grains. Once these elements reach a grain surface, they migrate to the surface of the meteoroid and escape into space in a geologically short period of time. The release of rare gases from grain interiors is generally attributed to diffusive loss at high temperatures, although mechanical fracture may also play a role in shock-produced loss.

The greatest body of K-Ar and U, Th-He age data is available for the chondrites, and especially for the ordinary chondrites. One of the most interesting facts revealed by these data is the different distribution of ages in the L group relative to the other groups of ordinary chondrites (ANDERS, 1964). Fig. X-2 shows plots of U, Th-He ages vs. K-Ar ages for the H, L, and LL groups. Most K-Ar ages fall between 3.5 and 4.6 Gyr; the U, Th-He ages for these chondrites show more spread, ranging from about 2.0 to 4.5 Gyr. This type of distribution is thought to reflect either long-term diffusive loss during the past 4.6 Gyr, or fairly recent mild metamorphic reheating. That the U, Th-He ages tend to be lower is attributed to one or both of the following effects; 1. diffusion coefficients for He are significantly higher than those for Ar; and 2. because the half-life of ^{40}K is smaller than that of ^{238}U and of ^{232}Th, equal loss of ^{4}He and ^{40}Ar results in a smaller reduction in the K-Ar age.

The L group also shows another type of age distribution, i.e., a cluster of nearly concordant ages between 0.4 and 1.3 Gyr. As noted by ANDERS (1964), meteorites falling in this trend exhibit more extensive evidence of shock than do the other L chondrites. HEYMANN (1967) carried out an extensive investigation of the rare gases of the shocked L chondrites and attributed the cluster of low concordant ages to a major collisional event about 520 Myr ago. Apparently the H and LL groups were not affected by this event, probably because they were stored in separate parent bodies.

The possibility of a severe metamorphism in the L group about 500 Myr ago was subjected to a Rb-Sr investigation by GOPALAN and WETHERILL (1971). They found evidence of severe metamorphism in the shocked L chondrites Orvinio and Farmington. Their data suggest lower ages (ca. 200—300 Myr) for the metamorphism event, but the individual points scatter and the errors are large enough to include HEYMANN'S estimate.

A more sophisticated technique for the determination of K-Ar ages was developed by MERRIHUE and TURNER (1966). WÄNKE and KÖNIG (1959) had shown that K and Ar could be determined simultaneously by first subjecting the meteorite sample to a neutron irradiation, thereby forming 325-yr ^{39}Ar by the reaction ^{39}K (n, p) ^{39}Ar. MERRIHUE and TURNER combined the neutron activation with controlled heating steps such that the Ar is released in a series of temperature fractions. A separate K-Ar is calculated for each fraction. This technique is generally called the $^{39}Ar-^{40}Ar$ method.

Figure X-3 shows $^{39}Ar-^{40}Ar$ age studies of six shocked L-group chondrites (TURNER, 1969). The low-temperature fractions tend to give plateaus corresponding to ages of about 500 Myr, except for Wittekrantz and Château-Renard, for which the plateau ages are 300 ± 30 Myr. These plateaus are interpreted to date major outgassing events and to support the contention that many L-group chondrites were produced by a single major breakup event. They indicate, however, that a single event cannot account for all the low ages. It is of interest to note that, although the high-temperature fractions give higher ages, enough gas loss has occurred to make them consistently lower than the 4.6 Gyr ages determined by Rb-Sr studies. In lunar crystalline rocks high-temperature fractions often form plateaus, with the plateau ages in excellent agreement with Rb-Sr ages (e.g., cf. TURNER, 1971). This presumably reflects lower degrees of reheating of lunar basalts relative to L chondrites.

Most of the recent $^{39}Ar-^{40}Ar$ ages have been reported by PODOSEK (1971, 1972, 1973; PODOSEK and HUNEKE, 1973), including some low ages [1.3 Gyr for the anomalous achondrites Nakhla and Lafayette

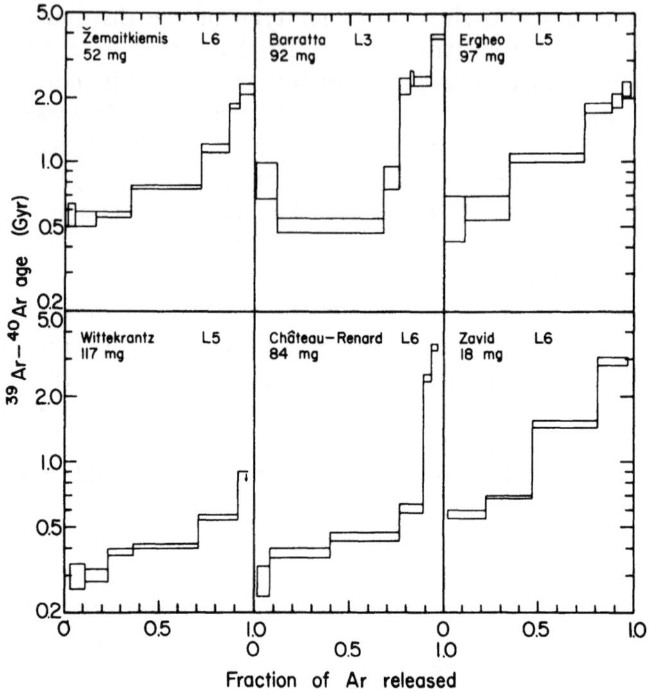

Fig. X-3. K-Ar ages based on gas fractions released during controlled heating experiments (TURNER, 1969). Neutron activation is used to determine K by the $^{39}K(n,p)^{39}Ar$ reaction. Plateaus on such a plot indicate either formation or metamorphism ages. Low temperature ages of six L-group chondrites are approximately constant, followed by sharp rises at higher temperatures. This is interpreted to indicate a reheating event at about 500 Myr (slightly lower for Wittekrantz and Château-Renard)

(which may be a mislabelled fragment of Nakhla), 4.4 Gyr for Petersburg and 3.5—3.9 Gyr for Stannern, both eucrites]. PODOSEK and HUNEKE (1973) suggest that these low ages reflect metamorphism, possibly during the parent-body breakup, but note that some of these events were more efficient in outgassing high-temperature sites than was the event which produced the L-group patterns shown in Fig. X-3; perhaps these really are formation ages.

C. Formation Intervals

The discovery by REYNOLDS (1960) that meteorites contain ^{129}Xe produced by decay of the extinct radionuclide ^{129}I brought a new degree of precision to the dating of events occurring during the early history of the solar system. Extinct radionuclides are those which have half-lives long enough that some activity survived the period between the last nucleosynthesis event and the formation of solid bodies in the solar system, but too short for survival until the present day of detectable activity (BROWN, 1947). Evidence of the former presence of these nuclides is found in the form of enrichments of their stable decay products, or as particle tracks if fission is the decay process. Decays involving radionuclides with appropriate half-lives include:

$$\text{7-Myr}\ ^{107}\text{Pd} \rightarrow {}^{107}\text{Ag}$$
$$\text{17-Myr}\ ^{129}\text{I} \rightarrow {}^{129}\text{Xe}$$
$$\text{30-Myr}\ ^{205}\text{Pb} \rightarrow {}^{205}\text{Tl}$$
$$\text{82-Myr}\ ^{244}\text{Pu} \rightarrow \text{fission products}$$
$$\text{superheavy nuclide} \rightarrow \text{fission products}$$

CHAKRABURTTY et al. (1964) and DEWS and NEWBURY (1966) were unable to confirm the enrichments of ^{107}Ag previously reported by MURTHY (1962). Searches for enrichments of ^{205}Tl by ANDERS and STEVENS (1960), OSTIC et al. (1969) and HUEY and KOHMAN (1972) have also been unsuccessful. There is definite evidence for the earlier presence of ^{129}I and ^{244}Pu, and marginal evidence for extinct superheavy nuclide(s).

It now appears that the bulk of nucleosynthesis occurred long before the formation of the solar system (HOHENBERG, 1969; WASSERBURG et al., 1969b). Thus, the decay products of extinct radionuclides can only be detected when the ratio of parent to daughter element is very high; to date sufficiently high ratios have been found only when the daughter element is a rare gas. As discussed earlier, rare gases are prone to escape, and the most precise studies of the decay products of extinct radionuclides are not of whole-rock samples, but of separated minerals (WASSERBURG et al., 1969a) or of the gas released during controlled heating experiments following neutron activation (HOHENBERG et al., 1967a; PODOSEK, 1970a) similar to those used for ^{39}Ar$-^{40}$Ar age determinations.

Figure X-4 is from PODOSEK (1970a) and shows the variation of ^{129}Xe/^{132}Xe and ^{128}Xe/^{132}Xe ratios in high-temperature gas-release fractions from the CO4 chondrite Karoonda and the aubrite Peña Blanca Spring. The variation in the ^{129}Xe/^{132}Xe ratio reflects differing amounts of ^{129r}Xe (radiogenic ^{129}Xe) added by the decay of ^{129}I to primordial Xe with a ^{129}Xe/^{132}Xe ratio of about 1.0. The variation in the ^{128}Xe/^{132}Xe ratio reflects differing amounts of ^{128a}Xe (^{128}Xe produced by neutron-activation of ^{127}I to ^{128}I followed by beta decay) added to primordial Xe with a ^{128}Xe/^{132}Xe ratio of about 0.08. The remarkable correlation between ^{129}Xe and ^{128}Xe indicates that the ^{129r}Xe/^{128a}Xe ratio is the same in all mineral sites independent of I content, and confirms the interpretation that ^{129r}Xe is produced by *in situ* decay of ^{129}I. The ^{129r}Xe content of the minerals which outgas during a given heating experiment is equal to the ^{129}I content of these minerals at the time Xe retention started. The ^{128a}Xe is proportional

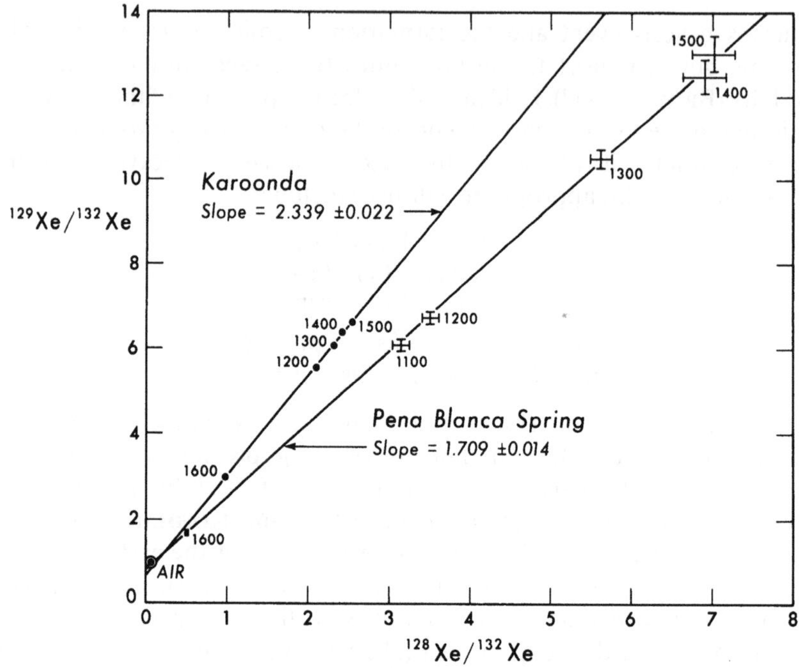

Fig. X-4. Plot of $^{129}Xe/^{132}Xe$ vs. $^{128}Xe/^{132}Xe$ in fractions of Xe released during the heating of neutron-activated meteorite samples through 100° temperature intervals (over periods of 20 min or longer). A strong correlation observed in temperature fractions above 1000° C indicates that excess ^{129}Xe resulting from the decay of ^{129}I resides in the same mineral sites as excess ^{128}Xe produced by neutron-capture and beta decay from ^{127}I. This indicates that the excess ^{129}Xe formed by *in situ* decay. (From PODOSEK, 1970a)

to the ^{127}I content of the same minerals; the proportionality constant is determined by measurement of the amount of ^{128}Xe produced in a known amount of ^{127}I irradiated with the meteorite samples. In this way one determines the $^{128}I/^{127}I$ ratio at the onset of Xe retention in the meteoritic material.

In Fig. X-5 are plotted $^{129}I/^{127}I$ ratios for most meteorites studied by this technique. The upper scale gives these ratios; the lower scale gives relative differences in formation ages (note that a ratio difference of a factor of 2 corresponds to an age difference of 17 Myr). In early papers such data were used to define model-dependent formation intervals. The current, and simpler practice is to discuss the data in terms of relative differences in formation ages or intervals. In calculating I-Xe ages, PODOSEK (1970a) normalized all ages to Bjurböle ≡ O.

Note that the total span of I-Xe ages illustrated in Fig. X-5 is only 14 Myr and that all samples are chondrites except Shallowater and Peña

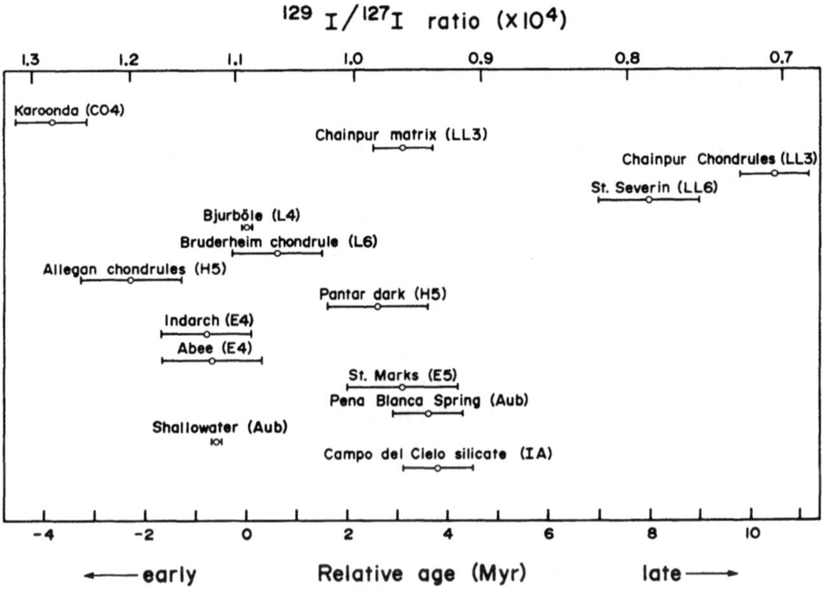

Fig. X-5. Relative ^{129}I-^{129}Xe formation times of meteorites (relative to Bjurböle) (PODOSEK, 1970a). All plotted meteorites began to retain Xe within a period of about 14 Myr very early in the history of the solar system. The aubrites Peña Blanca Spring and Shallowater and the IA iron El Taco (= Campo del Cielo) began retaining Xe at the same time as the chondrites

Blanca Spring which are aubrites, and the group-IA iron meteorite Campo del Cielo. Some unplotted data indicate formations substantially later than the chondrites; e.g., PODOSEK (1972) assigns an age of ≥66 Myr to the anomalous achondrite Lafayette.

KURODA (1960) proposed that fission products from extinct ^{244}Pu were responsible for differences in the abundance of heavy Xe isotopes between terrestrial and meteoritic samples. ROWE and KURODA (1965) isolated Xe from the Pasamonte eucrite which was very rich in a fissiogenic component. The isotopic composition of this component was further refined by ROWE and BOGARD (1966), EBERHARDT and GEISS (1966) and HOHENBERG et al. (1967b). Fission tracks in excess of those expected from the spontaneous or neutron-induced fission of U and Th isotopes have been observed in meteoritic minerals in various investigations; a status report is given by SHIRCK et al. (1969). WASSERBURG et al. (1969a) showed that Xe with the isotopic composition observed in Pasamonte was present in large amounts in whitlockite grains from St. Severin; CANTELAUBE et al. (1967) had found a large excess of fission tracks in the same material. ALEXANDER et al. (1971) studied Xe from synthetically produced ^{244}Pu and confirmed that Pasamonte-type fission Xe resulted from the decay of this nuclide.

There is no stable isotope of Pu, which seriously limits the precision with which relative formation ages can be assigned by the study of ^{244}Pu decay products. Both Pu and U are members of the actinide series of elements, and there is reason to believe that they have very similar geochemical properties. In most studies of ^{244}Pu decay features it is assumed that at any moment in time the Pu/U ratio was constant in meteorites, independent of meteorite class or mineral siting. If this assumption holds, measurement of the $^{244}Pu/^{238}U$ ratio at the time of meteorite formation provides a means for precise assignment of relative ages, a factor of 2 difference in this ratio corresponding to an age difference of 83 Myr (this includes an allowance for the concurrent decay of ^{238}U).

In principle, concordant ages should result from careful determinations of the $^{244}Pu/^{238}U$ and $^{129}I/^{127}I$ ratios in meteorites at the time of their formation (SABU and KURODA, 1967). REYNOLDS (1968) reviewed the evidence available at that time and concluded that there was too much scatter in the experimental data to allow any conclusions regarding such a concordancy.

Most studies of initial $^{244}Pu/^{238}U$ ratios have been based on whole-rock analyses of fission Xe and U in separate samples, which introduces uncertainties resulting from partial loss of Xe and from inhomogeneities in U distribution. PODOSEK (1970b, 1972) attempted to avoid these difficulties by applying the neutron-activation controlled-temperature-release method. This technique is significantly more difficult for Pu-Xe than for I-Xe, since ^{132}Xe, ^{134}Xe, and ^{136}Xe concentrations must be resolved into several fissiogenic components. PODOSEK (1972) was nonetheless able to estimate Pu-Xe ages for several meteorites (e.g., 146 Myr after St. Severin for the Petersburg eucrite, and ≥ 350 Myr after St. Severin for the Lafayette anomalous achondrite). In meteorites containing excess ^{129}Xe initial $^{244}Pu/^{238}U$ ratios were within a factor of 2 of those expected from the observed initial $^{129}I/^{127}I$ ratios. Thus it appears that the decay-corrected whole-rock Pu/U ratios in most meteorites were constant to within a factor of 2. Because of its potential for dating events substantially later than those datable by I-Xe, it seems likely that the Pu-Xe method will be developed further during the next few years.

The Xe in carbonaceous chondrites and type-3 ordinary chondrites contains a fission component which is distinct from that resulting from ^{244}Pu. The relative abundances of ^{244}Pu fissiogenic isotopes given by ALEXANDER et al. (1971) are:

$$^{131}Xe = 0.25;\ ^{132}Xe = 0.88;\ ^{134}Xe = 0.92;\ ^{136}Xe \equiv 1.00\,.$$

The spectrum of carbonaceous-chondrite fission Xe is not very well defined since its abundance is always small relative to trapped Xe. The relative isotopic abundances estimated by PEPIN (1967b) are:

$$^{131}Xe = 0.20;\ ^{132}Xe = 0.37;\ ^{134}Xe = 0.70;\ ^{136}Xe \equiv 1.0\,.$$

There seems to be little doubt that the fissioning species is an extinct transuranic nuclide. There is some evidence that it is a superheavy nuclide; i.e., one near the island of relatively stable nuclides thought to exist near atomic numbers 110—118 (ANDERS and HEYMANN, 1969; DAKOWSKI, 1969). This Xe component appears to correlate with volatile elements such as In, suggesting that the parent nuclide is volatile (ANDERS and LARIMER, 1971). SABU and MANUEL (1971) and MANUEL et al. (1972) challenged the evidence for superheavy-element fission products in carbonaceous chondrite Xe as well as that for the element being volatile, if it existed. Its association with the "primitive" carbonaceous chondrites suggests the alternative that it might be a relict contained in a presolar interstellar dust component in these meteorites. Exciting developments are to be expected in this research area.

XI. Breakup and Accretional History of Parent Bodies, Meteoroid Shape and Erosion, and Terrestrial Ages of Meteorites

A. Cosmic-Ray Interactions

Cosmic rays consist mainly of protons with mean energies of about one GeV. The mean absorption depth for such a particle and the energetic secondary particles it produces is about a meter. Thus, meteoroids with dimensions of the order of meters or less are subjected to continuous cosmic-ray irradiation in space.

Cosmic-ray primary and secondary particles chiefly undergo spallation-type nuclear interactions. After the fall of a meteorite, its cosmic-ray produced (cosmogenic) nuclides can be measured if they are radioactive, or if they are otherwise isotopically distinguishable from the inherent material. A typical cosmogenic radionuclide observed in recently recovered meteorites is 300 day ^{54}Mn, mainly produced by spallation of ^{56}Fe. The stable cosmogenic nuclides which have been resolved are always isotopes of rare elements — the rare gases, or, in some iron meteorites, lithophilic elements.

Production rates of cosmogenic nuclides are estimated by a combination of techniques. The production cross sections of nuclides can be determined by bombarding appropriate target materials with high-energy protons from terrestrial accelerators. Examples of such studies are those of FUNK and ROWE (1967), FUNK et al. (1967), MATSUDA et al. (1971), SHEDLOVSKY and RAYUDA (1964) and TRIVEDI and GOEL (1969). Others, such as ARMSTRONG (1969) and KOHMAN and BENDER (1967) have fitted bombardment data with equations based on nuclear systematics. Last, and probably best, components of differing composition have been separated from meteorites, the cosmogenic nuclide contents of these different components measured, and empirical production rates for each nuclide from each major element calculated. Examples of such studies are those of BEGEMANN (1965); BEGEMANN et al. (1967); BOCHSLER et al. (1969); BOGARD and CRESSY (1973); HINTENBERGER et al. (1964); HOHENBERG et al. (1967b); KRUGER and HEYMANN (1968); NYQUIST et al. (1967); PODOSEK and HUNEKE (1971); and SCHULTZ et al. (1971b). Recent reviews of production rates for large numbers of cosmogenic nuclides are by KOHMAN and BENDER (1967); HONDA and AR-

NOLD (1967); VOSHAGE (1968b); KIRSTEN and SCHAEFFER (1971); and REEDY and ARNOLD (1972). Production rates of the most commonly studied nuclides, ^3He, ^{21}Ne and ^{38}Ar are reviewed by HERZOG and ANDERS (1971a) and BOGARD and CRESSY (1973).

If the flux of cosmic rays is constant with time, as it now appears (see VOSHAGE, 1967, for a discussion of meteorite research applied to this problem), and the shape of the meteoroid remains unaltered, the amount of a stable cosmogenic nuclide divided by its production rate is an age, generally called the cosmic-ray exposure age. Since shielding affects the production rate, a change in shape of the meteoroid due to fragmentation or erosion changes the production rate.

The production rates of light spallation fragments such as ^3He and of nuclides which differ by only a few mass units from the target nuclide generally increase with depth for about the first 10 cm below the surface, then decrease with further increase in depth. The initial increase with depth parallels the increase in the flux of cosmic-ray secondary particles. Production rates of nuclides widely differing in mass from the target nuclide decrease monotonically with depth. Production rate ratios can show substantial variations with depth; this offers a means for assessing the preatmospheric depth of an individual sample. For example, the $^{38}Ar/^{21}Ne$ production ratio increases by a factor of about 1.5 between the surface and center of a m-sized iron meteorite (BEGEMANN, 1965; SCHULTZ and HINTENBERGER, 1967). An especially large production-rate/depth gradient is found for products of thermal-neutron-induced reactions (EBERHARDT et al., 1963).

B. Erosion and Preatmospheric Shapes

The preatmospheric shapes of meteorites have been investigated by FIREMAN (1958) and HOFFMAN and NIER (1958, 1959). Studies of neutron-induced reaction products have been used to estimate preatmospheric masses (e.g., BEGEMANN and VILCSEK, 1966; MARTI et al., 1966). These investigations indicate that mass losses of a factor of two during atmospheric passage are typical. In some cases much larger losses are indicated, but part of this missing mass may be material which survived atmospheric ablation but which was never recovered from the earth's surface. Densities of tracks produced by cosmic-ray Fe-group nuclei have been used to establish preatmospheric sizes of a few meteorites (PRICE et al., 1967; CANTELAUBE et al., 1969; MAURETTE et al., 1969), and in some cases indicate ablation losses of less than 1 cm.

It is reasonable to assume that the surfaces of meteoroids are slowly eroded by collisions with interplanetary dust or by sputtering by solar wind ions (WHIPPLE and FIREMAN, 1959). Whether space erosion is rapid enough to produce an observable effect on the distribution of cosmogenic nuclides is an unsettled question. The most sensitive indicators are ratios of cosmogenic nuclides, one of which is produced mainly by thermal neutrons (EBERHARDT et al., 1963). To date the magnitude of space erosion remains uncertain and can only be given as upper limits of about 5×10^{-6} cm/yr for stone meteorites (BEGEMANN and VILCSEK, 1966; FIREMAN, 1966) and about 10^{-7} cm/yr for iron meteorites (WHIPPLE and FIREMAN, 1959).

Present evidence favors the view that the destruction of meteoroids results chiefly from collisions with interplanetary objects large enough to destroy the meteoroid, as opposed to dust or ionic erosion (EBERHARDT and HESS, 1960; FISHER, 1966; WETHERILL, 1967; DOHNANYI, 1969). In other words, destruction is not a slow, steady-state process, but a stochastic event resulting in major alteration in the meteoroid dimensions. Most of the destruction appears to result from collisions with debris having dimensions ranging from 0.01—1 R, where R is the radius of the meteoroid (WETHERILL, 1967). The combined effects of space erosion and destructive fragmentation are called mass wastage.

The absence of a relationship between the duration of cosmic-ray exposure and friability indicates either that mass wastage is not a dominant effect in the determination of cosmic-ray age, or that such effects are very sensitive to variations in orbital parameters. For example, the aubrites are among the most friable meteorites, but have the highest mean cosmic-ray age known in a group of stony meteorites (EBERHARDT et al., 1965b). Mass-wastage rates for lunar rocks have been estimated by particle-track techniques and appear to lie in the range 10^{-8}—10^{-7} cm yr^{-1} (CROZAZ et al., 1972; FLEISCHER et al., 1971; BARBER et al., 1971).

C. Exposure-Age Distributions and Their Significance

Accurate cosmic-ray ages can only be obtained if the concentration of a stable cosmogenic nuclide and its production rate can be determined in the same sample. In general, this amounts to coordinated measurement of cosmogenic stable and radioactive nuclides. The production rate of the stable nuclide can be calculated from that of the radionuclide by use of the ratio in their respective production rates determined from bombardment studies and from studies on separated components, as discussed earlier. The determination of radioactive and stable isotopes of the same element should yield the most accurate cosmic-ray ages, since 1. the variation in production-rate ratio with depth or composition will generally not be as large as for more heterogeneous pairs; and 2. inaccuracies in determining the fractional recovery of the element do

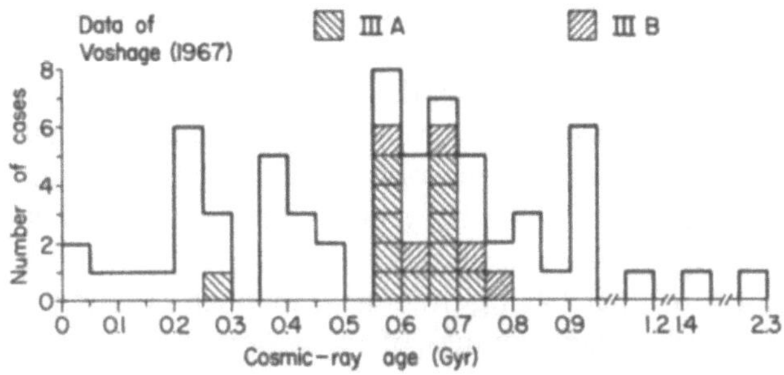

Fig. XI-1. Cosmic-ray age distribution in iron meteorites determined by the ^{40}K-^{41}K method (VOSHAGE, 1967). Seventeen of 18 group-III AB irons fall in a cluster at 650 ± 100 Myr

not affect the exposure age. Where high accuracy is sought, the production-rate ratio can be corrected for depth effects by the measurement of two stable cosmogenic nuclides with widely differing production rate depth dependencies. The outstanding studies in this regard are the iron meteorite investigations of VOSHAGE (summarized in his 1967 paper), in which he measures stable cosmogenic ^{41}K and radioactive ^{40}K and uses the $^{4}He/^{21}Ne$ ratio to correct for the change in the $^{40}K/^{41}K$ production ratio.

VOSHAGE'S (1967) exposure ages for 60 iron meteorites are shown in Fig. XI-1. The highest observed age is 2.3 Gyr in the anomalous ataxite Deep Springs. The absence of meteorites with ages less than 0.1 Gyr is an artifact; because of the long half-life of ^{40}K, VOSHAGE has not been able to obtain accurate results on meteorites with such low ages. If his preliminary studies showed that an iron had a low age, he eliminated it from his series of analyses. Studies by others indicate that a small but important fraction (perhaps 10%) of the irons have ages lower than 0.1 Gyr (COBB, 1966; VILCSEK and WÄNKE, 1963; BEGEMANN et al., 1970). A number of II A irons have very low ages.

VOSHAGE claims a relative error of ±100 Myr in his exposure ages. It is therefore not possible to say whether the clumping of ages of III AB irons near 650 Myr represents one or two peaks. For purposes of discussion we will consider it to represent a single peak.

What interpretation is to be attached to a peaking such as that observed for the III AB irons? Why should their cosmic-ray clocks have all been turned on 650 Myr ago, whereas those of structurally similar meteorites belonging to other groups were turned on at quite different times? The only reasonable explanation seems to be that they were all stored in a single parent body which was disrupted 650 Myr ago. Clustering of cosmic-ray ages appears to be the best evidence currently available in support of the idea that members of certain classes of meteorites were at one time stored in the same parent body.

As will be discussed in more detail in Chapter XII, the mean length of time a meteoroid spends in orbit before capture by a planet is a function of its orbital elements. Meteoroids in earth-crossing orbits have mean lifetimes of the order of 4×10^7 years (WETHERILL and WILLIAMS, 1968). The much higher mean cosmic-ray ages for the iron meteorites are inconsistent with their having originated in earth-crossing orbits, but are consistent with mean lifetimes of the order of 10^9 years expected for Mars-crossing asteroids (ARNOLD, 1964).

EBERHARDT and GEISS (1964) showed that the minimum size of the parent body for a given class of meteorites can be estimated by a knowledge of the mass influx dM/dt of that class and an estimate of the inverse lifetime λ, based on cosmic-ray ages. The equation, which is identical in form to the radioactive decay law, is

$$dM/dt = -\lambda M .$$

WASSON (1972b) has shown that a value dM/dt is about -9×10^5 g yr^{-1} for group IIIA—IIIB irons. If λ is taken to be about 10^{-9} yr^{-1}(consistent with the cosmic-ray ages), the total mass of IIIAB material in Mars-crossing orbits is 9×10^{14} g today, or about 1.8×10^{15} g 650 Myr ago. At a density of 7.9 g cm^{-3}, the minimum radius of a IIIAB parent body disrupted 0.7 Gyr ago is 0.3 km. This lower limit is much smaller than parent body sizes necessary to account for estimated cooling rates (WOOD, 1964; FRICKER et al., 1970).

As noted above, the use of a stable and radioactive nuclide from the same element should provide the most accurate exposure-age determinations. One such technique involving mass-spectrometric determination of stable ^{38}Ar and 270-yr ^{39}Ar has been used (e.g., FIREMAN and GOEBEL, 1970), but is limited to recent falls by the half-life of ^{39}Ar. A very promising development has been the use of 0.21-Myr ^{81}Kr and stable ^{83}Kr to determine exposure ages (EUGSTER et al., 1967c; MARTI, 1967c). This technique offers two additional advantages relative to those involving ^{39}Ar: 1. the ^{81}Kr is not counted, but determined in the same mass-spectrometer run as the other Kr isotopes; and 2. as a result of its longer half-life, it integrates the mean production rate during a larger fraction of the exposure age of the meteorite. Unfortunately, ^{81}Kr$-$^{83}Kr exposure ages have been reported for relatively few meteorites. It is to be hoped that this technique will be applied to many more in the future.

In a few studies a number of stable and radioactive cosmogenic nuclides have been measured in the same meteorite. Among these, the study of the St. Severin LL chondrite by MARTI et al. (1969) can be cited as especially comprehensive; exposure ages based on six different methods are reported. The study of Lost City and other meteorites by BOGARD et al. (1971) is nearly as comprehensive.

Most cosmic-ray age studies of silicate-rich meteorites consist of measurements of two or more of the stable cosmogenic nuclides ^{3}He, ^{21}Ne or ^{38}Ar without concurrent measurement of a radioactive nuclide. These nuclides are measured as part of the studies of the isotopic compositions of He, Ne and Ar. That these nuclides are good cosmogenic monitors results from their low abundances in primordial rare gas and the fact that they are not produced by the decay of long-lived radionuclides. The most precise data are generally those based on ^{21}Ne, since 1. Ne is less prone to diffusive loss than He, and 2. the cosmogenic and primordial contributions in Ne are easier to separate than in He and Ar, since in the former all three isotopes are non-radiogenic; He and Ar have only one and two nonradiogenic isotopes, respectively. Cosmic-ray ages based on ^{21}Ne include random errors of up to 40% if no allowance is made for the depth dependence of the production rate.

Most ordinary chondrite ^{3}He ages in the literature are based on the production rate of 2.00×10^{-14}cm^{3} ^{3}He g^{-1} yr^{-1} published by KIRSTEN et al. (1963), and ^{21}Ne ages on a production rate of 3.77×10^{-15}cm^{3} ^{21}Ne g^{-1} yr^{-1} calculated by dividing the ^{3}He rate by the mean ^{3}He/^{21}Ne ratio of 5.3 observed in ordinary chondrites (See HEYMANN, 1967)[10]. KRUGER and HEYMANN (1968) estimated a maximum variation in the ^{3}He production rate of 45% among stony meteorites, with the lowest values in CI chondrites, the highest in the aubrites. An equation given by HERZOG and ANDERS (1971a) and data by WIIK (1969) indicate that these range from 0.299 cm^{3} g^{-1} yr^{-1} in CI chondrites to 0.688 cm^{3} g^{-1} yr^{-1} in aubrites, a range amounting to a factor of 2.3. At depths up to a few tens of cm, ^{3}He production rates are relatively depth-independent, whereas ^{21}Ne rates are much more sensitive to depth (EBERHARDT et al., 1966). The values of HERZOG and ANDERS (1971a) hold only when the ^{3}He/^{21}Ne ratio is 5.3 (after correction for possible diffusion losses).

Figure XI-2 illustrates ^{21}Ne exposure-age data for ordinary chondrites (compiled by L. SCHULTZ, private communication) and for carbonaceous chondrites (MAZOR et al., 1970). Corrections have been made for primordial ^{21}Ne contributions. Cosmogenic ^{21}Ne concentrations are shown for the ordinary chondrites only; the conversion to exposure ages was obtained by dividing by 4.66×10^{-15} cm^{3} g^{-1} yr^{-1}. The carbonaceous chondrite ages are as given by MAZOR et al. (1970); no correction has been made to allow for the higher production rates of HERZOG and ANDERS (1971 a). Since the uniformly low K-Ar ages indicate that CI and

[10] All gas volumes are measured at 273 °K and 1 atm.

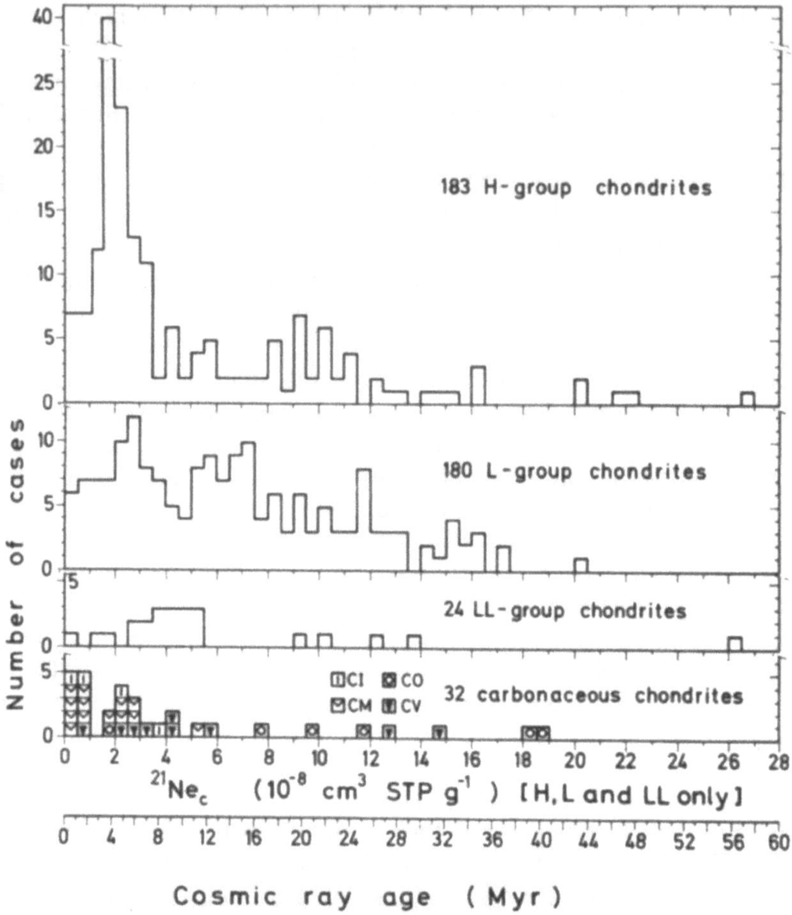

Fig. XI-2. Cosmic-ray age distribution in ordinary chondrites (MAZOR et al., 1970) based on their ^{21}Ne contents. Carbonaceous chondrite ages decrease with increasing friability. Cosmic-ray ages of chondrites tend to be lower than those of iron meteorites by factors of about 20—100

CM chondrites have experienced extensive gas loss (requiring a correction in the opposite direction), it appeared simplest to make no correction at all, but only to note that there is a large amount of uncertainty connected with these exposure ages.

As discussed for the iron meteorites, clusters of cosmic-ray ages are generally interpreted as major parent-body disruption events. At various times, clusters have been reported for the L-group (EBERHARDT and GEISS, 1964) and H-group (ANDERS, 1964; WÄNKE, 1966). TANENBAUM (1967) argues that the L-group clusters are not statistically significant.

The data in Fig. XI-2 seem to confirm the H-group cluster at 4 Myr. It appears that one or more major breakup events released H-group meteoroids into space at about this time. The absence of significant L-group clusters indicates either that these meteoroids were injected into space in a series of small events, or that the cosmic-ray ages are mainly recording the effects of some other process, such as fragmentation by collisions with smaller meteoroids. Note that the L-group exposure ages are all an order of magnitude or more lower than the ca. 0.5 Gyr event recorded by K-Ar and U, Th-He ages. The latter is interpreted as recording a major event which involved disruption of the L-group parent body. The LL group seems to give a cluster at bout 8—10 Myr, and it will be interesting to see whether this is confirmed by future studies.

No peaks are evident among the carbonaceous chondrites. A trend is observed for cosmic-ray age to decrease with increasing petrologic type number, or through the sequence

$$CI < CM < CV = CO\,.$$

To some degree this probably reflects an artificial effect resulting from decreasing gas loss with increasing mean grain size as petrologic group increases, and to some degree a real effect related to decreasing friability through the same sequence.

A few other meteorite classes show clustering of cosmic-ray ages. The most pronounced of these is observed among the aubrites, where six of nine members have ages of about 40 Myr. One of the three remaining aubrites, Norton County, has an age of about 100 Myr (HERZOG and ANDERS, 1971 b), the highest cosmic-ray age known in a stony meteorite. It is curious that the highly friable aubrites tend to have ages substantially greater than those observed in other stony meteorites, such as the ordinary chondrites (Fig. XI-2). This could indicate that the aubrite orbits avoid debris-rich portions of the solar system. The few diogenites which have been studied have cosmic-ray ages near 15—20 Myr (ZÄHRINGER, 1968; EBERHARDT et al., 1966). Some meteorites show correlations between cosmic-ray ages and compositional parameters; the enstatite chondrites tend to show decreasing age with increasing Si content of the metal (WASSON and WAI, 1970) and the IIAB irons seem to have cosmic-ray ages which decrease with increasing Ir content of the metal (WASSON, 1969).

As the studies of cosmogenic effects become more sophisticated, increasing evidence is found for two-stage or multistage irradiations. That different portions of large iron meteoroids such as Sikhote-Alin (KOLESNIKOV et al., 1972) or Canyon Diablo (HEYMANN et al., 1966) have experienced minor fragmentations is not surprising in view of their large size. More interesting is the occasional evidence of a two-stage irradiation in small and relatively friable stony meteorites such as Ivuna (CI 1), Mighei (CM 2) or Serra de Magé (eucrite) (FUSE and ANDERS, 1969). A common speculation regarding the origin of the CI and CM chondrites is that they originate in comets. The surface layers of a comet would seem to be ideally suited for providing a location in which an early irradiation could occur. The evaporation of ices could then result in the "gentle" release of the meteoroid into space without excessive fragmentation. Thermal neutron effects should be prominent if the early irradiation were in the presence of ices.

D. Accretional History

As discussed in Chapters VI and IX, certain brecciated silicate-rich meteorites consist of light clasts in dark matrices. Often the matrices are rich in solar-type rare gases. PELLAS (1972) and co-workers (SCHULTZ et al., 1972) studied two of these meteorites (Weston, Djermaia) in great detail; both cosmic-ray tracks and rare-gas isotopes were investigated. They found some xenoliths with ^{21}Ne ages or heavy-particle track ages greater than those observed in the matrix of the meteorite, and interpreted their evidence to indicate that the xenoliths of these meteorites have been subjected to different degrees (and types) of preirradiation for periods of the order of 10 Myr prior to the final accumulation of the breccia into the form in which it is today. It thus appears that some xenoliths existed as m-sized bodies in space for at least 10 Myr prior to the final accumulation of material to the parent body. This point is also discussed in Chapter XVIII.

E. Terrestrial Ages

After a meteorite falls, it is shielded from cosmic rays by the earth's atmosphere. If a series of radionuclides of differing half-lives are measured in a meteorite find of unknown age, those with half-lives much shorter than the terrestrial age will not be detectable and those with half-lives substantially greater than the terrestrial age will have undergone negligible decay. The activities of cosmogenic radionuclides with half-lives comparable to the terrestrial age will have undergone substantial reduction, but still be measurable with fair accuracy. The decrease in activity provides a direct measure of the terrestrial age. The most common nuclides employed in terrestrial age studies are 270-yr ^{39}Ar, 5700-yr ^{14}C, and 3×10^5-yr ^{36}Cl.

Table XI-1. Terrestrial ages of meteorites

Meteorite	Group	Terrestrial age (yr)	Isotope	Reference
Chondrites				
Brownfield (1937)	H3	12700	^{14}C	BOECKL (1972)
Estacado	H6	6800	^{14}C	BOECKL (1972)
Faucett	H	$\leqq$ 2000[a]	^{14}C	BOECKL (1972)
McKinney	L4	10200	^{14}C	BOECKL (1972)
Plainview (1917)	H5	$\leqq$ 2000	^{14}C	SUESS and WÄNKE (1962)
Potter	L6	$\geqq$ 20000	^{14}C	SUESS and WÄNKE (1962) GOEL and KOHMAN (1962)
Selma	H4	6500	^{14}C	GOEL and KOHMAN (1962)
Woodward County	H	14000	^{14}C	SUESS and WÄNKE (1962)
Irons				
Clark County	IIIF	600000	^{36}Cl	VILCSEK and WÄNKE (1963)
Henbury	IIIA	$\leqq$ 7000	^{14}C	GOEL and KOHMAN (1962)
Keen Mountain	IIA	1290	^{39}Ar	VILCSEK and WÄNKE (1963)
Odessa	IA	$\geqq$ 11000	^{14}C	GOEL and KOHMAN (1962)
Sardis	IA	$\geqq$ 16000	^{14}C	GOEL and KOHMAN (1962)
Tamarugal	IIIA	1500000	^{36}Cl	VILCSEK and WÄNKE (1963)
Washington County	Anom	1400	^{39}Ar	FIREMAN and DEFELICE (1960)

[a] Reported to be 2000 ± 2000 yr, but the high ^{14}C counting of 58 dpm, similar to levels observed in fresh falls, indicates that the correct age is $\leqq$ 2000 yr.

Since cosmic ray ages are generally long relative to the half-lives of these nuclides, the activities at time of fall are equal to the production rates, best estimated by measuring a series of cosmogenic nuclides and using these to determine the appropriate parameters (such as burial depth). The more common practice is to use average activities observed for the same nuclide in observed falls of the same class of meteorite. As a result of inherent experimental difficulties in determining the very low levels of radioactivity and resulting from uncertainties in the estimation of the initial decay rate, terrestrial age estimates are generally accurate to only about ±25% of the reported value.

Table XI-1 lists a sample of terrestrial-age data. My selection is slightly biased in terms of older ages and greatly biased in terms of samples from dry climates (especially the southwestern portion of the United States). However, some of the oldest irons (Clark County, Sardis) are from areas of moderate rainfall (ca. 100 cm/yr). The data tabulated and other data by BOECKL (1972) and GOEL and KOHMAN (1962) indicate that there is no evidence among ordinary chondrites of a dependence of the rate of weathering on petrologic type. BOECKL (1972) estimates a terrestrial half-life of 3600 yr for the weathering of ordinary chondrites in the southwestern United States.

Exact terrestrial age data are generally not accurate enough to confirm that a meteorite find is associated with a historic fireball. However, they can sometimes eliminate potential pairings. For example, O. A. SCHAEFFER (private communication) has shown by ^{39}Ar measurements that the Benthullen chondrite has been on the earth for less than 200 years and is therefore unrelated to the Oldenburg fall of 1368 mentioned in HEY (1966).

XII. Orbits

Heliocentric orbits are elliptical, with the sun occupying one focus of the ellipse. The long axis of the ellipse is called the major axis; half this distance is a, the semimajor axis. Half of the short axis is b, the semiminor axis. The distance of closest approach to the sun is q, the perihelion, and the distance of greatest recession is Q, the aphelion. The eccentricity of the ellipse, e, is defined by the following relationships:

$$e = \left(1 - \frac{b^2}{a^2}\right)^{\frac{1}{2}} = 1 - q/a\,.$$

A heliocentric orbit is normally defined by the five elements a, e, i, Ω, and ω. The latter three quantities give the orientation of the orbit in space and are illustrated in Fig. XII-1. The inclination, i, is the angle subtended between the plane of the orbit and the plane of the earth's orbit (the ecliptic plane). More exactly, the inclination is the angle between the north polar axial vectors of the orbits, where north is defined as the direction the thumb points when the fingers of the right hand extend in the direction of the motion of an object in its orbit. All planets, asteroids, and all but four periodic comets are in direct orbits with inclinations less than 90°. Orbits with inclinations greater than 90° are called retrograde orbits. The intersection of the orbital and ecliptic planes defines a line passing through the sun. The orbit intersects this

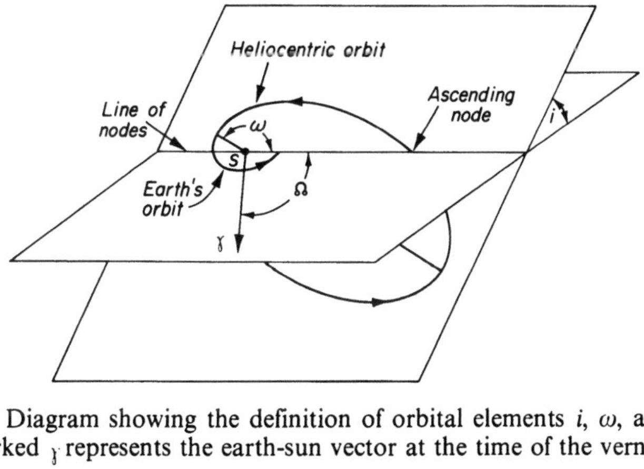

Fig. XII-1. Diagram showing the definition of orbital elements i, ω, and Ω. The arrow marked γ represents the earth-sun vector at the time of the vernal equinox

Table XII-I. Orbital elements of asteroids, comets and meteorites. The last column lists geocentric velocities for the meteorites and estimates of the radii of the asteroids

Object	q (AU)	Q (AU)	a (AU)	e	i (°)	ω (°)	Ω (°)	R (km) or V_g (km/sec)
Asteroid								
1 Ceres	2.55	2.98	2.77	0.079	10.6	70	80	550
2 Pallas	2.11	3.42	2.77	0.236	34.8	310	173	310
4 Vesta	2.19	2.53	2.36	0.071	7.1	150	104	280
433 Eros	1.13	1.79	1.46	0.223	10.8	178	304	8.5
1362 Griqua	2.16	4.40	3.28	0.342	24.1	263	121	12.9
1685 Toro	0.77	1.97	1.37	0.436	9.4	127	274	2.8
Comet								
Encke	0.34	4.10	2.22	0.847	12.0	186	334	—
Grigg-Skjellerup	0.99	4.93	2.96	0.665	21.1	359	213	—
Meteorite								
Pribram	0.79	4.05	2.42	0.674	10.4	242	17	17.7
Lost City	0.97	2.35	1.66	0.417	12.0	161	283	8.8

"line of nodes" at two points: at the ascending node it passes from south to north, at the descending node from north to south of the ecliptic plane. The longitude of the ascending node, Ω, is the angular distance in the ecliptic plane from the direction of the earth's vernal equinox to the ascending node, measured in the same direction as the earth moves in its orbit. The argument of perihelion, ω, is the angle between the ascending node and the perihelion direction, measured in the orbital plane and in the direction of motion of the object in the orbit. Note that the sun is at the apex of the latter two angles. In order to locate the position of an object within its orbit we must know a sixth quantity, a time. This is usually expressed as the time of perihelion passage.

Orbital elements for several meteorites and asteroids are listed in Table XII-1. Asteroid elements are from the *Ephemerides* (CHEBOTAREV, 1971) and comet orbits from MARSDEN (1972). Perihelia and aphelia are also given, since these are useful in envisioning the dimensions of the region swept out by the orbit.

The geocentric velocity, V_g, of the two meteorites is also listed in Table XII-1. This is the relative velocity between the object and the earth corrected for the effect of acceleration by the earth's gravitational field. The initial atmospheric velocity V_∞ of the meteorite is related to the geocentric velocity V_g as follows:

$$V_\infty^2 = V_g^2 + v^2$$

Where v is the escape velocity of the earth, 11.2 km sec^{-1}. Radii are listed for the asteroids. Those for the three largest are taken from the summary

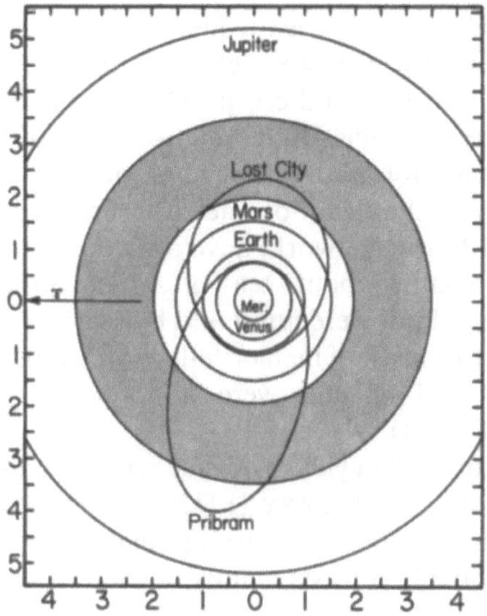

Fig. XII-2. Orbits of Pribram and Lost City relative to those of the planets and to the asteroid belt. All orbits are drawn in the same plane (i.e., as if $i = 0$)

of MORRISON (1973). They are subject to errors of the order of 10—20%. The radii of the smaller asteroids are calculated from the absolute magnitudes given in the *Ephemerides* (CHERBOTAREV, 1971), using a relationship quoted by WETHERILL and WILLIAMS (1968):

$$\log R = 3.39 - 0.2\,g$$

where R is the radius and g the absolute magnitude. This formula is derived by assuming a mean albedo of 0.07, the same as that of the moon.

Accurate orbital elements for meteorites can only be determined by simultaneous photographic observations from two or more locations. Such orbits are available for two recovered meteorites, the H5 chondrites Pribram and Lost City. Their orbits are shown in Fig. XII-2. Note that the aphelion of each is in or slightly beyond the asteroid belt. The perihelion of Lost City barely crosses 1 AU, whereas that of Pribram extends nearly to Venus.

Pribram was accidentally photographed by cameras installed in Czechoslovakia to observe artificial satellites (CEPLECHA, 1961). Partly as a result of its observation, two networks of all-sky cameras have been established to obtain additional orbital data on meteorites, as well as on large meteors and fireballs. The Prairie Network consists of 16 stations in a roughly circular area with a radius of approximately 500 km centered near southeastern Nebraska, USA (MCCROSKY and CEPLECHA,

1969; WOOD, 1968). Lost City is the first meteorite photographed by the Prairie Network (MCCROSKY et al., 1971). A smaller area (Czechoslovakia and the southern portion of the Federal Republic of Germany) is covered by the All-Sky Network. As of July, 1973, the All-Sky Network had not photographed a meteorite which had been recovered. Despite the lack of meteorite recoveries, the networks (particularly the Prairie Network) have provided orbital information on hundreds of fireballs, about ten of which probably resulted in recoverable material but were not found by search parties. A number of others appeared to consist of meteorite-like materials, but had estimated terminal masses <1 kg, and were not sought by ground parties (MCCROSKY et al., 1971).

The photographic data yield evidence related to the properties of the infalling meteoroids; MCCROSKY (1972) groups the objects into three broad categories: 1. compact, strong material—based on recovery statistics, this probably consists mainly of chondrites of petrologic types 3—6; 2. compact but friable material, probably CI or CM carbonaceous chondrites; and 3. low-density, weak material which may be of cometary origin. The relative fall abundances of these three categories MCCROSKY estimates at 30%, 60% and 10%, respectively. The abundance of friable material is much higher than expected from recovery statistics according to Table II-8 (CI + CM = 3% of observed falls), as had been proposed earlier by SHOEMAKER and LOWERY (1967) from airwave data.

Although visual observations are incapable of yielding orbits, they nonetheless yield valuable information bearing on the meteorite's orbit. The time of day when the meteorite falls is generally known to within one-half hour. To a rough approximation, meteorites which fall in the afternoon have perihelia near 1 AU, those which fall in the morning have aphelia near 1 AU [this statement is based on the assumption that the meteorites are in direct orbits lying in the ecliptic plane, and ignores effects resulting from the fact that the earth's capture radius is considerably greater than its geometrical radius—see WETHERILL (1968 a) for a detailed discussion].

More information is available from an accurate determination of the radiant of the meteorite, i.e., the apparent point of origin on the celestial sphere. According to LEVIN and SIMONENKO (1969), good radiant data are available for 13 meteorites in addition to Lost City and Pribram. Individual radiants can be used to search for associations between the meteorites and astronomically observed members of the solar system. For example, meteorites coming from comet Encke should have radiants which are the same as the Taurid meteors. In fact, no meteorites are definitely associated with meteor showers, although large fireballs are often associated with the Taurids (WHIPPLE, 1950; MCCROSKY, 1968),

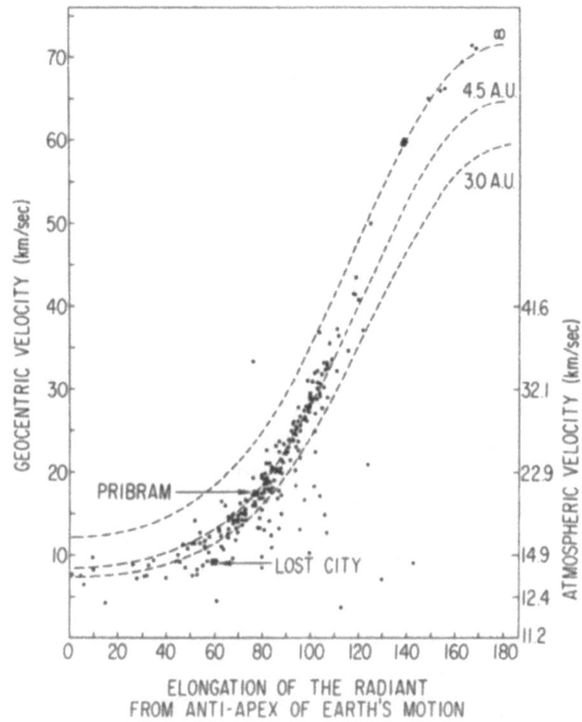

Fig. XII-3. This plot of geocentric velocity versus elongation of the radiant shows data for the two photographed meteorites as well as for 200 Prairie Network fireballs. The curves show the loci of points expected for meteorites with aphelia at 3.5, 4.6 and ∞ AU. See text for details. (From WETHERILL, 1974)

and there is marginal evidence that the H chondrite Blansko is associated with this shower (ÖPIK, 1966).

Radiant information can also be combined to investigate classes of orbits. Figure XII-3 is a plot (provided by G.W. WETHERILL) of geocentric velocity versus elongation (i.e., angle) of the radiant from the antiapex of the earth. The apex of the earth is the direction its orbital velocity vector points to any given moment; the antiapex is the opposite direction. Plotted are 200 major fireballs observed by the Prairie Network and the two meteorites with accurately determined orbits. The curves show the loci of direct low inclination orbits with aphelia at 3.5 and 4.5 AU and at infinity (i.e., parabolic orbits). Note that most fireballs have aphelia between 3.5 and 4.5 AU. The absence of orbits with large elongations and low geocentric velocities indicates that most meteorites make very few close approaches to the earth prior to capture (WETHERILL, 1971).

Another parameter closely related to the orbits of meteoroids is the cosmic-ray age (Chapter XI). The cosmic-ray clock is started by a fragmentation event which reduces the meteoroid to dimensions such that most or all of its volume is subjected to cosmic-ray interactions. Processes that remove the meteorite from interplanetary space determine its survival lifetime, which in most cases must be comparable to its cosmic-ray age. These processes include mass wastage (destruction by particle erosion and fragmentation) and removal from interplanetary space as a result of capture by a planet, or gravitational interaction with a planet resulting in ejection from the solar system.

Rates of mass wastage in the solar system are estimated by WETHERILL (1967), DOHNANYI (1969) and GAULT and WEDEKIND (1969) (see GAULT et al., 1972 for estimations of these processes on the lunar surface). Estimates vary considerably, but indicate a mean mass-wastage lifetime of stony material in meteoritic orbits of about 10^7—10^8 years. Lifetimes with respect to short-range interactions with planets range from 10^7 to $>10^{10}$ years (ÖPIK, 1951; ARNOLD, 1964). Because of the difficulty in resolving mass-wastage and planetary-interaction effects, it is difficult to use cosmic-ray ages to define meteorite orbits. On the other hand, these ages can provide valuable boundary conditions. For example, some orbits yield planet-interaction lifetimes substantially shorter than iron-meteorite cosmic-ray ages. It is not reasonable to suppose that the III A—III B iron meteorites (with mean cosmic-ray ages of 650 Myr) were in earth-crossing orbits (with planet-interaction lifetimes of about 10 Myr) at the time their cosmic-ray clocks were started. Stated in more general terms, the mean cosmic-ray age of a meteorite class tends to set a lower limit to the dynamic lifetime associated with the orbit that the material occupied immediately following the fragmentation event which started the cosmic-ray clock. Note, however, that the orbit of a meteoroid at the time of earth capture may differ substantially from that initial orbit.

A number of telescopically observable solar system objects have been cited as possible meteorite parent bodies: the moon, Mars, ring asteroids, asteroids near Kirkwood gaps, Mars-crossing asteroids, earth-crossing (Apollo) asteroids, and periodic comets, particularly those with aphelia well inside Jupiter's orbit ($Q < 4.6$ AU). The samples returned from the moon differ in many respects from known meteorites; it has not been a source. The high escape velocity of Mars (5 km/sec) makes it extremely unlikely that meter-sized objects could survive acceleration from its surface (particularly since lunar meteorites are not known). We are left with the latter five sources.

Before discussing the features of these sources, it is necessary to discuss effects which bring about changes in their orbits and in those of meteoroids. Because of its large mass, Jupiter has a perturbing influence on all objects in the inner solar system. As a result, e and i tend to vary out of phase (WETHERILL, 1969), and the longitude of the node and the argument of perihelion undergo precessional motion. The angular momentum of the object tends to remain constant, and thus a does as well. These effects of Jupiter are called secular perturbations and occur on a time scale of about 10^4—10^6 years. As a result, fragments initially in nearly identical orbits will become widely separated in periods of this magnitude.

Despite these secular effects, HIRAYAMA (1918) was able to recognize nine families of asteroids having similar orbital elements, each of which was presumed to have formed by the breakup of a larger object. BROUWER (1951) used proper elements (obtained by calculating the means of the ranges which the orbital elements assume as a result of secular perturbations) to search for asteroid families and listed a total of 29 families including confirmation of the original nine families of Hirayama. ANDERS (1965) discussed the fragmentation of the asteroids which formed the families. Improved calculations by ARNOLD (1969) and LINDBLAD and SOUTHWORTH (1971) increased the number of Hirayama families to about 40. WILLIAMS (1971) used improved procedures to calculate proper elements which will result in further refinements of the families; he notes that some families result from orbital selection as a result of Jupiter resonances, rather than from breakup of a larger body. There is marginal evidence for the existence of jetstreams (ALFVÉN, 1969)—groups of asteroids in which all five proper elements are similar (see DANIELSSON, 1971, for a discussion of the evidence).

Close approaches to the planets cause major changes in all orbital elements. ÖPIK (1951) was the first to attempt an evaluation of this effect. ARNOLD (1964, 1965a, b) used Monte Carlo methods and an electronic computer to make a more quantitative study. Further calculations have been made by ANDERS (1971c), ANDERS and MELLICK (1969) and WETHERILL (1967, 1968a, b, 1969, 1971).

Objects which are in Jupiter-crossing orbits have lifetimes of only about 10^5 years before undergoing one of various fates, the most likely being ejection from the solar system. In contrast, objects which cross only the orbit of Mars will have mean lifetimes of about 10^9 years, since Mars' mass is small. The earth is about 10 times more massive than Mars, and objects crossing the orbit of the earth have mean lifetimes of the order of 10^7—10^8 years. Earth-crossing objects are generally captured by the earth or perturbed by the earth into Jupiter-crossing orbits, where their lifetimes are short.

Figure XII-4 is a histogram showing asteroid frequency as a function of semimajor axis. Kirkwood gaps occur at values of a such that the ratio of the period of the asteroid to Jupiter's period is equal to the ratio of two small integers. The semimajor axes corresponding to these ratios are listed at the top of the graph. Peaks appear at the 2/3 (Hilda family) and 3/4 (Thule) resonances. Asteroids with values of a in the resonance

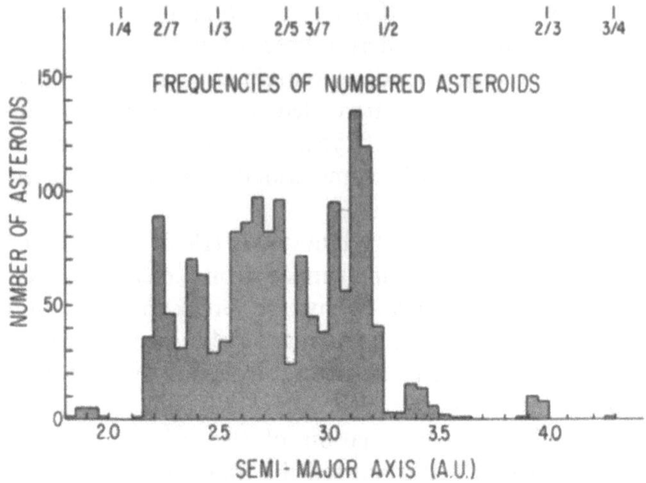

Fig. XII-4. Asteroid frequency as a function of semimajor axis. The fractions along the top are the ratios of the period of the asteroid relative to that of Jupiter. Peaks occur at the two largest ratios and minima called Kirkwood gaps at the lower values

region are perturbed by Jupiter into orbits which have much larger values of e and i than the ring asteroids, but which cannot make close approaches to Jupiter, and appear to be stable over periods of time comparable to the age of the solar system. The precise reason why gaps in the asteroid distribution occur at the 1/2 and smaller resonances is not known, although it may be that the high encounter velocities which these objects have relative to ring asteroids lead to their rapid destruction; the Hildas and Thule may have survived because the large semimajor axes of their orbits make them less likely to intersect those of the ring asteroids.

Programs of increasing sophistication have been developed to search for parent-body orbital elements consistent with the orbital properties and cosmic-ray ages observed in meteorites. Orbital data are essentially available only for ordinary chondrites, which account for about 79% of observed meteorite falls, including the two photographically determined orbits. WETHERILL (1968a, 1969, 1971) has made the most extensive investigations of the orbits of these meteorites.[11] He shows that radiant

[11] Although WETHERILL includes all chondrites in his plots (e.g., of exposure age or time of fall), the L- and H-group chondrites account for most of the data. WÄNKE (1966) showed that the time-of-fall distributions of these groups are essentially the same, and that their cosmic-ray age distributions differ mainly as a result of the 4-Myr peak in the age group (see Fig. XI-1). Thus WETHERILL's basic conclusions should be valid for either of these groups, although the reduced populations cause some weakening in the statistical strength of the conclusions.

and time-of-fall data can be understood only in terms of relatively unique types of initial earth-crossing orbits, with low inclinations, perihelia near 1 AU and aphelia of about 4.5 AU. Presumably objects move into such orbits as a result of perturbations by Jupiter.

P. ZIMMERMANN and G.W. WETHERILL (unpublished results, 1972) think that it may be possible to get asteroidal material into such orbits by taking advantage of the increased mean eccentricity which is associated with the Jupiter resonances. If a km-sized asteroidal fragment is jostled from the highly populated region at about 3.2 AU into the 1:2 Kirkwood gap at 3.3 AU, Jupiter will quickly pump it into a highly eccentric orbit. Cratering events remove meteorite-sized debris from the asteroid. Some of this debris will be removed with relative velocities as high as 0.2 km sec^{-1}, and as a result will be placed in an orbit which is still highly eccentric, but not in resonance with Jupiter. Jupiter's perturbations will then remove most of these meteoroids from the solar system, but some will become earth-crossing and will be captured by the earth in a period of the order of 10^7 years, comparable to the cosmic-ray ages of meteorites.

ANDERS (1971c) argues that the Mars-crossing asteroids are the meteorite parent bodies. The iron-meteorite classes do have ages comparable to the dynamic lifetimes of Mars-crossing asteroids, which tends to support this viewpoint. The chondrites have much shorter ages, however; ANDERS (1971c) argues that this is no problem, provided that the mean cosmic-ray age of these objects is determined by mass wastage rather than by planetary interactions. Although this is true, the yield would be greatly reduced by the mass-wastage process. A more serious difficulty is associated with the discovery that secular variations in the orbital elements occur in such a way that the argument of perihelion is near 90° or 270° when the eccentricity is at a maximum (WETHERILL, 1969). As a result, most asteroids with present day values of q which are less than Mars' a or Q (1.5 or 1.7 AU) never make close approaches to Mars. For this reason, the amount of material entering earth-crossing orbits by this mechanism may be too low to provide the observed meteorite flux at the earth's orbit.

ÖPIK (1966, 1968) and WETHERILL (1969) suggested that meteorites (more specifically, chondrites) come from earth-crossing comets, particularly those with aphelia less than about 4.6 AU. ÖPIK pointed out that there is one live comet, Encke, and its associated meteor stream (the Taurids) in such an orbit now, and that the Geminid meteors, which have a mean aphelion distance of 2.66 AU, are probably associated with a comet whose ices have evaporated. The dynamic lifetime of a cometary body such as Encke is about 10^8 years, whereas the evaporation lifetime of its ices is only about 10^4 years; from this ÖPIK extrapolated a total population of 10^4 burned out comets in the inner solar system. Some comets have quite large nuclei; ÖPIK (1963) and ROEMER (1966) estimated a radius of 10—56 km for Comet Humason. No evidence is available which can be used to estimate the mass of the nucleus after the ices have evaporated. There seems to be no question about the general magnitude of these numbers, and a substantial fraction of the interplanetary debris captured by the earth must be cometary in origin. On the other hand, the high temperature metamorphism ($T \geq 1000\,°$ K) observed in many meteorites seems easier to rationalize for asteroids than for comets. In the absence of orbital data, a cometary origin for unmetamorphosed CI and CM chondrites seems a reasonable hypothesis, however.

It had generally been concluded that the only way to remove material from the bulk of the asteroidal belt was by a series of mutual collisions, a mechanism which WETHERILL (1968c) showed to be inadequate to explain the observed flux of meteorites. A surprising development is the discovery of LOWERY (1971) and J. G. WILLIAMS (unpublished results, 1972) that the orbits of Lost City and Pribram undergo rapid variations in eccentricity when extrapolated backwards in time. WILLIAMS finds that the variations of Lost City result from a resonance in the rate of its nodal precession with that of Jupiter, while that of Pribram results from a resonance between the arguments of perihelion of this object and Jupiter. It is not known whether such resonances can remove material from the asteroidal belt at a rate sufficient to explain the observed meteorite flux, but it seems more than coincidence that each of these photographed falls should have a Jupiter resonance.

Because of the difficulty in allowing for the combined gravitational perturbations of the sun, Jupiter, and one or more of the terrestrial planets, all calculations to date are approximate. It seems likely that improved techniques and additional orbit determinations will result in substantial gains in our understanding of meteorite orbital problems during the next few years.

As discussed in Chapter XVII, telescopic spectrophotometric studies can also help to identify solar system objects which seem to be appropriate parent bodies for certain classes of meteorites (CHAPMAN et al., 1971). Unfortunately there are relatively few well-defined absorption bands in the observed spectra, and unambiguous association of most asteroids with meteorite classes on the basis of telescopic observations does not appear possible in the near future. Rendezvous or landing missions by spacecraft seem necessary in order to obtain unambiguous compositional information on asteroids.

XIII. Fall and Recovery

Meteorites enter the earth's atmosphere with velocities of about 11—30 km sec^{-1} (see Fig. XII-3). Meteoroids with higher initial velocities rarely reach the earth's surface. At altitudes of about 100 km the air density becomes great enough to create appreciable friction and deceleration begins, accompanied by an increase in the surface temperature. The surface becomes hot enough to emit visible light and to melt. The melted material is blown away.

The initial velocity is far in excess of the speed of sound (~0.3 km sec^{-1}), and the meteorite is subjected to substantial shock pressure, which often causes it to be disrupted into fragments. According to KRINOV (1960), disruption normally occurs between 12 and 30 km. Photographs show that the Lost City H-group chondrite broke up at 32 km (MCCROSKY et al., 1971). HELLYER (1969) surveyed fall statistics and found that 1. about half the meteorite falls are multiple; 2. multiple falls are more common for carbonaceous chondrites and less common for irons than for other classes; and 3. the larger the object, the more likely is a multiple fall.

Air resistance slows down most meteorites to a uniform terminal velocity which is generally of the order of 100—300 m sec^{-1}. The average altitude at which the meteorite reaches terminal velocity is 12.5 km (KRINOV, 1960). Calculations show that a meteorite must have a preatmospheric mass greater than 10 t (and must avoid appreciable fragmentation) in order to reach the earth's surface at a velocity greater than terminal velocity (HEIDE, 1957). At terminal velocity a meteorite is no longer hot enough to emit light, and it no longer suffers appreciable ablation. Falling meteorites look dark to observers near the impact site (HEIDE, 1957).

The interiors of meteorites remain cool during atmospheric passage. They are usually cool enough to touch immediately after striking the earth's surface (WATSON, 1962). There are some reports of frost forming on the interior surfaces of stony meteorites which broke on impact. Iron meteorite falls have a well-defined heat-altered zone immediately below the fusion crust. Thermal-alteration effects can be observed in this zone, which is seldom as wide as 1 cm. Meteorite interiors remain cool because conduction is a relatively slow process, unable to transport appreciable heat during the interval of about 10 sec when the surface is hot. Because

Table XIII-1. Classification of meteorites found near explosion (hypervelocity) craters

Meteorite	Group	Crater diameter (m)
Boxhole	IIIA	160
Campo del Cielo[a]	IA	56
Canyon Diablo	IA	1200
Dalgaranga	Mes	70
Henbury	IIIA	200
Kaalijarv	IA[b]	90
Odessa	IA	150
Wolf Creek	IIIB	840

[a] Possibly not an explosion crater (see CASSIDY, 1968).
[b] Tentative identification by V. F. BUCHWALD (private communication).

of their lower thermal conductivities, the interiors of stony meteorites are subjected to still less heat alteration than are irons.

Meteorites experience ablation during atmospheric passage. The amount of material removed can be estimated by detailed studies of cosmic-ray-produced nuclides or particle tracks (see Chapter XI). BALDWIN and SHAEFFER (1971) report calculations of ablation and breakup effects on meteoroids. They find that the ratio of final to initial mass decreases exponentially with increasing initial atmospheric velocity, decreasing from about 0.9 at 11 km sec^{-1} to about 0.1 at 20 km sec^{-1} for H-group chondrites.

On rare occasions a meteorite strikes the earth with a velocity near its initial atmospheric velocity, and an explosion crater is produced. A combination of aerial mapping and the development of methods for recognizing the effects of shock metamorphism has led to the recognition of a large number of meteorite craters ("astroblemes") on the earth's surface. Reviews of recent studies of meteorite impact structures are given by DENCE (1972), VON ENGELHARDT (1972), DIETZ (1972) and SHORT and BUNCH (1968). Only eight explosion (hypervelocity) craters are associated with recognizable meteoritic debris; these are listed in Table XIII-1, along with the classification of the meteorite. It is probably of some significance that seven of these features are associated with iron meteorites and that these irons belong to only two groups. Four are members of the IA portion of group IAB, two of the IIIA portion and one of the IIIB portion of group IIIAB; these are the two largest groups of iron meteorites. Although it is true that iron meteorites survive weathering better than stones, there is only one additional meteorites crater (Aouelloul) with a diameter ≥ 100 m and a freshness comparable to those listed in Table XIII-1. We must conclude that either large metallic meteoroids are more abundant than large stony meteoroids or, more likely, that stony meteoroids large enough to produce craters ≥ 100 m in diameter undergo fragmentation during atmospheric passage.

No hypervelocity impacts of meteorites with the earth's surface have occurred during historic time. During hypervelocity (in excess of about 2.5 km sec^{-1}) impacts, the response of both target and projectile is essentially fluid. The Sikhote-Alin (II B-An) iron, which fell in Siberia in 1947, while not a hypervelocity object, made craters with diameters up to 28 m. Sikhote-Alin, at 23 t, is the largest recovered fall. The largest recovered fragment is about 1.7 t, and the estimated mass of the material entering the atmosphere is 70 t (KRINOV, 1960).

The largest extraterrestrial object to fall during historic time fell on the Siberian *taiga* near the Podkamennaya (Stony) Tunguska River in 1908. This object created an air wave recorded all around the world (WHIPPLE, 1930, 1934; ASTOPOVICH, 1934). Unfortunately, because of the remoteness of the site and the unstable political situation, the first scientific expedition to the site was in 1927. No crater was found, but the epicenter of the explosion was defined by the direction of fall of uprooted trees. Trees were uprooted and seared to a distance of 15 km, uprooted without appreciable searing to 20 km, and occasionally uprooted or topped as far away as 50 km. No meteoritic specimens have been recovered, although FLORENSKI et al. (1968a, b) found a high concentration of magnetic spherules in the region.

The explosion appears to have occurred at an altitude of about 10 km (FESENKOV, 1968). The energy estimated from airwave data is about $4 \pm 2 \times 10^{23}$ ergs, equivalent to about 10 Mt of TNT (HUNT et al., 1960). KAULA (1968) reports that the relationship between crater diameter D (in m) and energy W (in kt TNT) is:

$$D = 49\,W^{0.294}\,.$$

This formula yields an energy of 53 Mt TNT for the event which produced the Arizona (Canyon Diablo) crater. If the Tunguska object had impacted with the earth as a single mass, a crater with a diameter of about 735 m would have resulted.

It is generally believed that Tunguska was a comet. According to WHIPPLE'S (1950) model, a comet consists of solid mineral grains in an icy matrix ("dirty snowball"). LA PAZ (1941) proposed that Tunguska consisted of antimatter. However, the available evidence, such as the absence of an appreciable increase in ^{14}C activity (COWAN et al., 1965; GENTRY, 1966; MARSHALL, 1966) does not support this hypothesis. An icy comet would be expected to have a low resistance to fracturing, and seems in keeping with the high detonation altitude.

If we assign a nominal atmospheric velocity of 30 km sec^{-1} and densities of 1.0 and 7.9 g cm^{-3} respectively to Tunguska and Canyon Diablo, their preterrestrial radii were about 28.1 and 24.5 m. If the velocity were 20 km sec^{-1}, the radii were higher by a factor of 1.3.

According to SHOEMAKER and LOWERY (1967) a number of very large meteoroids produce signals on sensitive microbarographs but do not reach the earth's surface as appreciable specimens. One such event near Revelstoke, British Columbia, produced only 1 g of recovered material, which FOLINSBEE et al. (1967) identified as a CI chondrite. The most accurate microbarograph data are not published, but kept in classified status because of their bearing on the detection of atmospheric testing of nuclear weapons. Although some skepticism is warranted until they are published, they appear to provide the strongest support for the belief that the bulk of the meteoritic debris entering the earth's atmosphere is CI-like (but possibly still more friable) material which does not survive atmospheric passage. Numerous workers have suggested that the CI chondrites are of cometary origin.

As noted above, most large meteorites fall as showers. The individual fragments scatter over a roughly elliptical area known as a strewn field. Figure XIII-1 (adapted from CLARKE et al., 1970) shows the strewn field of the CV3 chondrite Allende. Fragments with the largest mass/cross-sectional area ratio travel the greatest distance; thus the largest fragments tend to lie at the forward end of the strewn field, and the smallest fragments at the opposite end. This is the case at Allende, where the largest fragment (110 kg) defined the northeast end of the field and the smallest fragments (mean mass 12 g) at the southwest end. The axis of the Allende field points about 32° east of north. However, CLARKE et al. (1970) think that the smaller fragments may have suffered greater deflections by surface winds from the west, and they suggest that the true flight direction was 37° east of north.

The Allende meteorite was one of the largest meteorite falls in historic time—perhaps second only to Sikhote-Alin in terms of the mass of material which reached the surface of the earth in the form of recoverable fragments. CLARKE et al. (1970) estimate the mass to be about 4 t or greater. Its strewn field is among the largest known.

A number of prehistoric strewn fields are known. One of the largest and most thoroughly investigated is that at Campo del Cielo, Argentina (CASSIDY, 1968, 1971), which consists of craters linearly oriented over an area 18 km long and of meteoritic fragments (IA iron) over an area 75 km long and 4 km wide. The largest strewn field of all is that

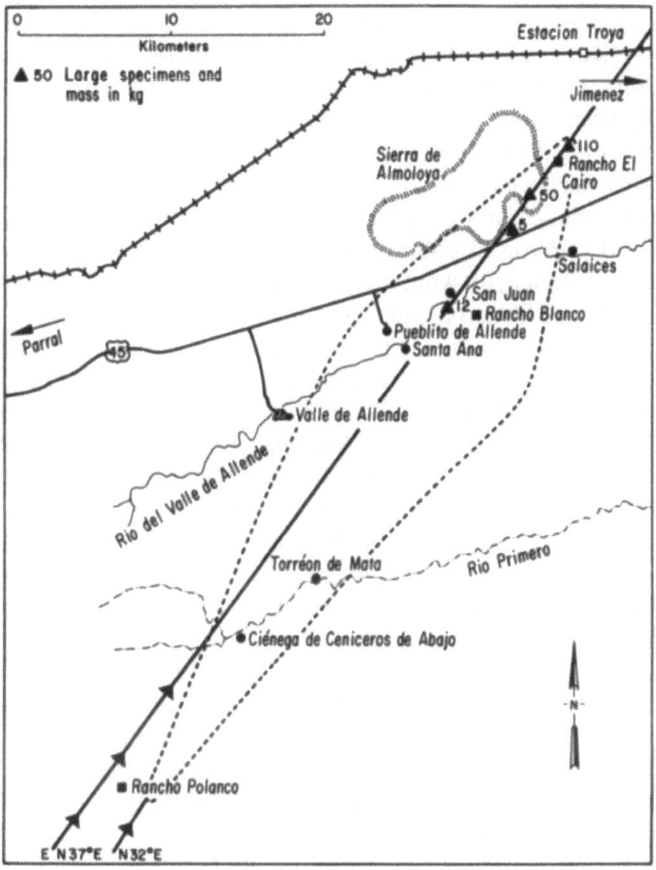

Fig. XIII-1. Strewn field of the CV3 chondrite Allende, which fell on 8 February 1969. The meteorite was moving from southwest towards northeast. The largest specimens are at the northeast end of the field and tend to line up along a vector pointed 37° east of north. (From Clarke et al., 1970)

connected with the Gibeon (IVA iron) shower in South-West Africa (Buchwald, 1969b), which is about 400 km long and 100 km wide.

Human transport has occasionally affected the field distribution of meteorite specimens. Perhaps the best documentation of such a factor is that of the Arizona octahedrites. Of 13 medium and coarse octahedrites recovered and catalogued from the state of Arizona, 11 have the same composition and appear to be fragments of Canyon Diablo, despite reported discovery sites which are as much as 350 km distant from the crater (Wasson, 1968). The pallasitic material from the Hopewell Indian burial mounds in Ohio appears to have been brought there from Brenham, Kansas, some 1400 km distant (Wasson and Sedwick, 1969).

Hey (1966) uses *synonym* to indicate a meteorite which is a fragment from a historically older meteorite but incorrectly treated as an independent fall. He uses *paired* to indicate meteorites which are possibly fragments of the same fall. In this book and in earlier papers (e.g., Wasson, 1969) I use *paired* in a more restrictive sense; i.e., to mean meteorites accorded separate catalog entries in Hey's (1966) catalog, but which on detailed study appear to be parts of the same fall (e.g., Camp Verde and Canyon Diablo). I use *synonym* to designate a second name applied to a meteorite where there is no serious doubt that only one fall is involved (e.g., Otumpa and Campo del Cielo).

Meteorites are still being found at a relatively rapid rate, particularly in Western Australia and the Plains States of the USA. Identification of a meteorite in the field is not easy, but a dedicated meteoritophile usually has a pocket magnet along on hikes, since most meteorites contain some metal. Fresh meteorites have glassy fusion crusts; the crusts are generally darker than the interior. A stone in an area where the fields are free of stones is always of interest. Persons having rocks which may be meteoritic should detach a single fragment weighing about 10—20 g and send it to a major natural history museum or the geology or mineralogy department of a major university.

The official name of a meteorite is that given in the latest edition of the British Museum catalog (Hey, 1966) or, if of recent vintage, in the *Meteoritical Bulletin* (edited by E.L. Krinov through No.49, March, 1970, and published by the Committee on Meteoritics, USSR Academy of Sciences, Moscow; currently edited by R.S. Clarke, Jr., and appearing periodically in *Meteoritics)*.

Meteorites are named after geographic landmarks. During the past few decades the general rule has been that this should be the nearest city or village large enough to have a post office. In a number of cases, however, confusingly similar names have arisen as a result of the discovery of several independent meteorites near a single village. As Hey (1966) emphasizes, the chief value of a meteorite name is as a label of a scientific sample. For this reason, Buchwald and Wasson (1972) recommend that meteorites be given unique names based on any permanent landmark in the neighborhood of the discovery site. They further recommend that these names be short (no longer than 25 spaces) and without diacritical marks.

XIV. Morphology and Macrostructure

The exterior morphology of meteorites recovered soon after fall is generally characterized by flight markings. Meteorites which have maintained a constant orientation during their fall have furrows on their forward surfaces radiating away from the apex. The rear surface is either unmarked or is furrowed, with the wavelength longer and amplitude smaller than on the forward surface. An excellent example of an oriented meteorite is the Cabin Creek iron (pp. 40—41 of MASON, 1962a).

Fusion crust forms during the period of deceleration. KRINOV (1960) gives an exhaustive discussion of different types of fusion crusts. Individual fragments recovered following stony meteorite showers show a variety of crusts ranging from a mature, thick (ca. 1 mm), black, glassy crust to a thin crust formed later in the braking period to no crust at all for surfaces formed by fractures occurring after the meteorite reached terminal velocity. Of course, some meteorites fragment when they strike the ground; most uncrusted surfaces result from this effect. Iron meteorites tend to have a very thin fusion crust and to have a polished appearance to the naked eye. Figure XIV-1 illustrates such a surface on the recently discovered meteorite Bushman Land IVA fine octahedrite from South-West Africa.

Weathering eventually destroys the original surfaces of meteorites. Chondrite finds are almost always covered with a patina of ferric oxide which hinders the naked-eye recognition of a fusion crust. Iron meteorites often weather to surfaces which at first glance appear to be flight marked, although metallurgical examination shows no trace of a heat-altered zone beneath the surface. Irons recovered from deserts sometimes have sandblasted surfaces; an example is Sierra Sandon (PALACHE, 1926). Sometimes weathering leads to quite deep pits on the lower surface of an iron. This seems to be the origin of the deep pits in the 15 t Willamette (IIIA) iron (see photo on p. 29 of MASON, 1962a).

Most types of meteorites have characteristic macrostructures. Perhaps the most striking is the pallasitic structure, which consists of roughly equal portions of olivine and Fe-Ni, with typical domain dimensions of about 1 cm. Figure XIV-2 shows a polished and etched section of the Salta pallasite. In this specimen the olivine crystals are angular and in some cases show parallel faces separated by thin strips of metal. Other pallasites (e.g., Brenham) have rounded olivine crystals.

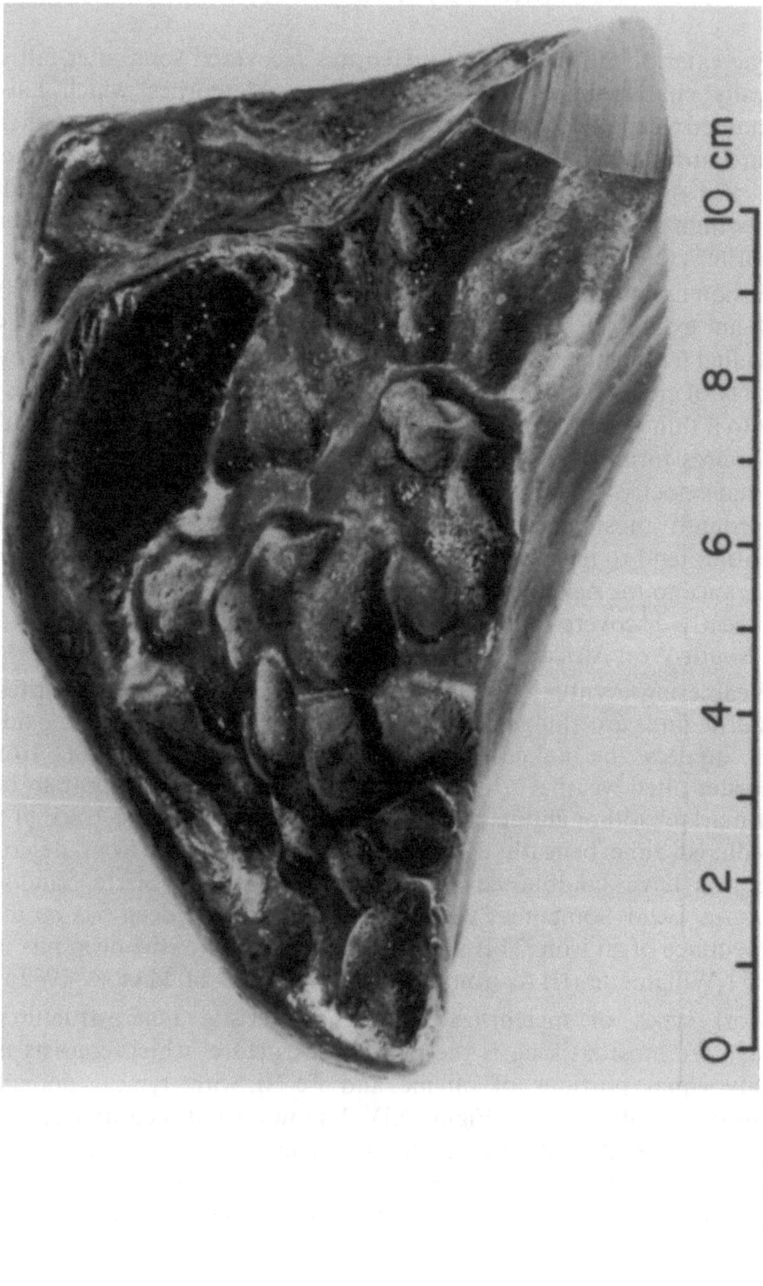

Fig. XIV-1. Exterior of the IV A iron, Bushman Land. The shallow depressions and polished appearance of the surface are typical of fresh iron meteorite falls. A small sample removed from the lower left corner shows a well-preserved heat-alteration zone and confirms that this meteorite find is a recent fall. (Smithsonian Institution photo)

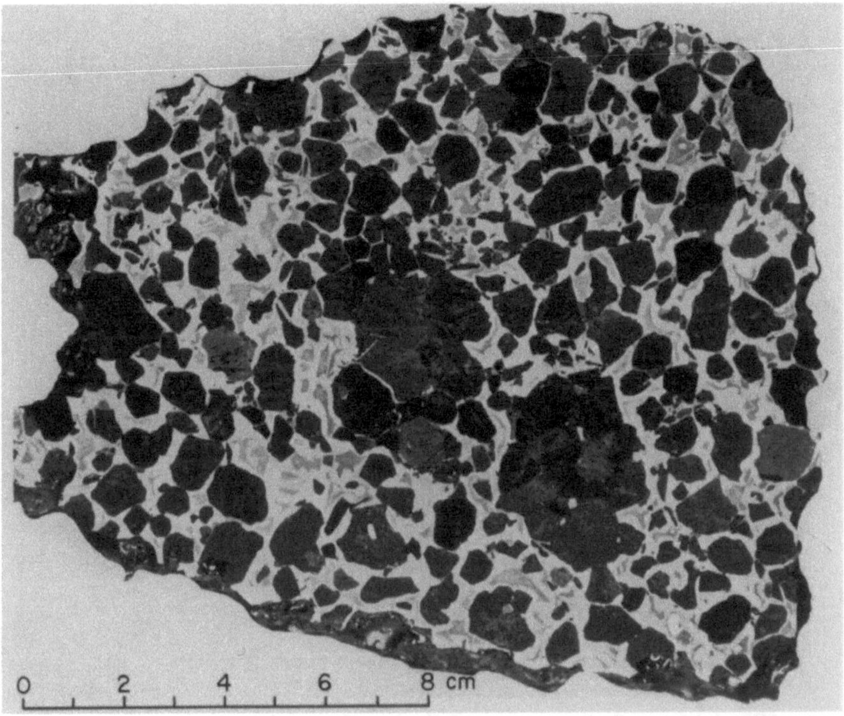

Fig. XIV-2. Polished and etched section of the Salta pallasite in reflected light. The dark fragments are olivine; the lighter matrix material is metal. The metal consists of kamacite (white and light gray) and plessite (dark gray); in some cases kamacite bars cross plessite fields. The olivine is clearly fragmental, with thin metal veins separating fragments with complementary surfaces. The large masses of olivine near the center of the section have atypically large areas. (Smithsonian Institution photo)

Note that the etching makes it possible to resolve the kamacite (white or light gray) portion of the metal from the plessite (dark gray). Taenite is not resolvable at the scale of this photo. It is easy to imagine why the pallasites were the first meteorites to be recognized as extraterrestrial (CHLADNI, 1794).

The mesosiderites, the other kind of "stony-iron", have macrostructures which are distinctly different from those of the pallasites. Figure XIV-3 shows the structure of the Mount Padbury mesosiderite. Note that most of the metal (white) is finely disseminated, although some also is in the form of cm-sized nuggets. It is evident that Mount Padbury is a breccia. The dark, angular fragments (which consist mainly of orthopyroxene) are clearly parts of a rock which predates the present struture. Further, a close inspection shows a number of light

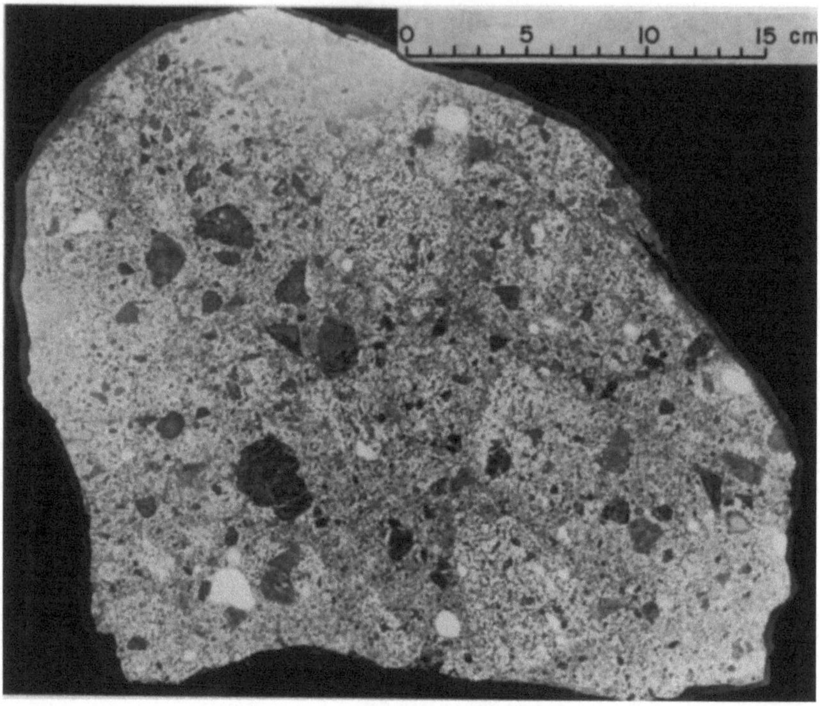

Fig. XIV-3. Mount Padbury mesosiderite in reflected light. Metal is white, silicates dark. At least two brecciation events are recorded in the structure, since certain well-defined metal-rich areas are clearly breccias, and these in turn are embedded in a slightly darker matrix. (Photo by V. F. BUCHWALD)

(richer in metal) breccia blocks (typical dimensions 5 cm) enclosed in a darker matrix. One triangular shaped block is located at the 5 o'clock position in Fig. XIV-3. Thus, in Mount Padbury, we have evidence for at least two brecciation events. The macrostructure of the mesosiderites is variable. Some, such as Pinnaroo or Bondoc, have the metal mainly in the form of large nuggets with diameters as great as 9 cm.

The howardites are also breccias. A small fragment of the Kapoeta howardite is illustrated in Fig. XIV-4. This is a good example of a light-dark structure. FREDRIKSSON and KEIL (1963) studied this fragment and found that the mineralogical compositions of the two areas are identical, but that the grain size is smaller in the dark portion. The light portion is also a breccia, but it is not clear whether more than one brecciation event is recorded in the structure. The dark portion is rich in solar-type primordial rare gas (SIGNER and SUESS, 1963) (see Chapter IX). The microstructure of Pasamonte, a member of the closely related class of eucrites, is illustrated in Fig. VI-5.

Fig. XIV-4. Kapoeta howardite in reflected light. The light area on the left is a xenolith in the darker matrix. The light and dark material have very similar bulk compositions, but the dark matrix is enriched in certain volatile elements, including solar-type rare gases. Howardites consist mainly of orthopyroxene and plagioclase. (Photo supplied by H. E. SUESS)

Figure VI-7 in the chapter on petrology shows a section through the remarkable Cumberland Falls meteorite. It illustrates how extreme the contrast between differing portions of a polymict breccia can be. The light material is typically aubritic. It has scattered mm-sized metal inclusions, one of which appears gray about 1 cm below the center of the top edge of the section. The dark fragments consist of a rare type of chondrite (BINNS, 1969). Note the angular nature of both the white and black xenoliths.

Figure XIV-5 shows a section of the Indarch E4 chondrite. The chondrules are abundant and sharply defined. The many small, irregular crystals are mainly (clino-) enstatite and may be fragments of chondrules. Small amounts of olivine and tridymite coexist in Indarch (BINNS, 1967b); if it were an equilibrium assemblage these would have reacted to form pyroxene.

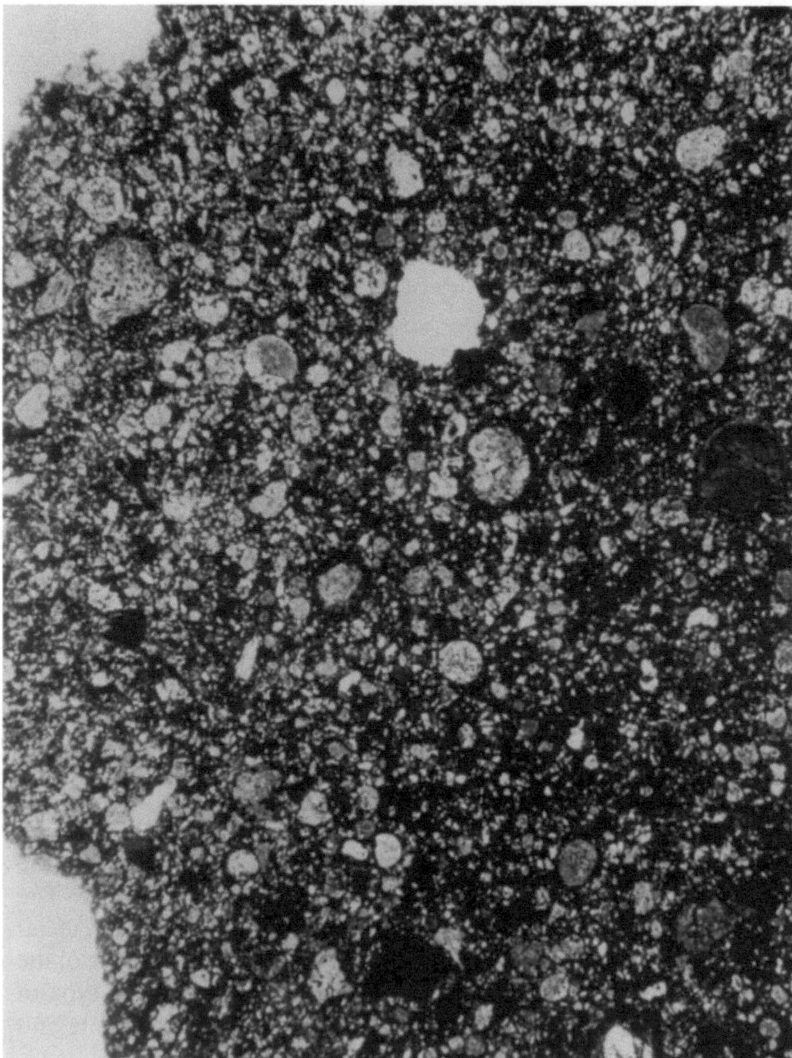

Fig. XIV-5. Section of Indarch E 4 chondrite in transmitted light. Main minerals are metal and troilite (dark) and enstatite (light). Chondrules are sharply defined. The enstatite chondrites are highly reduced, as evidenced by the fact that the metal contains 1—3% Si. White spot in right center is a hole in the section. (Smithsonian Institution photo)

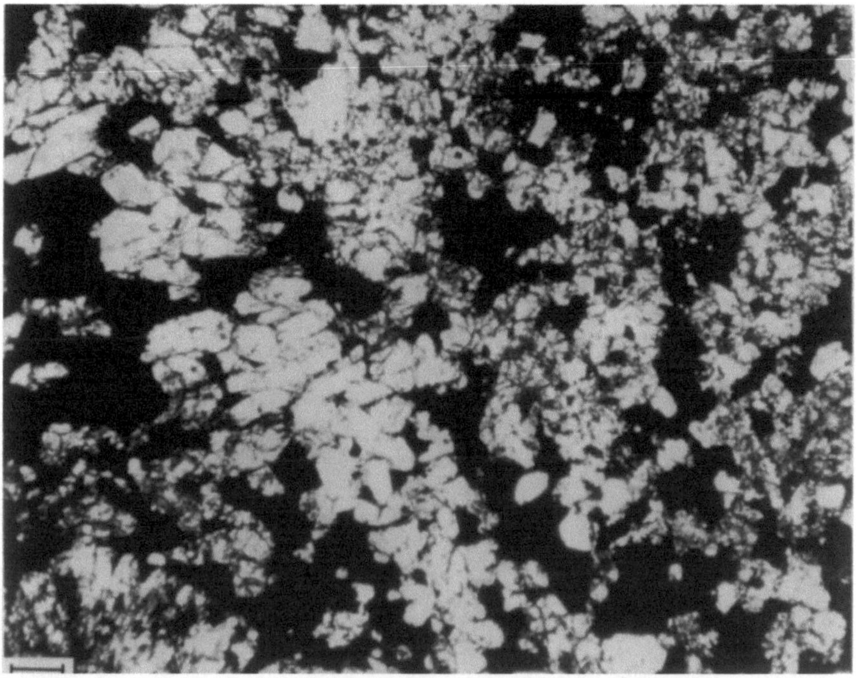

Fig. XIV-6. Thin section of Khairpur E6 chondrite photographed in transmitted light. Main minerals are metal and troilite (dark) and enstatite (white). Differences in bulk composition show that E6 chondrites cannot be made from E4 chondrites by metamorphism. Width of section about 4 mm. (Smithsonian Institution photo)

Figure XIV-6 is a thin section of the Khairpur E6 chondrite. Like Fig. VI-3 of Peace River, it illustrates the integration which is present in the higher petrologic types of chondrites. The light-colored grains are orthoenstatite. The E4 and E6 chondrites have distinctly different bulk compositions (higher Fe/Si, lower Mg/Si in E4) (MASON, 1966; LARIMER, 1968b); it is therefore impossible to convert a E4 chondrite into an E6 chondrite by metamorphic recrystallization.

Figure XIV-7 shows a thin section of the Bremervörde, a xenolithic or genomict H3 chondrite. The upper left portion is similar to the Bishunpur L3 chondrite (Fig. VI-1), with very sharply defined chondrules. In contrast, chondrules can scarcely be resolved in the lower right portion of the section, which appears to consist of type 5 material. It is probable that the volatile-element content differs greatly between these two portions; bulk analyses of xenolithic chondrites are difficult to interpret if not combined with petrographic information.

Because of its great mass (≥ 4 t) and interesting composition, the Allende CV3 chondrite has been studied more thoroughly than any

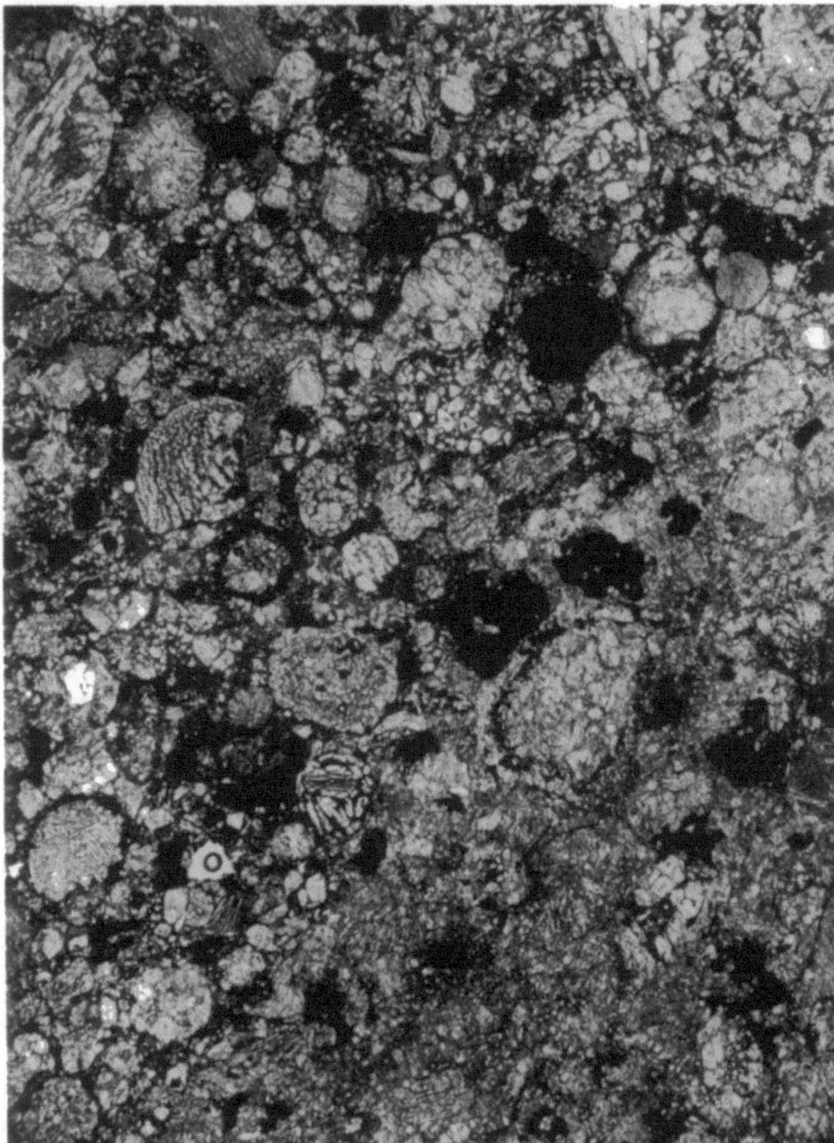

Fig. XIV-7. Thin section of Bremervörde H 3 chondrite in transmitted light. Upper left portion shows sharply defined chondrules and chondrule fragments characteristic of a type 3 chondrite. Lower right portion is an indurated xenolith of type 5 material. Width of section about 6 mm. (Photo by F. WLOTZKA)

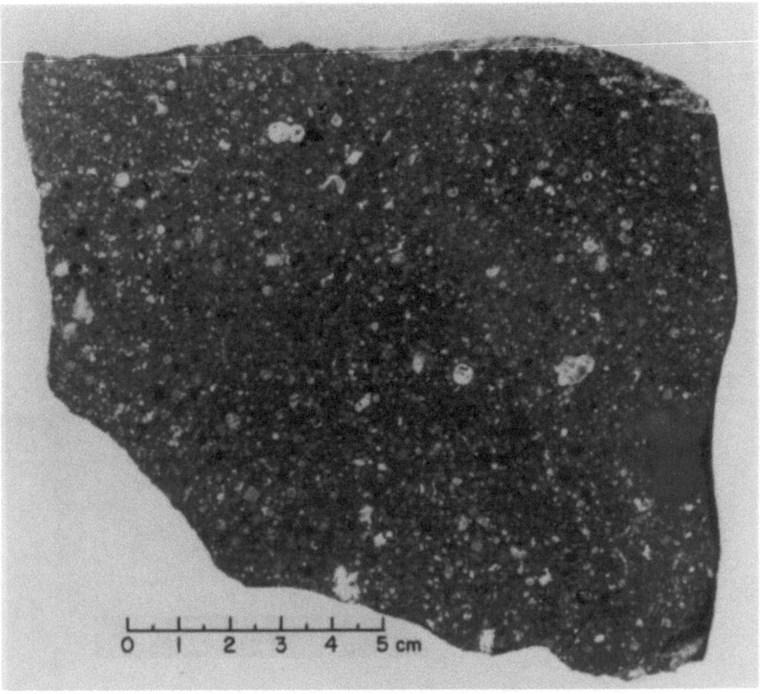

Fig. XIV-8. Polished face of Allende chondrite showing a typical CV3 structure. The irregular white inclusions are the high-temperature Ca-Al rich materials discussed in Chapter VI; normal chondrules are abundant, and a dark xenolith which is nearly chondrule-free is present. It seems reasonably certain that each of the different materials in Allende originated by fractionation processes occuring in the solar nebula. (Smithsonian Institution photo)

other chondrite. As discussed in Section VI D, a sizable fraction of these studies was performed on the remarkable white inclusions which Allende bears. These inclusions are enriched in refractory elements such as Ca, Al, and Ti, and appear to be high temperature condensates or residues of nebular origin. Figure VI-4 illustrates a rough specimen of Allende which contains white inclusions with diameters up to 2 cm. Figure XIV-8 shows a polished section of Allende which has a more typical structure. The white inclusions are smaller and less abundant. A dark xenolith which appears to contain no chondrules or white inclusions is located along the left edge of the specimen. Petrographic inspection generally shows that such inclusions are similar to CM chondrites.

The group IAB irons are very rich in inclusions. In contrast to those in silicate-rich meteorites, most of these inclusions achieved their morphological form by *in situ* growth (inclusion is a misnomer, but too

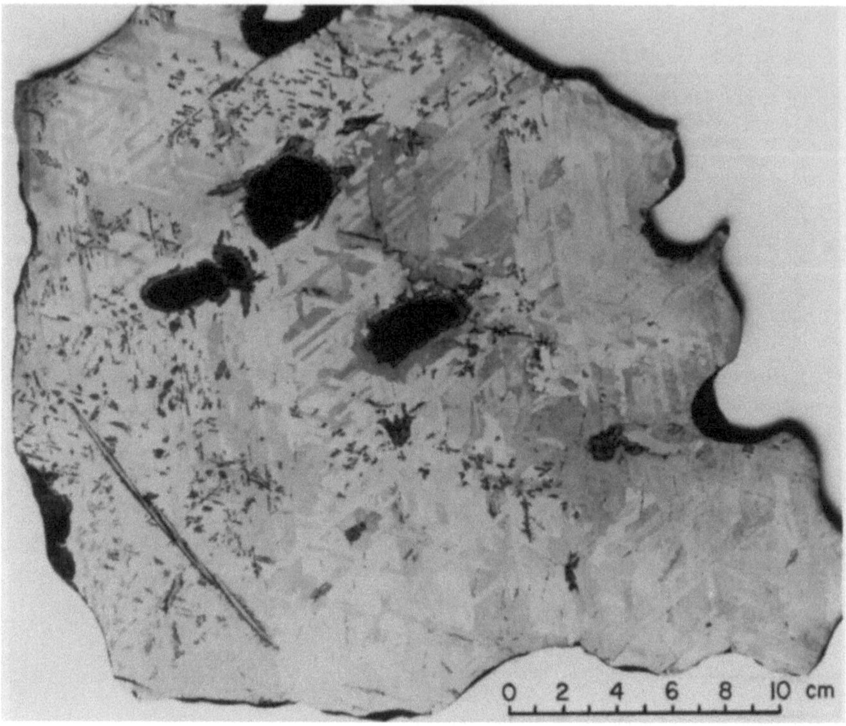

Fig. XIV-9. Polished and etched section of Canyon Diablo IA coarse octahedrite (reflected light). Three sets of kamacite bands are easily recognized, forming 60° angles with one another. The large inclusions are troilite-graphite surrounded by schreibersite. The long inclusion is schreibersite. The small dark inclusions near the centers of some kamacite bands are cohenite (Fe_3C). The embayments on the upper and left sides are caused by the mechanical loss of troilite-graphite nodules as a result of the weathering away of adjacent metal. (Smithsonian Institution photo)

entrenched to alter). Figure XIV-9 illustrates the structure of the typical IA iron Canyon Diablo. The large, dark, irregular inclusions consist of troilite (dark gray) and graphite (black), surrounded by a rim of schreibersite (medium gray). The 20 cm-long inclusion on the lower right side of the specimen is also schreibersite. Carbon is also present in the form of metastable cohenite, which occurs mainly in the centers of kamacite lamellae. Some metallic areas are cohenite-rich (e.g., the area near the long schreibersite inclusion); others are nearly cohenite-free (most of the right portion of this specimen). Not visible on the photo are minor amounts of silicates, mainly found in the troilite-graphite nodules. Some members of group IAB contain very large amounts of silicates—about 15% by weight in Woodbine, for example (MASON, 1967e).

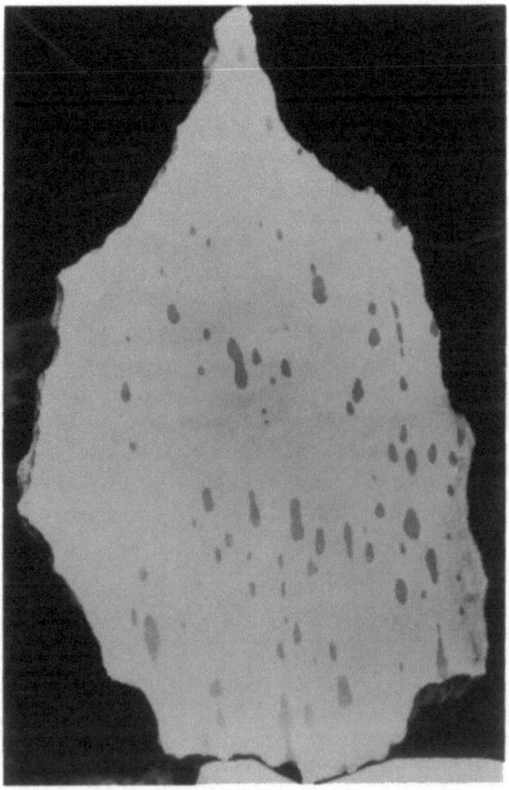

Fig. XIV-10. Section of Cape York IIIA iron, with dimensions about 1.3 × 1.8 m. The inclusions consist of troilite and are parallel to a 100 plane of the parent taenite crystal. The location of phosphates and metal inclusions shows that the long axes of the troilite nodules were parallel to the gravitational field. The specimen was sliced with a wire saw using a slurry of SiC as the abrasive. (Photo by V. F. BUCHWALD)

In the low-Ni, IIIA portion of iron-meteorite group IIIAB the inclusions are mainly troilite. Figure XIV-10 shows the 2.4 m^2 slab removed from the 20 t Agpalilik specimen of the Cape York IIIA shower by BUCHWALD (1971a). This section provides several pieces of important evidence regarding the origin of the IIIA irons. It (and smaller sections removed from other locations) shows that the Widmanstätten pattern is continuous throughout the entire mass, and thus, that the parent taenite crystal was originally at least 2 m across. Further, as seen in Fig. XIV-10, the ellipsoidal troilite nodules are oriented with their long axes parallel to one another. They are also parallel to a 100 plane of the parent taenite crystal. BUCHWALD (1971b) found that phosphate crystals are

Fig. XIV-11. Polished and etched section of Edmonton (Canada) hexahedrite, like all large-crystal hexahedrites, a member of group II A. The fine lines are called Neumann lines and are mechanical twins produced by low-pressure shock. The irregularly round inclusions are troilite. Small, white needles are schreibersite crystals in the rhabdite form. The density of rhabdites is quite variable. The specimen is about 9 cm wide. (Smithsonian Institution photo)

concentrated at one end of the nodules and metal inclusions at the other end. Since phosphates have densities less than, and Fe-Ni a density greater than FeS, the direction of the gravitational field in Cape York is uniquely established.

In most cases the low-Ni, II A meteorites of group II AB consist of single crystals of kamacite. Perhaps because they cleave relatively easily along cubic planes none of the largest specimens of iron meteorites are

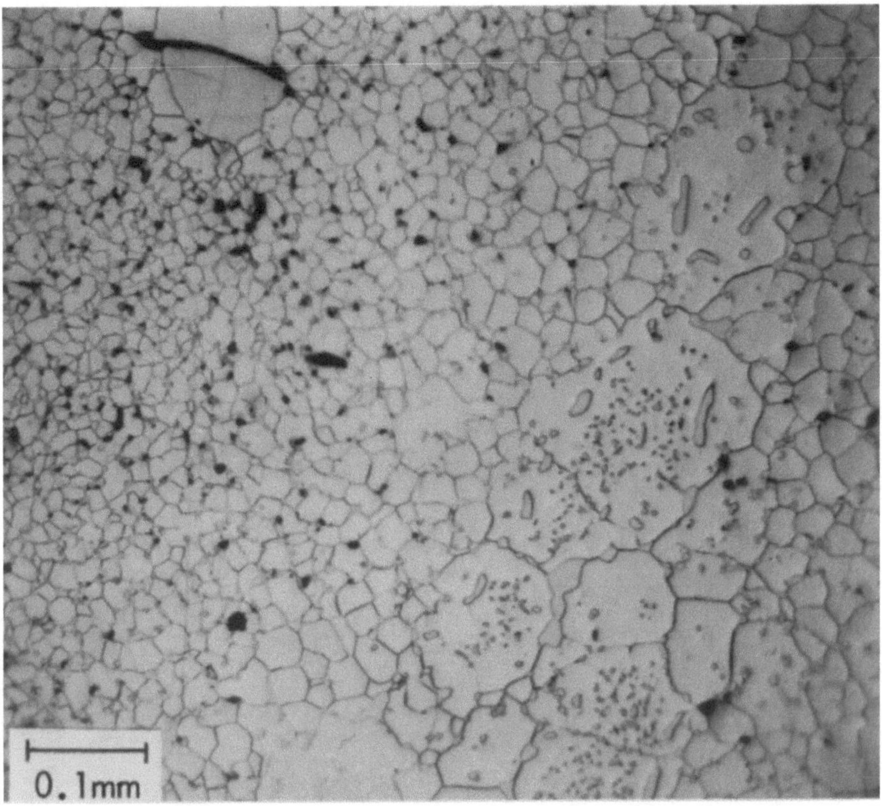

Fig. XIV-12. Polished and etched section of Mejillones (1905) II A recrystallized hexahedrite. Schreibersite, $(Fe, Ni)_3P$, outlines the new kamacite grains. This structure is produced by the cosmic reheating of a structure similar to that of Edmonton, Canada. (Photo by J. L. GOLDSTEIN)

members of this group. The largest known hexahedrite specimen is a member of the Coahuila shower and has a reported mass of about 200 kg (the II A member Chico Mountains was reported to weigh 2000 kg, but a) it is cosmically recrystallized to 0.5 mm crystallites, and b) the reported mass is suspect, since only about 1 kg is in museum collections). Edmonton (Canada), a typical hexahedrite, is shown in Fig. XIV-11. The fine parallel lines which cross the specimen are Neumann lines. These are shock-produced mechanical twinning lamellae which form along the 211 planes of the kamacite at shock pressures of the order of 10 kbar (see Section VI F). There are twelve 211 plane orientations, and thus twelve possible sets of Neumann lines (UHLIG, 1955). The larger inclusions in the Edmonton (Canada) section are troilite; fine schreibersite inclusions

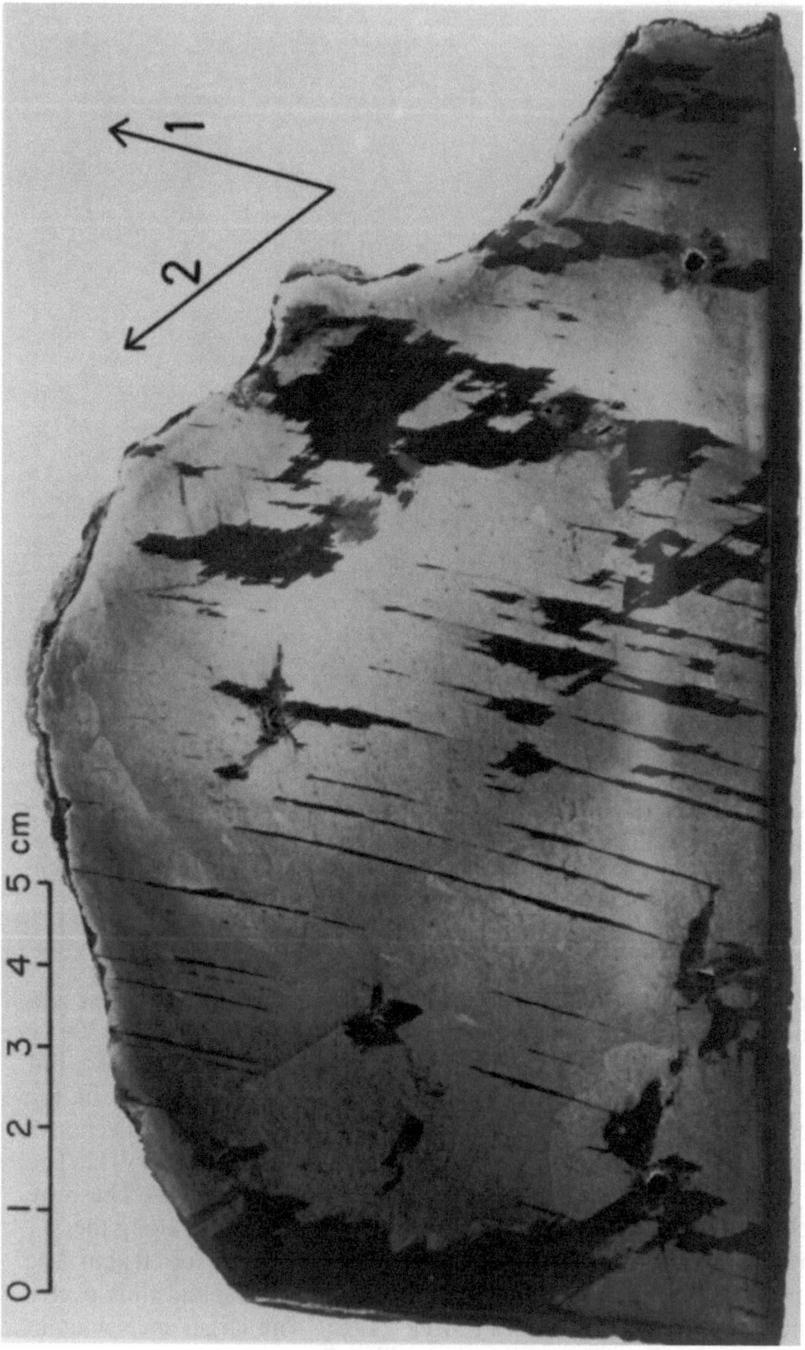

Fig. XIV-13. Polished and etched section of Hoba IV B ataxite. The dark bands are probably of shock origin; those in direction 1 seem to predate those in direction 2. The inclusions are troilite. The splotchy, light gray areas are etching artifacts. (Smithsonian Institution photo)

(called rhabdites) can be recognized as a background of short needles parallel to some sets of Neumann lines.

In the earlier meteorite literature one often finds irons with Ni contents less than about 9% classified as "Ni-poor ataxites". Analyses of Ge, Ni, and other elements by LOVERING et al. (1957) and by my co-workers and me (cf. WASSON, 1969) have shown that most "Ni-poor ataxites" are members of group II A and have Ni contents of $5.5 \pm 0.1\%$, Ge contents of 180 ± 10 ppm, and are chemically members of group II A. BUCHWALD (1974) argues that the name ataxite (from the Greek meaning "without structure") should only be used for iron meteorites which appear to have experienced a typical cooling history, i.e., have cooled from high to low temperatures at a rate similar to or less than that recorded in the most rapidly cooled groups of irons, group IV A. If he can surmise the original structure, he assigns the meteorite that structure preceded by the word "recrystallized". Otherwise he recommends that the structure be designated anomalous.

The structure of the Mejillones, recrystallized hexahedrite, is illustrated in Fig. XIV-12. The original structure, which consisted of a single crystal of kamacite tens of cms across, has been replaced by crystals 0.02—0.2 mm across. The sheared schreibersite crystal on the upper left may indicate that a shock event is responsible for the reheating recorded in this structure.

The (normal) ataxites are not nearly as structureless as their name implies. Figure XIV-13 shows a polished and etched section of the IV B iron Hoba (at 60 t, the largest meteorite known). Dark, mm-sized troilite inclusions can be seen. Most impressive, however, is the dark banding, which occurs in two prominent directions indicated by the arrows. The origin of this banding is not known, but is probably related to shock. In this section the bands in the second direction seem to have caused small offsets in those in the first direction. Hoba, like most ataxites, also has a few narrow spindles of kamacite oriented octahedrally with respect to one another. These are too fine to be visible in this photo.

XV. Organic Matter

Many studies of meteoritic carbonaceous compounds have been published during the past decade; the review of HAYES (1967) contains 130 references, and many more papers have been published in the interim. In most cases the studies involved carbonaceous chondrites, although minor amounts of organics are also found in other meteorite classes; of these, the highest are in type-3 ordinary chondrites and in ureilites. Detailed investigations of organic matter in the latter are lacking. Studies of graphite, cohenite and diamond in ureilites (VINOGRADOV et al., 1967; VDOVYKIN, 1969, 1970) and in group-IAB irons (e.g., DEINES and WICKMAN, 1973) may be mentioned, but are not truly germane to this chapter.

Investigations of organic matter have focussed on two interrelated problems: 1. the search for evidence of extraterrestrial life; and 2. the attempt to define abiogenic processes for producing complex organic compounds. The evidence to date (KVENHOLDEN et al., 1970; see also HAYES, 1967) seems overwhelmingly in favor of an abiogenic origin for all organic constituents. As a result much recent research has concentrated on the question of nebular synthesis of the observed mixtures of organic compounds.

STUDIER et al. (1968, 1972), ANDERS et al. (1973), and GELPI et al. (1970) favor a Fischer-Tropsch-type synthesis, in which a mixture of CO, H_2O and NH_3 is heated to a moderate temperature (ca. 700° C) in the presence of a catalyst. Under such circumstances the products consist of a mixture of stable and metastable organics. The relative amount of each product depends on both thermodynamics and kinetics, and thus on the details of the reaction conditions. A critical evaluation of some aspects of Fischer-Tropsch-type processes is given by BELSKY and KAPLAN (1970).

The other main contender for a source of the organics is a Miller-Urey-type synthesis in which a mixture of simple molecules (CH_4, NH_3, H_2O, etc.) is subjected to a brief burst of energy at relatively low temperatures [e.g., those appropriate to a terrestrial-type atmosphere (MILLER and UREY, 1959; UREY and LEWIS, 1966)]. The reactions in this case are hot-atom reactions, and the composition of the product mixture is controlled mainly by kinetic processes rather than by an approach to equilibrium.

The Fischer-Tropsch-type processes appear to produce mixtures which more nearly approximate the composition of the organic matter observed in carbonaceous chondrites. It is necessary, however, to cook the products for an additional period at lower temperatures (ca. 200° C) in order to make the spectrum of products match the meteoritic spectrum reasonably well. Further, as STUDIER et al. (1972) point out, studies of terrestrial shales by FRIEDEL and SHARKEY (1968) suggest that metamorphic diagenesis (partial equilibration) of carbonaceous matter will produce a Fischer-Tropsch-like distribution of compounds, in which case the details of the precursor gas-phase synthesis may be relatively unimportant. Definition of the synthetic processes which produced the meteorite organics is a difficult problem, and one in which much research is still needed.

More pertinent to the theme of this book is the bearing that organic geochemical evidence has on the classification and origin of meteorites. Unfortunately, there is little progress to report in this direction.

Data on hydrocarbons appear to be the most reliable and relevant. Figure XV-1 is a plot of isoprenoid[12] hydrocarbon content in carbonaceous chondrites versus that of straight-chain hydrocarbons and is modelled after a similar plot in HAYES (1967). The data are from GELPI and ORÓ (1970) for the four meteorites Grosnaja, Mokoia, Murray and Vigarano, and from NOONER and ORÓ (1967) for the remainder of the meteorites. Averages are plotted for all meteorites except Orgueil, where replicates are plotted in order to illustrate the great degree of scatter in the data determined on this meteorite. For convenience in plotting, samples reported by NOONER and ORÓ (1967) to have isoprenoid contents of 0.00 ppm are plotted at 0.012 ppm on the log-log diagram.

The only clear trend which can be read from Fig. XV-1 is that the lowest n-alkane concentrations are found in Orgueil. Except for Kainsaz, the CO3 meteorites have the highest n-alkane/isoprenoid ratios (mass ratio > 10). The CM group is the most homogeneous, with concentration ranges of both types of alkanes which are each smaller than a factor of 10. The four CV2 meteorites separate into two widely divergent pairs—Mokoia and Al Rais are rich in alkanes and Renazzo and Kaba have low concentrations, lower than those in the two CV3 meteorites studied.

The variations in CV chondrites may not be significant, since these are among the most heterogeneous chondrites in terms of structure. Figure XV-2 illustrates the structure of the Mokoia CV2 chondrite. Light xenoliths are embedded in a darker matrix. Light-colored chondrules (or inclusions) in the xenoliths are reminiscent of the light inclusions in the Allende CV3 chondrite (see Figs. VI-4, XIV-9). A wavy, 0.1 mm-thick dark vein crosses the specimen from bottom to top slightly left of center.

---

[12] Isoprenoids are 2,6,10-trimethyl and 2,6,10,14-tretamethyl alkanes.

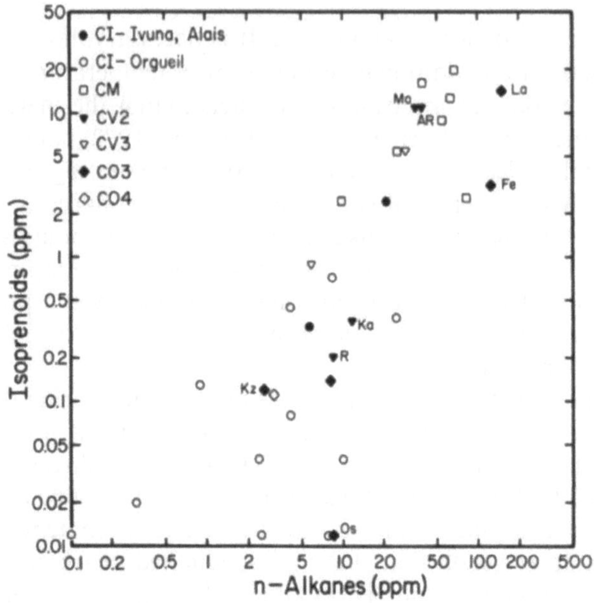

Fig. XV-1. Concentration of isoprenoid hydrocarbons plotted against n-alkane concentration. The two type of alkanes correlate. The lowest n-alkane concentrations are observed in the Orgueil CI chondrite, the highest in the Lancé and Felix CO3 chondrites. Abbreviations: AR, Al Rais; Fe, Felix; Ka, Kaba; Kz, Kainsaz; La, Lancé; Mo, Mokoia; Os, Ornans; R, Renazzo. (Data taken from GELPI and ORÓ, 1970; NONNER and ORÓ, 1967)

As summarized in Table II-5 and Fig. VIII-1, the four carbonaceous chondrite groups can be dinstinguished on the basis of bulk oxygen isotope data. The $^{13}C/^{14}C$ data of SMITH and KAPLAN (1970) suggest that each group also has a characteristic distribution of C isotopes among the various compounds, but the number of investigated meteorites must be increased before this can be confirmed. Additional discussion of stable isotope variations is given in Chapter VIII.

Perhaps the most promising evidence of a systematic trend observed in recent studies of organic matter is in the relative distribution of alkanes. Figure XV-3 is a plot of relative alkane abundance versus number of C atoms in normal alkanes, with the two main isoprenoids pristane (C_{19}) and phytane (C_{20}) plotted separately to the right. The CI chondrites have broad n-alkane distributions with double peaks in the range C_{17}—C_{23}; they have minor isoprenoid contents. The typical CM 2 chondrites Cold Bokkeveld, Mighei, and Erakot show moderately wide n-alkane distributions with single maxima in the range C_{17}—C_{19}; they have moderate isoprenoid contents. The n-alkane distribution in the CV 2 chon-

Fig. XV-2. Section of Mokoia CV 2 chondrite (about 2 cm long). Irregular fragments of a lighter rock are imbedded in a dark matrix; the structure is clearly that of a breccia. A dark, fragment-free vein crosses the specimen from bottom to top. The larger light xenoliths (especially that in the top left) contain white inclusions which appear similar to those in Allende (Fig. XIV-8), a type-3 member of the same group. (Smithsonian Institution photo)

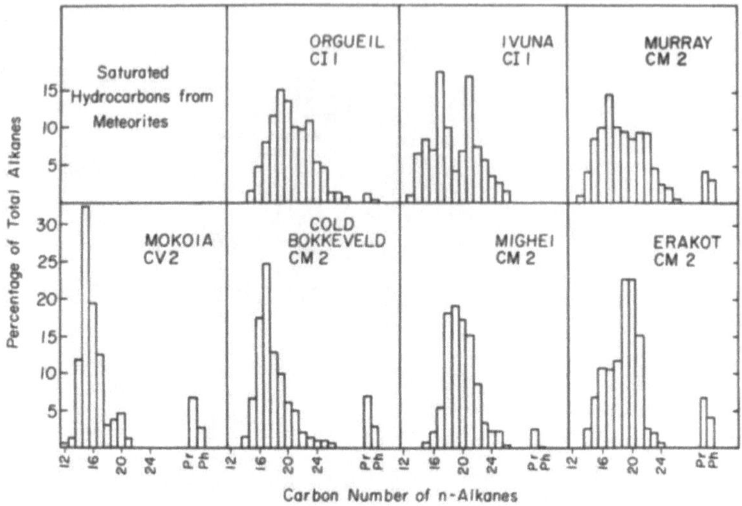

Fig. XV-3. Distribution of straight-chain and isoprenoid (Pr, pristane and Ph, phytane) hydrocarbons in seven type 1 and type 2 carbonaceous chondrites. The width and mode of the n-alkane distribution decreases through the series CI 1 > CM 2 > CV 2. (From SMITH and KAPLAN, 1970)

drite Mokoia is narrower than in the CM 2 chondrites, and its single maximum occurs at C_{15}; Mokoia also has a moderate isoprenoid content. Thus, there appears to be a systematic trend for the width and mode of n-alkane distributions to decrease in the order CI > CM > CV. Data are not available on CO chondrites. Murray shows a distinctly wider alkane distribution than those of other CM chondrites; it is also somewhat anomalous isotopically (see Chapter VIII).

It is to be hoped that more attention will be paid to possible petrographic and bulk chemical correlations during future studies of organic constituents of meteorites.

XVI. Magnetic Properties

The magnetization of a meteorite is the vector sum of magnetization induced by an external field and the natural remanent magnetization (NRM) which is independent of field (see general discussions in LINDSLEY et al., 1966; STACEY, 1969, Chapter 6; the best and most extensive review of meteorite magnetism is BRECHER's 1972a thesis). The induced magnetization is dependent on the magnetic susceptibility, the ratio of the intensity of magnetization I to the intensity of the magnetic field H. This susceptibility is usually divided by the mass to give the specific magnetic susceptibility χ.

The types of magnetic behavior found in meteoritic minerals are summarized in Table XVI-1. Most minerals are paramagnetic or diamagnetic and contribute very little to meteorite magnetization. The main carriers of magnetization are the ferromagnetic Fe-Ni minerals kamacite and taenite, and, in carbonaceous chondrites, the ferrimagnetic minerals pyrrhotite, $Fe_{0.87-1.0}S$ and magnetite, Fe_3O_4. Ferro- and ferrimagnetic minerals show similar behavior with respect to temperature, field intensity, etc., and are often referred to together as ferromagnetic.

Ferromagnetism disappears above the Curie temperature, which is characteristic for each mineral. If the basic mineralogy of a meteorite is known, the amount of magnetization associated with each ferromagnetic mineral can be determined by observing the drop in magnetization during carefully controlled heating past each Curie temperature. Generally these lie between about 240° C (Ni-rich taenite) and 750° C (kamacite). The analysis of the data is complicated if the grain size or mineral composition changes in the same heating interval. A list of the more common meteoritic magnetic phases is given in Table XVI-2.

The natural remanent magnetization in meteorites results from the fixing into position of magnetic domains which had become preferentially oriented in the presence of a magnetic field. Acquisition of remanence as a result of cooling past the Curie temperature is called thermal remanent magnetization (TRM); that which results from a phase change is called chemical remanent magnetization (CRM); a combination of these two (for example, the formation of kamacite from taenite after cooling across the $\gamma/\alpha+\gamma$ phase boundary) is called thermochemical remanence (TCRM); deposition or shock heating in a magnetic field leads to DRM or SRM; and lastly, remanence acquired at constant

Table XVI-1. Types of magnetism present in natural samples

Diamagnetism
No net magnetic moment present at the atomic level. An applied magnetic field induces a very small opposing field.

Paramagnetism
A magnetic moment is present at the atomic level. In the absence of a field, these moments are randomly oriented and no net moment is present. In the presence of a field, the individual moments tend to line up parallel to the applied field (although the net moment is generally small because thermal motions randomize all but a small fraction of the moments).

Ferromagnetism
Atomic magnetic moments act in concert to line up all moments parallel within a small volume called a domain. Above a maximum temperature (called the Curie temperature) thermal motions randomize the atomic moments and destroy the ferromagnetic behavior. In an applied field, ferromagnetic samples attempt to orient themselves with net moments parallel to the field.

Antiferromagnetism
Atomic magnetic moments act in concert to line up adjacent moments in opposite directions, the alternation continuing across a domain. Above a maximum temperature (called the Neel temperature) thermal motions randomize the atomic moments and destroy the antiferromagnetic behavior. Since the domain has no net moment, it does not respond to an applied magnetic field.

Ferrimagnetism
Atomic moments act in concert, and all moments within a domain are either parallel or opposite. Because of specific crystal factors, one direction is always dominant. The behavior in an applied field is essentially the same as that of ferromagnetic substances except for the small net moment in each domain. The maximum temperature at which the moment orientation can be maintained is called the Curie (or by some authors, the Neel) temperature.

temperature in a reasonably strong field is called isothermal remanent magnetization (IRM), and a time-dependent form of IRM is called viscous remanent magnetization (VRM). Part of the remanent magnetization in each meteorite is an IRM acquired in the earth's field. Such effects can be removed by "cleaning", i.e., tumbling the sample in a moderate alternating field of decreasing frequency.

A measure of the amount of magnetization which can be induced in a particular phase is the saturation magnetization (M_s) induced by a very strong field, typically of several thousand Oe. Estimates of M_s for meteorite phases are listed in Table XVI-1.

The earliest modern studies of NRM in meteorites are those of ALEXEYEVA (1958) and LOVERING and co-workers (LOVERING, 1959; STACEY et al., 1961; LOVERING and PARRY, 1962). The most extensive studies are those of GUSKOVA and POCHTAREV (1967, 1969), GUSKOVA (1972),

Table XVI-2. Magnetic phases in meteorites (BRECHER, 1972a). T_C is the Curie temperature, T_N the Neel temperature, and M_s the maximum remanence which can be induced into a phase at high field strengths

Formula	Name	Type	T_C or T_N (°C)	M_s (emu/g)
Fe–Ni α-iron	kamacite	ferromagnetic	~750	90—140[a]
γ-iron	taenite	paramagnetic or ferromagnetic	0—600[a]	180—220[a]
Fe[b]	iron	ferromagnetic	770	~218
Ni[b]	nickel	ferromagnetic	358	~ 54.4
Fe_3O_4	magnetite	ferrimagnetic	580	92
$NiFe_2O_4$	trevorite	ferrimagnetic	585	51
$Fe_{0.87-1.00}S$	pyrrhotite	ferrimagnetic	≲320	< 19.5—62
FeS	troilite	antiferromagnetic	320	
$(Fe, Ni)_3P$	schreibersite	ferromagnetic	≦420	
$(Fe, Ni)_3C$	cohenite	ferromagnetic	≦215	
Fe_2SiO_4[c]	fayalite	antiferromagnetic	−147	
$FeSiO_3$[c]	ferrosilite	antiferromagnetic	−233	

[a] Properties depend on Ni content.
[b] Only for reference: present in meteorites only as Ni-Fe alloys.
[c] Not magnetic at ordinary temperatures; pure $FeSiO_3$ is not stable.

and GORSHKOV et al. (1972). HERNDON et al. (1972) list data for 106 chondrites, 64 irons and 27 other differentiated meteorites studied by GUSKOVA and co-workers. Detailed studies of carbonaceous chondrites have recently been published by BANERJEE and HARGRAVES (1971, 1972), BRECHER (1972a, b) and BUTLER (1972).

These studies show that the remanent magnetizations of meteoritic irons, pallasites and mesosiderites generally lie in the range 0.01—0.3 emu g^{-1} (or gauss cm^3 g^{-1}, or cgsm = cgs magnetic units), those of metal-bearing chondrites fall in the range 10^{-4}-0.06 emu g^{-1}, and those of achondrites and chondrites with <0.5% metal fall in the range $5\times10^{-5}-2\times10^{-3}$ emu g^{-1} (BANERJEE and HARGRAVES, 1972; HERNDON et al., 1972; BRECHER, 1972a).

In principle, remanent magnetizations can be used to infer the strength of the magnetic field in which the magnetization was acquired. In fact, however, it is first necessary to determine how the magnetization was acquired, since mechanisms for acquiring remanents are not all equally efficient. One of the most efficient mechanisms is thermomagnetic (or thermochemical magnetic) acquisition, and the efficiency of this process can be studied in straightforward fashion by heating the sample above the Curie point for the ferromagnetic minerals and then inducing a new remanence in a field of known strength (BANERJEE and HARGRAVES, 1972; BRECHER, 1972b; BUTLER, 1972). An alternative but less

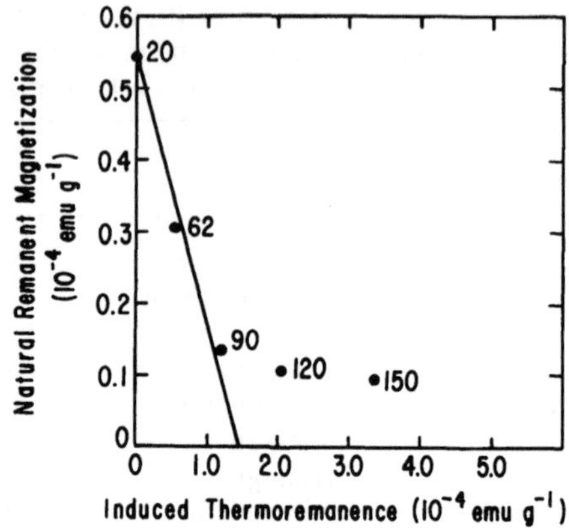

Fig. XVI-1. Plot of the remanent magnetization in the Murchison CM2 chondrite remaining after heating to a fixed temperature versus the magnetization induced by an 0.52 Oe field at the same temperature in a previously demagnetized sample of Murchison: if the magnetization in Murchison is a thermal remanent, the intensity of the paleofield is the slope times −0.52. (From BANERJEE and HARGRAVES, 1972)

satisfactory approach is to estimate the fields from the known magnetic properties of the carrier minerals (GUSKOVA and POCHTAREV, 1967; STACEY et al., 1961). Figure XVI-1 shows the application of the former technique to estimate the intensity of the field in the Murchison CM2 meteorite (BANERJEE and HARGRAVES, 1972). The slope of the 20, 62, and 90° C points indicates a field of 0.20 Oe. BANERJEE and HARGRAVES (1972) do not consider the data found at higher temperatures reliable enough to be used for paleofield estimates.

Recently, MEADOWS (1972) suggested that remanence induced by shock heating may be even more efficient than TRM or TCRM, since some ordering of domains by the shock wave may occur. Until this higher efficiency is confirmed experimentally, however, we should probably assume that the magnitudes of the paleofields estimated by TRM are correct.

The intensities of the ancient fields estimated by TRM techniques are consistently in the range 0.2—1.2 Oe, independent of the meteorite class studied. If the remanence was induced by a less efficient process (e.g., as depositional remanence during accretion), the fields were higher yet. The field at the surface of the earth varies from about 0.3 Oe at the equator to 0.6 Oe at the magnetic poles. The interplanetary magnetic field is normally about 3—6 γ (one $\gamma = 10^{-5}$ Oe), increasing by about an order of magnitude during magnetic storms (NESS, 1968). The upper limit on a dipole field at the surface of the moon ranges from 1.5—3.0 γ (COLEMAN et al., 1972); upper limits on the surface magnetic fields of Mars and

Venus are also low, $\leq 60\,\gamma$ at their magnetic equators (DOLGINOV et al., 1973; BRIDGE et al., 1967). Remanent magnetizations of lunar rocks vary widely, and although the more stable remanences are consistent with a strength of about $1000\,\gamma$ (10^{-2} Oe) at the time at which they formed 3—4 Gyr ago (NAGATA et al., 1972; PEARCE et al., 1972), some softer remanences suggest field strengths of several Oe (DUNN and FULLER, 1972). With the exception of the latter, the paleofields recorded in the meteorites are remarkably higher than those found in the moon or neighboring planets. It is generally believed that the earth's strong field is associated with the combination of a liquid core and a relatively high rate of axial rotation. If the meteorites record the paleofields of their asteroidal or cometary parent bodies, it is surprising that they all had such strong fields and, presumably, liquid cores. A liquid core demands a minimum temperature of about 1750 °K, the melting point of Fe-Ni. The process for heating of meteorite parent bodies is an unsolved problem (see Chapter XVIII), and the presence of a core seems unlikely in the parent body of CI1 or CM2 chondrites, which do not appear to have been at a temperature greater than about 450 °K since accretion. In fact, BANERJEE and HARGRAVES (1972) interpret their data to indicate that such meteorites acquired their remanence below 410 °K. Although field strength increases with depth within a planet, temperature also increases if there is a liquid core. As a result, temperatures low enough to freeze in magnetic remanence are found only in an exterior shell (RUNCORN, 1972).

An alternative explanation of the apparent uniformity in field intensity, independent of meteorite class, is that the solar magnetic field was about 10^4 times stronger 4.6 Gyr ago. Some theories of solar system origin call for very strong early fields (e.g., ALFVÉN, 1954; HOYLE, 1960). SONETT (1971) suggests that the heating of meteorite parent bodies could be produced with a minimum disturbance field of 0.3 Oe. Although this explanation may be found satisfactory for very primitive meteoritc types, it is difficult to envision how iron meteorites with low cooling rates in the taenite field (e.g., group IIIAB—GOLDSTEIN and SHORT, 1967b) could have cooled to their Curie temperatures before this high field had decayed to an intensity approaching the present-day value.

A second significant result of magnetic studies of meteorites is the discovery that their susceptibility varies with field direction (STACEY et al., 1961; WEAVING, 1962; BRECHER, 1972b). The degree of anisotropy is expressed by $P = \chi_{max}/\chi_{min}$. For terrestrial rocks the value of P is generally between 1.0 and 1.5 (NAGATA, 1961), whereas in chondrites a typical value is about 1.5. The orientation of the direction of maximum susceptibility is related to foliation fabrics (STACEY et al., 1961; WEAVING, 1962); the degree of foliation is independent of degree of recrystallization (i.e., of petrologic type) and thus appears to have been established during accretion. Mechanisms by which the fabric could have been established are reviewed by DODD (1965) and BRECHER (1972a) and include sedimentation from a fluid medium onto a planetesimal or parent body, alignment by solar fields during aggregation, thermal metamorphism in a hot parent body, and shock metamorphism; a choice among these is not yet possible.

As noted above, bulk susceptibility measurements reflect mainly the fraction of ferromagnetic phases in the meteorite sample. A somewhat surprising development, therefore, is the discovery of GORSHKOV et al. (1972) that the mean value of χ increases with increasing petrologic type within the H and L groups, even though the amount of metal in these groups is essentially independent of type.

Magnetic properties seem capable of providing important constraints on models for meteorite and solar system origin. It is to be hoped that detailed studies such as those of BRECHER (1972a, b) and BANERJEE and HARGRAVES (1972) will be produced at an accelerated pace during the coming years.

XVII. Miscellaneous Physical Studies

There are a number of physical techniques which have only rarely been applied to meteorites. This section is a summary of techniques which have been the subject of publications during the past two decades. Data on these properties in terrestrial rocks are tabulated in CLARK'S (1966) very useful compilation.

A. Density and Porosity

The density of meteorites is often reported in papers describing petrographic and bulk chemical investigations (e.g., CLARKE et al., 1970;

Table XVII-1. Density ranges observed for unweathered members of several meteorite classes

Group	Density ($g\ cm^{-3}$)	Source
E4	3.6—3.8	MASON (1966)
E6	3.5—3.7	MASON (1966)
H	3.6—3.8	MASON (1962a)
L	3.5—3.6	MASON (1962a)
LL	3.4—3.5	JAROSEWICH (1966) [b]
CV, CO	3.3—3.6	MASON (1963c)
CM	2.6—2.9	MASON (1963c)
CI	2.2—2.3	MASON (1963c)
Aub	3.2	(Density of enstatite)
Dio	3.3—3.4	MASON (1963b)
Ure	ca. 3.3	VDOVYKIN (1970)
Euc	3.1—3.2	CHIRVINSKY (1941) [c]
How	3.2—3.3	JAROSEWICH (1967) [d]
Mes	ca. 5	[e]
Pal	4.3—5.8	CHIRVINSKY (1949)
IIA, IIIA [a]	7.8—8.0	HENDERSON and PERRY (1954)

[a] And other iron meteorites which contain <2 wt % inclusions.
[b] Also MASON and WIIK (1961), KEIL et al., (1964)
[c] Also ENGELHARDT (1963).
[d] Also MASON and WIIK (1966a, b).
[e] Silicate density about 3.2 $g\ cm^{-3}$ (as in howardites), metal density about 7.9 $g\ cm^{-3}$ (as in IIA, IIIA).

Table XVII-2. Porosity of meteorite stones

Meteorite	Class	Micro-porosity (%)	Macro-porosity (%)	Total porosity (%)	Ref.[a]
Krymka	L3	3.7	3.0	6.7	1
Barrata	L4	—	—	0.7	2
Saratov	L4	5.9	12.4	18.3	1
Bjurböle	L4	—	—	16.7	2
Farmington	L5	—	—	2.9	2
Kunashak	L6	3.9	3.2	7.1	1
Mount Browne	L6	—	—	6.8	2
Elenovka	L	2.4	8.1	10.5	1
Forest Vale	H4	—	—	18.1	2
Zhovtnevyi	H	2.9	10.1	13.1	1
Mokoia	CM2	—	—	24.4	2
Cumberland Falls	Aub	—	—	4.3	2
Pesyanoe	Aub	2.7	12.4	15.1	1
Stannern	Euc	—	—	14.8	3

[a] References: 1. ALEXEYEVA (1958); 2. STACEY et al. (1961); 3. ENGELHARDT (1963).

MASON and WIIK, 1961). Table XVII-1 gives density ranges for unweathered specimens of the different classes.

The porosity of meteoritic stones is high and variable; data for 14 stones are tabulated in Table XVII-2. Chondrite porosities average about 11%. HOLMES (1965) notes that the porosity of plutonic rocks is generally less than 1%, that of sandstones about 14%, and that of loose sand or gravel about 35%. Thus, the porosity of the chondrites is in keeping with an origin by accretion or sedimentation on a parent body surface followed by compaction at shallow depths. Interestingly, there is no correlation between porosity and petrologic type among the seven L-group chondrites. The solar gas-rich brecciated aubrite Pesyanoe probably also accumulated in its present form on a parent-body surface, and its porosity is comparable to the highest values observed in chondrites. In contrast, the high porosity of Stannern probably reflects an origin as a surface lava (ENGELHARDT, 1963). The microporosity of iron meteorites is about an order of magnitude lower than that of chondrites (SHUR et al., 1970).

B. Mechanical Properties

Compressive strengths of eight ordinary chondrites were reported by BUDDHUE (1942); the values ranged from 0.062 to 3.7 kb and appear

Table XVII-3. Seismic velocities in ordinary chondrites (ALEXEYEVA, 1960)

Meteorite	Group and type	Longitudinal velocity ($km\,sec^{-1}$)	Transverse velocity ($km\,sec^{-1}$)
Bielokrynitschie	H4	3660	1220
Bjelaja Zerkov	H6	3990	—
Farmington	L5	3850	1000
Krymka	L3	2760	830
Kunashak	L6	3570	—
Misshof	H4, 5	2050	900
Saratov	L4	3300	1130
Sevrukovo	L	4200	600

unrelated to petrologic type (although only types 5 and 6 were included in his study). Except for Holbrook (which showed the least strength), all samples were weathered finds; it is probable that tests of fresh materials including unweathered enstatite chondrites and shock-hardened ordinary chondrites would yield still higher values. BUDDHUE (1942) reported a compressional strength of 3.7 kb for the IIIA iron Descubridora. KNOX (1970) found compressional strengths ranging from 3.2 to 4.4 kb for three irons, the highest being in shock-hardened Canyon Diablo. GORDON (1970) determined the tensile strength of the IVA iron Gibeon to be 3.2 kb; he reported that Gibeon specimens elongated about 20% prior to failure, indicating a ductility similar to 2024 Al alloy, but a factor of 2 less than that for pure Fe. The reduction of ductility in Gibeon resulting from a decrease in temperature to 100° K was only a factor of 2, whereas pure Fe became brittle at this temperature.

Seismic velocities in eight chondrites measured by ALEXEYEVA (1960) are tabulated in Table XVII-3. Both the longitudinal (P-wave) and transverse (S-wave) velocities are substantially smaller than those observed in terrestrial igneous rocks and are more similar to values for sedimentary rocks or volcanic tuffs (CLARK, 1966). One expects the lowest seismic velocities in the least consolidated materials, and might on this basis predict a correlation with petrologic type. No such correlation is observed.

C. Thermal Properties

ALEXEYEVA (1958) reported that the heat capacity at 0° C of five ordinary chondrites ranged from 0.166 to 0.182 cal g^{-1}, with higher values in more metal-rich samples. A sixth ordinary chondrite (Kukschin) inexplicably gave a widely divergent result (0.24 cal g^{-1}).

ALEXEYEVA (1958) found that the L-chondrite Elenovka began to melt at about 1180° C and was completely molten at 1350° C. A typical melting range for an igneous rock is much narrower, generally of the order of 50° C. VOLAROVICH and LEONTIEVA (1941) reported that the L4 chondrite Saratov melted at 1320° C and had a viscosity of about 1.3×10^4 poises at 1400° C. IVANOVA et al. (1968) observed that the Saratov and Elenovka chondrites become molten in the range 1360—1410° C; they also measured the rate of vaporization of meteorites over the range 1900—2200° C and found that the fraction of the starting mass which vaporized in four hours rose from 0.2 at 1900° C to 0.6 at 2200° C.

Thermal conductivity of three ordinary chondrites and the Norton County aubrite were determined by ALEXEYEVA (1960) to lie in the range $(3.6-5.8) \times 10^{-3}$ cal sec^{-1} cm^{-1} $°C^{-1}$ at 50° C. If one disregards the data on the Orlovka find, thermal conductivity increases monotonically with increasing metal content.

D. Electrical Properties

Metallic minerals in meteorites behave as conductors; most oxygen-bearing minerals behave as insulators. Thus, the electrical properties of meteorites are strongly dependent on the composition of their major minerals and on textural factors, especially whether metal grains in silicate-rich meteorites are in contact with one another.

The DC resistivity of metallic Fe is about 10^{-8} ohm-m; typical values for Precambrian plutonic rocks are about 5000—20000. It is not surprising that stony meteorites fall nearer to the latter than the former. Data of various authors summarized by WOOD (1963b) show values ranging from about 0.1 to 5×10^6 ohm-m. If we eliminate data on finds and on Rose City, which has large metallic veins of shock origin, the lower limit is raised to about 1200. There is no systematic decrease in resistivity with increasing metal content in the investigated samples, but the highest values are observed in the LL-chondrite Kelly and the aubrite Norton County, both of which contain less than 3% metal.

FENSLER et al. (1962) reported dielectric constants at a frequency of 420 MHz of about 37 and 12 for the H-chondrite Plainview and the L-chondrite Leedey; the dielectric constants decrease 5—20% as the frequency is increased to 1800 MHz. The conductivity of ordinary chondrites is too high to permit the determination of low frequency dielectric constants. Additional dielectric constant measurements are reported by ALEXEYEVA and TOVARENKO (1961) and ALEXEYEVA and GUSKOVA (1969).

E. Optical Properties

Optical properties of meteorites have been studied chiefly in order to compare meteorites with planets, satellites, and asteroids. Reflectances of several stony meteorites and one iron measured by WATSON (1938) at three wavelengths in the visible spectrum are listed in WOOD (1963b). The reflectances are 5—30% higher in the red than in the blue; the brightest objects were the Allegan H4 and Drake Creek L6 chondrites, which reflected about 30% of the incident light. The shocked black L5 chondrite Farmington and the CM 2 carbonaceous chondrite Mighei reflected only about 8% of incident light. Asteroid albedos estimated by MATSON (1971) and MORRISON (1973) range from 3% for Bamberga to 30% for Vesta, more or less in the same range. The two sets of data are not strictly comparable because the meteorite reflectances are measured on flat rather than spherical surfaces; the albedos, in turn, are somewhat dependent on assumptions regarding rotation, surface roughness, etc., and may also include significant systematic errors.

An interesting recent development is the comparison of spectral reflectivities of asteroids with those of meteorites. MCCORD et al. (1970) showed that the reflection spectrum of Vesta is very similar in shape to that of the Nuevo Laredo eucrite (Fig. XVII-1) and distinctly different from those of ordinary chondrites and an Apollo-11 basalt. The absorption minimum near 0.9 μm indicates the presence of pyroxene; this is magnesian pigeonite in Nuevo Laredo and, by analogy, in

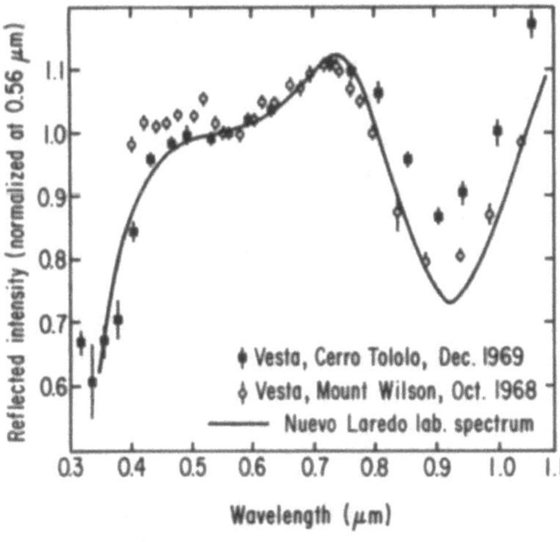

Fig. XVII-1. Telescopically determined spectral reflectivity of Vesta compared to laboratory spectrum of Nuevo Laredo eucrite. Reflectivities are normalized to 1.0 at 0.56 μm. (From MCCORD et al., 1970)

Vesta. A comparison of the spectral reflectivity of 41 meteorites and 36 asteroids is reported by CHAPMAN and SALISBURY (1973). They find that the depth of the pyroxene absorption line increases with increasing petrologic type number in the chondrites. The spectra of many asteroids are basically similar to those of ordinary chondrites, but there are differences in detail which exceed the quoted experimental errors. A study by JOHNSON and FANALE (1974) shows that the reflection spectra of carbonaceous chondrites are quite flat and featureless, presumably because of the high content of opaque minerals.

Infrared reflection spectra of meteorites have been studied mainly to investigate the use of these wavelengths for remote reconnaissance of planets and asteroids. A chondrite spectrum for the wavenumber range 180—670 cm^{-1} is reported by ARONSON et al. (1967). Visible and near infrared spectra for a number of stony meteorites are given in GAFFEY (1974).

Recent interest in the polarimetric properties of meteorites has been stimulated by investigations which show that asteroid albedos can be determined from the linear portion of plots of polarization versus the earth-asteroid-sun angle (VEVERKA, 1971a, b). The calibration of the relationships is based on studies of polarization produced by meteorite or lunar samples (VEVERKA and NOLAND, 1973). EGAN et al. (1973) report a detailed study of reflectance and polarization as a function of meteorite grain size and use these results to conclude that main-belt asteroids differ in composition from ordinary chondrites.

The observation of luminescence on the lunar surface (KOPAL and RACKHAM, 1963; MIDDLEHURST, 1967) stimulated a series of studies of the luminescence of various types of meteorites. Luminescence induced in meteorite samples by proton and ultraviolet irradiation was studied by GEAKE and WALKER (1967) and BLAIR and EDGINTON (1968). Aubrites are the only meteorite class which converts the incident energy to visible radiation with an efficiency as high as 1%; other investigated classes (ureilites, eucrites, ordinary chondrites) have efficiencies of $\leq 0.07\%$. Now, as a result of the study of samples returned by the Apollo and Luna missions, we know that the lunar surface is not aubritic and that the luminescent efficiency of returned lunar materials is indeed too low to allow solar-powered luminescence to account for the observed emissions from the moon (GREENMAN and GROSS, 1972).

As discussed in Chapter X, thermoluminescence records solid-state damage produced by cosmic rays or radioactive decays. Unfortunately, thermoluminescent intensity is not capable of yielding accurate formation or cosmic-ray ages (LIENER and GEISS, 1968).

XVIII. An Example of the Interpretation of Meteorite Properties: The Origin of Ordinary Chondrites[13]

A. Introduction

The ordinary chondrites account for 79% of observed meteorite falls. The relative abundances of the three ordinary chondrite groups (H, L, and LL) are 42, 49, and 9%, respectively. Each group includes representatives of VAN SCHMUS-WOOD (1967) petrologic types 3—6; the distribution among these types is shown in Table II-4. Type 6 is the most abundant type in the L and LL groups, type 5 in the H group.

The chief distinctions between the three classes are summarized in Table II-5: the most common indices for distinguishing between them are total Fe and metallic Fe contents which decrease, and oxidized Fe content (often measured as the Fe/(Fe+Mg) ratio in pyroxene or olivine), which increases through the sequence H-L-LL. Figure II-3 illustrates this relationship.

The distinctions between ordinary chondrite and other chondrite groups are also summarized in Table II-5. The variations in total Fe and its distribution between silicate and reduced phases (including FeS) are apparent in Fig. II-3. As shown by Fig. II-2, the ordinary chondrite groups have very similar refractory element contents, intermediate between the higher values in carbonaceous chondrites and the lower values in enstatite chondrites.

The Fe/(Fe+Mg) ratios measured in ferromagnesian minerals by KEIL and FREDRIKSSON (1964) and FREDRIKSSON et al. (1968) are plotted in Fig. XVIII-1. Distinct hiatus are found between individual groups. Largely as a result of these hiatus the three groups are often considered to be separate populations, each with a characteristic mean composition, with the range in any given property reflecting random scatter about this mean. The alternative viewpoint, that we may be dealing with a single population which has been incompletely sampled, has been stressed by MÜLLER et al. (1971). These alternatives will be discussed further in Section H.

---

[13] This chapter is a shortened and updated version of the author's review paper on the same subject (WASSON, 1972a).

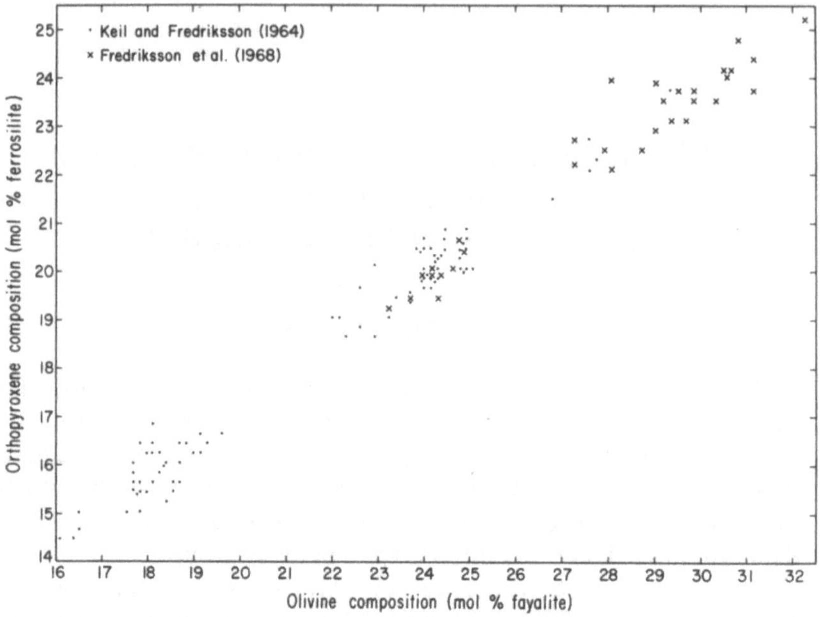

Fig. XVIII-1. Distribution of orthopyroxene and olivine compositions among ordinary chondrites of petrologic types 4—6 (Keil and Fredriksson, 1964; Fredriksson et al., 1968). Pronounced hiatus are observed between adjacent groups

Four seemingly independent fractionations are recorded in the compositions of the ordinary chondrites. Two are observable as trends within the ordinary chondrites. First, as discussed previously, the abundance of siderophilic elements decreases and the degree of oxidation increases from H to LL; the former trend was first recognized by Urey and Craig (1953), the latter by Prior (1916). This fractionation will be discussed in Section H.

The second fractionation observed entirely within the ordinary chondrites is an increase between type 6 and type 3 in the concentration of certain highly volatile elements. This increase is as much as a factor of 1000 within the L group; slightly smaller fractionation factors are observed within the other groups. This discovery, which was made by Zähringer (1966a), Marti (1967a) and Heymann and Mazor (1967, 1968) for the planetary rare gases and by Tandon and Wasson (1967, 1968) for In, was stimulated by the petrologic classification of Van Schmus and Wood (1967). This fractionation is discussed in Section E.

The other fractionations reflect systematic differences in abundance between the ordinary chondrites and the CI chondrites, which appear to be most representative of mean solarsystem material. A comparison shows that the ordinary chondrites have lower abundances of Mg, Ca, Ti, Al, Sc, and the rare-earth elements, which appear to have only one property in common—they are the elements which are concentrated in the phases which are stable at high solar nebular temperatures. The fractionation of Mg was first noted by UREY (1961), DUFRESNE and ANDERS (1962) and AHRENS (1964). The recognition that refractory elements in general were fractionated was made by LARIMER and ANDERS (1970). This fractionation will be discussed in Section K.

The ordinary chondrites also have systematically lower abundances of moderately volatile elements, as was first recognized as a general pattern by ANDERS (1964) and LARIMER and ANDERS (1967). Alternative fractionation processes which could produce the observed patterns are discussed in Section J.

B. Formation Ages and Intervals

All accurate age data indicate that the ordinary chondrites formed within a span of a few tens of Myr 4.6 Gyr ago. GOPALAN and WETHERILL (1969, 1971) and KAUSHAL and WETHERILL (1969) determined Rb-Sr isochrons based on whole-rock studies of suites of meteorites from the LL, L, and H chondrites, respectively. Well-defined ages between 4.54 and 4.69 Gyr were obtained and support the underlying assumption that entire groups formed simultaneously, which is necessary for whole-rock based isochrons.

By separating minerals or components, ages can be obtained for individual chondrites. Ages of 4.56, 4.63, and 4.70 Gyr for Guareña (H), Olivenza (LL) and Krähenberg (LL) were found by WASSERBURG et al. (1969b), SANZ and WASSERBURG (1969) and KEMPE and MÜLLER (1969), respectively. The Guareña isochron is shown in Fig. XVIII-2; the $(^{87}Sr/^{86}Sr)_I$ value of 0.69995 is distinctly higher than BABI (see Chapter X) and is most readily interpreted as indicating formation or metamorphism about 74 Myr after the time of the Rb-Sr fractionation observed in the basaltic achondrites.

Ages of 4.5—4.6 Gyr for several ordinary chondrites have been determined by Pb-Pb techniques (TATSUMOTO et al., 1973; HUEY and KOHMAN, 1973). PODOSEK (1971) reported ^{39}Ar—^{40}Ar ages of 4.6 ± 0.1 Gyr for Allegan (H5) and St. Severin (LL6). Conventional K-Ar and U, Th-He ages are discussed in Sections XB and XVIIIC.

Relative formation intervals by PODOSEK (1970a) and HOHENBERG et al. (1967a) indicate that the high-temperature minerals in ordinary chondrites began retaining Xe within a period of about 10 Myr for the six ordinary chondrites studied. Formation intervals for ordinary chondrites do not differ significantly from those for other chondrites. A study of the Manych L3 chondrite gave a poorly defined I-Xe isochron with a low $^{129r}Xe/^{127}I$ ratio (PODOSEK and HOHENBERG, 1970). Although the au-

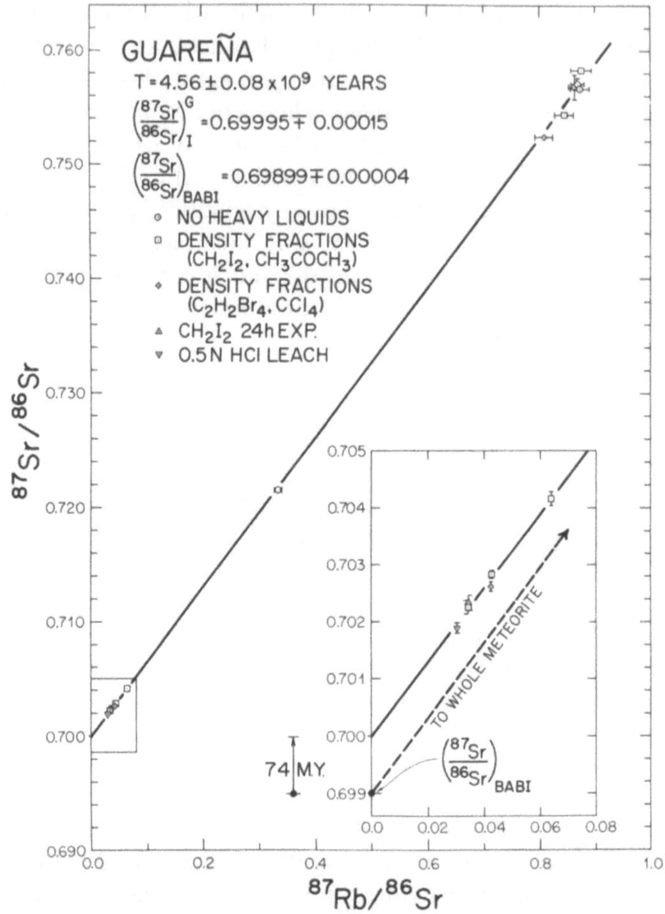

Fig. XVIII-2. Rb-Sr isochron obtained by WASSERBURG et al. (1969b) for the Guareña H6 chondrite. The $(^{87}Sr/^{86}Sr)_I$, value is distinctly higher than the BABI value determined on eucrites. If the Rb/Sr ratio was not changed during metamorphism, the period of metamorphism ended about 74 Myr after the fractionation of the eucrites

thors interpret this as evidence of cold assembly of Manych, it may more plausibly indicate partial loss of Xe from the poorly crystallized minerals in this meteorite.

C. On the Number, Size, and Location of the Parent Bodies

Clusters in cosmic-ray or metamorphism ages are perhaps the most direct evidence for storage in single parent bodies. Thus, the 4-Myr peak in the H-group cosmic-ray age distribution (Fig. XI-2) and the ca. 0.5-Gyr-old cluster of concordant U, Th-He and K-Ar ages in the L group (but not in H and LL, Fig. X-2) are indicative of the fact that these fractions of the H and L groups were stored in the same place when the clock-starting event occurred. Calculations by EBERHARDT and GEISS (1964) showed that minimal volumes of tens of km^3 of material must have been reduced to debris in order to account for such clusters.

The compositional hiatus observed between the ordinary chondrite groups (cf. Figs. II-3, VII-5, XVIII-1) and the differences in age distributions suggest that each group originated in separate parent bodies. The only evidence suggesting more than one parent body for a group is the bimodal distribution in the histogram of olivine composition in KEIL and FREDRIKSSON (1964, Fig. 7), and possibly the difference in siderophile element contents in H 3 relative to H 4—6 chondrites (C.-L. CHOU, unpublished data). There cannot have been too many parent bodies involved in providing ordinary chondrites to the earth, otherwise the compositional hiatus would surely have disappeared. It is possible and even likely that additional ordinary chondrite parent bodies exist which are not in meteorite-yielding orbits. Some of these probably have compositions which either fall in the hiatus or extend the sequence past the present extremes of the H or LL groups.

As noted above, the minimum radius of the parent bodies responsible for age clusters is of the order of 2—3 km. The actual size could be estimated if metastable phases produced by static pressures (see Chapter V) could be found in the ordinary chondrites. None are known, however; the high-pressure phases which are known (such as ringwoodite—BINNS et al., 1969) are of shock origin.

Sizes of parent bodies can be inferred from cooling rates. WOOD (1967b) estimated cooling rates at 770 °K of 2—10 °K Myr^{-1} for ordinary chondrites based on measurements of the central Ni contents of taenite grains. These imply burial depths of the order of 50 km, or radii > 50 km. Care must be exercised in the interpretation of cooling rates in absolute terms, however; diffusion coefficients are obtained by extrapolation over several hundred° K, and the kinetics of the γ—α phase transformation are not well known. It seems possible that the absolute cooling rates estimated by WOOD (1967b) are too low by a factor of 10, and as a result, that the minimum parent-body radius is reduced by a factor of 3. In fact, it is also possible that the cooling rate is not determined by the thermal inertia of the parent body, but rather by a decrease in the rate of heat transferred from an external source, such as a highly luminous phase of the sun (see Section F), or even that the interpretation of a cooling rate is incorrect, since the same characteristics could be produced by prolonged heating at a constant temperature of about 400° K.

CHAPMAN et al. (1973) found that the surface reflectivity of the earth-crossing Toro asteroid is very similar to those of certain L-group chondrites. The present radius of Toro is about 2 km, and the light curve shows considerable variation, indicating an irregular shape and suggesting that it is a fragment of a larger body of unknown size (DUNLAP et al., 1973). Thus, although it is of interest to know that Toro (and by implication, some other Apollo-type asteroids) may be a source of ordinary chondrites, these observations do not tell us the original size of the parent bodies.

A lunar origin for chondrites was proposed prior to the first sample-return missions. Studies of returned lunar samples, however, show that they have undergone chemical differentiation, and thus that the ordinary chondrites are not from the moon. Because of the difficulty of removing samples and perturbing them to earth-crossing orbits, it is also certain that the ordinary chondrites have not originated on any other solar-system object of lunar size or larger.

The two meteorites with accurately determined orbits are H 5 chondrites (see Chapter XII). These orbits extend out to the asteroid belt and provide evidence that ordinary chondrite parent bodies are presently located in this region. Further, the bulk of non-cometary debris is in this region, and it seems reasonable that the most abundant class of compact meteorites should originate here. Some belt asteroids have reflection spectra which are similar to those of ordinary chondrites (CHAPMAN and SALISBURY, 1973), although none provide as good a match as Toro (CHAPMAN et al., 1973—see preceding discussion). Thus, available evidence favors parent-body orbital semimajor axes in the range of 1.4 (Toro)-3.5 (outer edge of belt—see Fig. XII-4 and ZIMMERMAN and WETHERILL, 1973) AU for the ordinary chondrites.

In summary, each group of ordinary chondrites appears to have originated mainly in one or two parent bodies. Those parent bodies providing the age clusters had minimum radii of 2—3 km. The actual radii are not known, but there is some evidence that these were about 20—50 km or larger. The original orbital semimajor axes were probably between 1.4 and 3.5 AU.

D. Evidence for Metamorphism

VAN SCHMUS and WOOD (1967) showed that the petrographic texture of ordinary chondrites can vary from those in which chondrules are abundant and sharply defined (as in Bishunpur, Fig. VI-1) to those in which chondrules are scarcely recognizable and in which individual minerals extend across the original boundaries of the chondrules (as in Peace

River, Fig. VI-3). As discussed in greater detail in Chapters II and VI, textures such as those observed in Bishunpur are associated with the presence of non-equilibrium assemblages such as isotropic glass and ferromagnesian minerals with Fe/(Fe+Mg) ratios ranging from 0.00 to 0.40 and occasionally to 0.90 (KEIL et al., 1964; DODD and VAN SCHMUS, 1965; DODD et al., 1967; VAN SCHMUS, 1967). At the other extreme of the spectrum, glass (other than some of shock origin) is absent, olivine and pyroxene have constant compositions, and feldspar crystals have grown from submicroscopic size to dimensions of the order of 10 μm.

As discussed in Section VI C, there seems to be general agreement that petrologic types 4—6 represent a metamorphic sequence. It is my opinion that the arguments against the possibility of forming type 4 chondrites such as Bjurböle from type 3 chondrites such as Bishunpur are not as strong as the counter-arguments, and that the entire type 3—6 sequence is of metamorphic origin.

E. Fractionation of Highly Volatile Elements

As discussed in Section VII C, the highly volatile elements are highly fractionated among the petrologic types, and their concentrations generally decrease monotonically through the sequence 3-4-5=6. In Fig. VIII-4, ^{132}Xe is plotted against In for a petrologic suite of L-group chondrites (TANDON and WASSON, 1968; KEAYS et al., 1971). A strong correlation is observed. The In concentration varies by a factor of 400, that of ^{132}Xe by a factor of 50. Other volatiles behave similarly; ^{36}Ar (ZÄHRINGER, 1966a), Cs, Bi, and Tl (KEAYS et al., 1971) are fractionated by factors similar to that for In; other volatiles which generally correlate but are fractionated by factors of about 50 or less include C (MOORE and LEWIS, 1967), Br (KEAYS et al., 1971), and Cd (C.-L. CHOU, unpublished data).

The two major alternative hypotheses to explain the fractionation of highly volatile elements are: 1. differing degrees of partial condensation-agglomeration of the solar nebula (KEAYS et al., 1971; LAUL et al., 1973; LARIMER, 1973); and 2. differing degrees of metamorphic loss from a common starting material (WOOD, 1967b; DODD, 1969; WASSON, 1972a).

According to the LARIMER-ANDERS (1967) condensation model, the ordinary chondrites consist of 75% high-temperature fraction and 25% low-temperature fraction. The low-temperature fraction carries the entirety of the highly volatile elements. KEAYS et al. (1971) and LARIMER (1973) propose that the large variation in the content of these elements in the different petrologic types indicates that the low-temperature fraction was isolated (by agglomeration) from the nebula at successively lower temperatures, which they call *accretion temperatures.* The bulk of the condensation is thought to involve dissolution of metals (Bi, Tl) in Fe-Ni or sulfides (In, Cd) in FeS, but these elements condense as compounds after the solubility in the substrate is exceeded. The correlation between volatile content and mineral recrystallization is considered to be a secondary effect. Material condensed at high temperatures accretes to the interior, that condensed at low temperatures to the exterior of parent bodies; if the parent body is heated internally, the deepest material will experience the greatest intensity of heating (ANDERS, 1968).

According to a recent version of the metamorphic model (WASSON, 1972a), the volatile trace elements adsorbed from the nebula onto a variety of mineral phases. Further growth of these phases trapped previously adsorbed material and created new surfaces for further adsorption. The material from each group which agglomerated and accreted from the nebula was of a uniform composition similar to that in the more unequilibrated type 3 chondrites. During metamorphic reheating volatiles were not lost by diffusion, which is a slow process, but rather, they were lost completely whenever phase changes occurred which resulted in major lattice modifications. Trace elements may have been transported to the exterior of planetesimals or parent bodies by carrier gases such as H_2O, CO_2 or CO.

The main arguments in favor of the condensation model are: 1. accretion temperatures estimated from the observed abundance of different elements generally agree to within about 10° K; and 2. the strong degree of correlation between geochemically dissimilar elements (such as $^{36}Ar_p$ and In) is *a priori* unlikely to be produced by metamorphic loss, since these elements should be concentrated in different phases.

The main arguments in favor of the metamorphic model are: 1. a single process explains both the variation in degree of recrystallization and the volatile element content; 2. partial agglomeration should result in a fractionation of minerals depending on their tendency to stick together; 3. the probability that three elements (In, Bi, Tl) should independently undergo the same degree of condensation (e.g., be 10% condensed) within a range of only 25° (435—460°, LARIMER, 1973) is *a priori* low, as is the probability that five (the preceding plus Cd and Pb) should condense within a 70° span, 435—505 °K; 4. despite the fact that equilibration at low temperatures should result in the further oxidation of Fe, evidence of progressively higher degrees of Fe oxidation through the type 6—3 sequence is not found (WASSON, 1972a); 5.a) because of the low diffusion coefficients expected for large atoms such as Bi, Tl, and Pb at temperatures of ca. 450 °K, it is doubtful that appreciable dissolution in pre-existing metal grains could have occurred (BLANDER, 1974); and 5.b) if diffusional transport was possible, reaction with H_2O should have

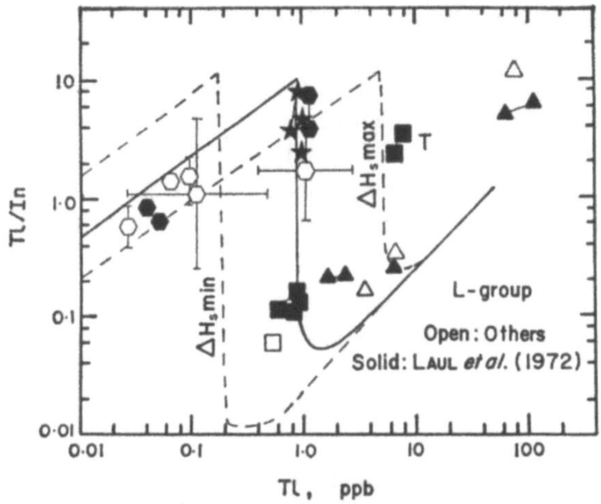

Fig. XVIII-3. Relationship between the Tl/In ratio and Tl in L-Group chondrites (LARIMER, 1973). The solid curve is calculated by a condensation model; the dashed curves show the effect of varying the ΔH_s (heat of solution) value by ± 1.5 kcal. The sharp break in the theoretical curve at a Tl content of 1 ppb results from In reaching its solubility limit in FeS and starting to condense as pure InS

resulted in the formation of an oxide layer on all metal grains, thus preventing further interaction with Bi, Tl, and Pb in the gas phase.

A recent development in support of the partial condensation model is the discovery by LARIMER (1973) that with the exception of one meteorite, he was able to fit a plot of Tl/In vs. Tl in the L group with his condensation equations (Fig. XVIII-3). The sudden rise in In content at a Tl content of about 1 ppb occurs because In has just exceeded the solubility in its substrate and begun to condense as a pure phase. This fit is quite impressive, and it will be interesting to see whether the behavior of other pairs of elements can be modeled so well by the partial condensation model. A curious aspect of the plot is that it shows that chondrites of all four types, L3—6, are found with Tl concentrations of 0.8—2 ppb, corresponding to essentially identical accretion temperatures of 458 ± 5 °K.

ONUMA et al. (1972b) derived a formula for the calculation of nebular equilibration temperatures from O-isotope data. Their temperatures fell in the same range as those calculated by LARIMER (1973) and LAUL et al. (1973) from the concentrations of highly volatile trace elements. In a more recent study ONUMA et al. (1974) note that their earlier value for the mean O-isotope composition of the solar nebula was in error, and indicate that isotopic equilibration ceased at high temperatures (near the condensation temperatures of major silicate phases). Thus, the agreement of their earlier values with accretion temperatures estimated from trace elements seems to have been fortuitous.

The debate by proponents and critics of these alternative volatile-element models and their variations is active (BLANDER, 1974; DODD, 1973; ANDERS and LARIMER, 1974) and not likely to diminish in intensity in the near future. I find that the evidence favors metamorphism and will assume this to be correct during the remainder of this chapter.

F. Metamorphic Reheating

1. Temperatures

Mineralogic indicators of temperatures during metamorphism were discussed in Chapter V. The system showing the greatest promise (subject to caveats mentioned by BLANDER, 1972) involves the distribution of Fe and Mg between augite and orthopyroxene. VAN SCHMUS and KOFFMAN (1967) reported that this system yields metamorphic temperatures of about 1090 °K for type 6 chondrites if the calibration curves of KRETZ (1963) are applied. WASSON (1972a) argues that this should be considered an upper limit and that the maximum temperature was probably 100—200° less. ONUMA et al. (1972a) report O-isotopic mineral partitioning temperatures of about 1220 °K for types 5 and 6 ordinary chondrites. There is a discrepancy of about 100° between the temperatures derived from the plagioclase-pyroxene and pyroxene-olivine pairs.

In a recent reinvestigation of the pyroxene-pyroxene thermometer, BUNCH and OLSEN (1974) report more accurate data than that of VAN SCHMUS and KOFFMAN (1967). They show that the Blander model yields similar metamorphic temperatures for the three ordinary-chondrite groups, but that the actual temperature cannot be determined because the system is not calibrated.

As discussed by WASSON (1972a), the release of primordial rare gases as a function of temperature appears to be a suitable technique for assessing metamorphic temperatures. The data determined by MERRIHUE (1966) on the Bruderheim L6 chondrite allow a temperature of 950 °K to be estimated. WOOD (1967b) notes that Bruderheim is heavily shocked. It seems unlikely that this has affected the high-temperature release patterns, however, since MERRIHUE and TURNER (1966) showed that radiogenic ^{40}Ar has not been appreciably lost by the minerals that outgas at $T > 900$ °K.

The metamorphism temperatures recorded in the other extreme of the petrologic spectrum have not been directly investigated by partitioning studies. DODD (1969) suggests that the high Ni contents observed by

WOOD (1967b) in the taenite of Bishunpur and Krymka (both L3) reflect reheating to about 670 °K. Alternatively, the fact that highly volatile elements have not been appreciably depleted in type-3 chondrites suggests metamorphism temperatures which are not appreciably higher than the nebular condensation temperatures of about 450° estimated for these elements (LARIMER, 1967, 1973).

The internal atmosphere in a parent body probably consists mainly of H_2O vapor. It should be possible to study the release of volatiles during heating in such an atmosphere and thus directly determine the maximum metamorphism temperatures experienced by the different petrologic types (H. WÄNKE, private communication, 1971).

In summary, maximum temperatures during metamorphism are poorly determined, but probably lie in the ranges 450—650 °K for the type-3 and 950—1150 °K for the type-6 chondrites.

2. Heat Sources: Internal Versus External Heating

The heat sources available for heating planetary bodies fall into six general categories: 1. decay of long-lived radionuclides; 2. decay of short-lived radionuclides; 3. release of gravitational (accretional) energy; 4. electrical heating by a T-Tauri solar "storm"; 5. radiant heating by a highly luminous Hayashi phase of the evolving sun; and 6. impact heating during accretion. It is difficult to show that any of these could have provided the intensity needed to metamorphose the ordinary chondrites if, as seems likely, they originated in bodies the size of the smaller asteroids.

Long-lived radionuclides release their heat at too low a rate to heat small bodies. Extinct radionuclides, particularly ^{26}Al, release heat at a high rate for several half-lives following their production, but recent estimates of the amount of ^{26}Al remaining at the time of parent-body formation yield values too low to have provided appreciable heating (REEVES and AUDOUZE, 1968; SCHRAMM et al., 1970).

UREY (1962) proposed that meteorite bodies were heated by the release of gravitational energy during an early stage when the parent body retained the light volatiles (especially H_2 and He). However, OSTIC (1965) carried out computations showing that, for mixtures of the elements in mean solar-system proportions, the minimum mass necessary to produce metamorphic temperatures would include a mass of non-volatile silicates and metal equal to half that of the moon. Since there are no solar-system objects of this size which could have provided the ordinary chondrites (see Section C), they cannot have been heated by the compression of gas spheres.

SONETT et al. (1968, 1970) suggested that meteorite parent bodies were heated by electric currents resulting from fields induced by a very high intensity, T-Tauri-type solar wind. Calculations show that heating to metamorphic temperatures (i.e., ca. 1000 °K) can be achieved only if the starting temperature is at least 600 °K and if very high solar spin rates and solar magnetic fields are assumed. The theory has not allowed for the fact that the parent bodies were spinning; nor for the fact that the rate of T-Tauri mass loss appears to change sporadically (KUHI, 1964). These calculations seem promising but clearly need to be extended to include more effects. The high values (ca. 1 Oe) of ancient magnetic fields inferred from magnetic remanence determinations (e.g., BRECHER, 1972b; GUSKOVA and POCHTAREV, 1967) offer some support to models involving T-Tauri solar winds.

Possibly the most plausible heat source is provided by the energy flux from the sun during a brief period in which it was fully convective prior to evolving onto the main sequence. Calculations by EZER and CAMERON (1963, 1965) show that such a Hayashi phase should result in a solar luminosity 500 times greater than that at present for a period lasting about 10^3 years. This luminosity would result in a temperature of 800 °K in a rotating black body at 2.8 AU (near the center of the asteroid belt). If at this time the (pre-T-Tauri) mass of the sun were 1.5 times the present mass, orbital radii would have been 0.67 times the present value, and the black-body temperature in the asteroid belt about 970 °K. As noted in Chapter I, LARSON's (1972) calculations indicate lower Hayashi-phase luminosities, and the above temperatures should be considered upper limits.

Impact heating is of little importance for heating asteroids. An asteroid with a density of 3.6 g cm^{-1} and a 300-km radius has an escape velocity of 0.4 km sec^{-1}. Accretion of material from orbits having low velocities relative to the asteroid releases about 20 cal g^{-1}. Even if this were converted to heat with 100% efficiency, accretion would only raise the temperature of the accreting mass by 100 °K.

Electric and solar radiation heating are thus the only plausible heat sources, and one fears that their plausibility is in no small part due to the fact that their theoretical bases are not well developed.

3. Internal Versus External Heat Sources

If metamorphism occurred in internally heated bodies, some of the volatiles released would recondense in the cooler portions near the surface, particularly if the external temperature were similar to that of a black body at 2.8 AU at the present time (170 °K). That ordinary chondrites showing appreciable enrichments in Bi, Tl or In are not known is evidence against an internal heat source. On the other hand, as F. BEGEMANN (private communication, 1971) has pointed out, a thick insulating regolith (debris) layer could provide the cold trap but have too little coherency to provide meteorites which could survive atmospheric passage. A more direct explanation of the absence of chondrites with excess volatiles, however, is that the heating was external, in which case volatile escape is no problem and is probably enhanced by the presence of a carrier gas such as H_2O.

Another argument against internal heating is that water, which accounts for 1% of the mass of type 3 chondrites, would be trapped beneath a permafrost region in an internally heated body and would react with the metallic Fe to produce FeO (WASSON, 1972a). Complete reaction of 1% of H_2O would oxidize 3% Fe to FeO. This should be detectable as variable degrees of oxidation among members of individual groups, since the amount of alteration should have depended on metamorphic temperature as well as on random factors such as loss of water to higher layers. Evidence of such water alteration in ordinary chondrites has not been reported.

The only external heat source which has been proposed is solar radiation heating during a Hayashi phase. Since the highest luminosity is present for only about 10^3 years, parent bodies could be heated only to depths of about 100 m. That most ordinary chondrites show evidence of metamorphic reheating suggests that these were mainly present as planetesimals (bodies with radii between 1 and 100 m—WASSON, 1972a) at the time of the Hayashi phase. As will be discussed in Section G, a planetesimal stage also seems necessary to account for the properties of the xenolithic chondrites.

The choice between internal versus external heating is thus closely linked to the question of possible heat sources. This is a very important area of meteorite research, and one which can be attacked both theoretically and by careful gathering and interpretation of observations on petrologic suites of ordinary chondrites.

G. Genomict Chondrites and Solar-Type Rare Gases

Many ordinary chondrites are breccias in which light-colored, generally angular fragments (xenoliths) are embedded in a darker host, the two materials differing mainly in degree of recrystallization (WLOTZKA, 1963). A section of the St. Mesmin LL chondrite, a reasonably typical xenolithic chondrite, is shown in Fig. XVIII-4. BINNS (1967c) surveyed about 500 ordinary chondrites and found that about 20% (25% in H group, 10% in L group, 62% in LL group) contained such fragments. In

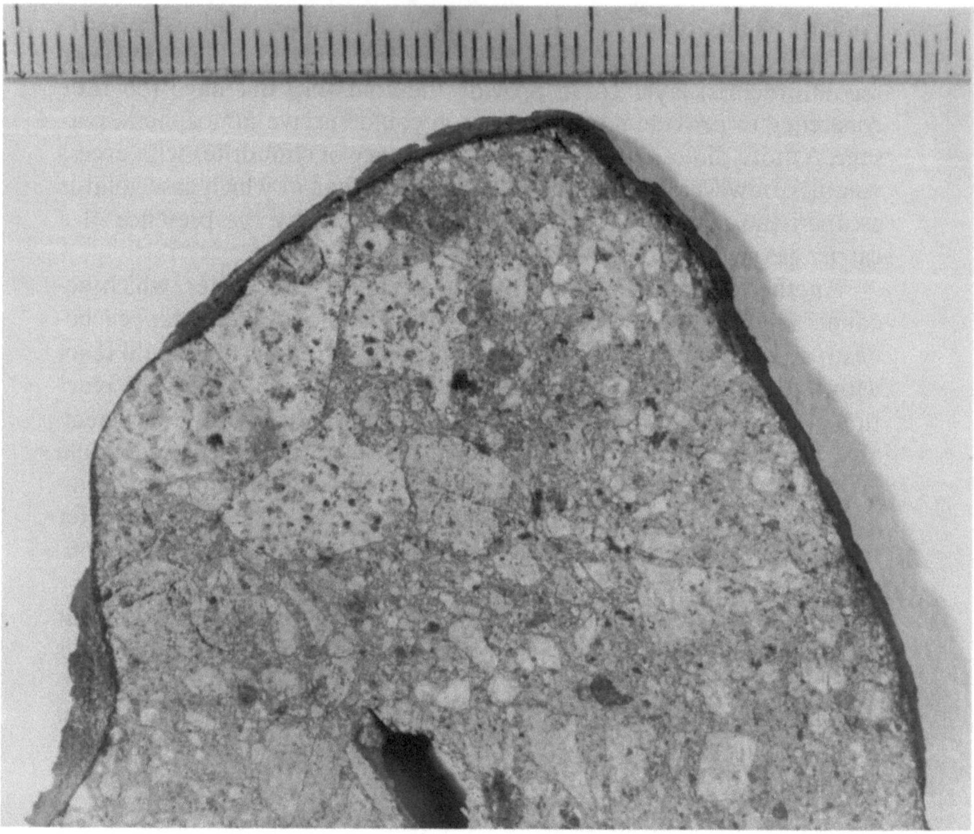

Fig. XVIII-4. St. Mesmin LL chondrite section showing genomict breccia structure. The various types of xenoliths and the host are compositionally LL-chondrite material, although another section of St. Mesmin includes an H-group clast. Note the angular nature of the xenoliths and their widely varying sizes and appearances. Xenolith material was removed from the hole near the bottom for study by a consortium of scientific groups. The host material in St. Mesmin contains solar-type rare gases. Smallest scale divisions are mm. (Photo by G. POUPEAU)

those cases in which the ferromagnesian minerals are equilibrated, the xenoliths and host have the same composition within experimental error (BINNS, 1968; KURAT et al., 1969). In a few cases, resolvable differences are found, particularly in the LL group; e.g., the Kelly LL4 chondrite olivine compositions of Fa 30.1 and Fa 28.0 in a xenolith and the host material, respectively (BINNS, 1968; see also FREDRIKSSON et al., 1968), but even in these cases the xenoliths and host have compositions within the limits found for that group. In Sharps (H 3) and Mezö-Madaras (L 3), xenoliths similar to CM chondrites have also been found (FREDRIKSSON et al., 1969; VAN SCHMUS, 1967), and in one section of the St. Mesmin LL chondrite PELLAS (1972) found a large (4-cm^2 area) dark angular fragment which appears to be H-group material. In all cases the xenoliths are of the same petrologic types as the host or are more recrystallized; in some cases individual xenoliths differ in petrologic type (e.g., in Ghubara (BINNS, 1968) and Castalia (VAN SCHMUS, 1969a).

Two important general conclusions can be drawn from petrologic studies of xenolithic chondrites (WLOTZKA, 1963; PELLAS, 1972; WASSON, 1972a). 1. the period of heating which resulted in the recrystallized textures of the xenoliths predated the final accumulation of the material into its present configuration; and 2. the similarity in composition (though differing in detail) between xenoliths and their hosts indicates that material from each group was isolated in a geographically distinct area, i.e., a narrow range of orbital radii.

As discussed in Chapter IX, high concentrations of solar-type rare gases are found in the dark (host) material of some xenolithic chondrites. Etching experiments (EBERHARDT et al., 1965a; HINTENBERGER et al., 1965b) show that these gases are surface correlated and suggest an origin by implantation of solar wind. As summarized in Table IX-1, rare-gas studies have uncovered 23, 5, and 2 solar gas-rich meteorites accounting for about 10, 2.5, and 8% of the analyzed chondrites in the H, L, and LL groups, respectively. When these data are combined with the preceding data of BINNS (1967c), it appears that about 40% of H-group xenolithic chondrites bear solar-type gas, in contrast with only 15—20% of those in the L and LL groups. These figures may be misleading, however, since a much larger proportion of L and LL chondrites belong to petrologic type-6 (solar rare gases are unknown in H6 chondrites—WÄNKE, 1966), and statistical fluctuations can be quite large when the number of observed cases is so small. Among the H3-5 chondrites, all petrologic types have about the same fraction (ca. 15%) of solar gas-rich members (WÄNKE, 1966).

Since the surficial distribution of solar-type rare gases would have been destroyed by metamorphic reheating, their presence confirms the conclusion based on petrographic evidence, viz., that the recrystallization of the xenoliths must have predated the final accumulation of the parent body. Other than the abundance of rare gases and of heavy-particle tracks (PELLAS et al., 1969; LAL and RAJAN, 1969), there appears to be no distinction between those xenolithic chondrites with solar-type rare gas and those without. It appears that the incorporation of the gas was an independent effect which only occasionally accompanied the accumulation process. Material which remained on the surface for long periods received large dosages of solar particles; material buried rapidly received negligible dosages.

The difference in petrologic grade between xenoliths and host may be related to their relative friabilities (H. WÄNKE, private communication, 1971). The material of lower petrologic type was more friable and was comminuted to form the host; the stronger materials of higher petrologic type were fragmented, but generally survived as larger clasts. If this explanation is correct, some cases should be found where the clast is of lower petrologic type than the host, since the correlation between friability and petrologic type is not a perfect one (K. FREDRIKSSON, private communication, 1971). That H6 chondrites do not contain solar-type rare gas may mean that this highly indurated material always formed xenoliths and never host (VAN SCHMUS and WOOD (1967) assigned petrologic type on basis of host texture).

H. Siderophilic-Element and Oxidation-State Fractionations

The related fractionation of Fe and O in the ordinary chondrites is illustrated in Fig. II-3 and discussed in section VD. Studies of other siderophilic elements invariably show decreasing abundances through the sequence H-L-LL, parallel to the trend observed for Fe. The fractionation of siderophilic elements must have occurred while the material was dispersed in the solar nebula. An alternative model involving partial gravitational separation of immiscible phases can be ruled out on the basis of the fact that the maximum temperatures recorded in the textures of ordinary chondrites (see Section F) are much lower than the temperature of melting of chondritic material of ca. 1600 °K (ALEXEYEVA, 1958—see Section XVII C).

Detailed recent studies show that the ratio of one siderophile to another does not remain constant during this fractionation (EHMANN et al., 1970; MÜLLER et al., 1971). For example, Fig. VII-5 illustrates the fact that the Ir/Ni ratio decreases by a factor of about 1.25 between the H and LL groups.

Before discussing possible fractionation mechanisms we should once again (see Section C) consider whether the individual groups represent homogenized material or whether intergroup fractionation trends are continuous within the groups as well. The manner in which MÜLLER et al. (1971) attempted to answer this question is shown in Fig. XVIII-5, in which the Ni/Si ratio is plotted against the Fe content of the olivine. The Ni data were obtained by neutron activation and the Fe in olivine by electron microprobe (KEIL and FREDRIKSSON, 1964; FREDRIKSSON et al., 1968) on separate samples. A dashed line drawn through the means of the three groups shows the approximate loci of the points expected if a single fractionation process produced both the intergroup and intra-

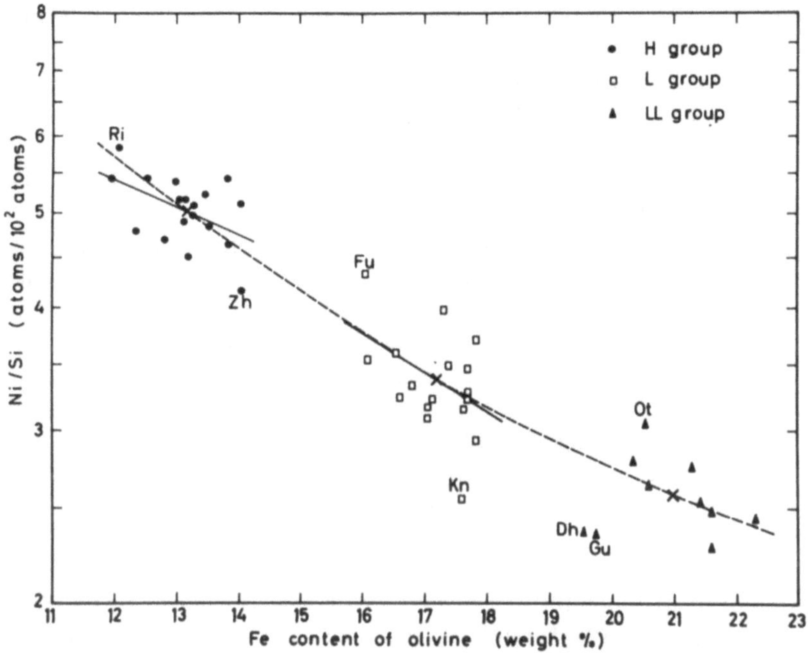

Fig. XVIII-5. Variation of the Ni/Si ratio as a function of the Fe content of the olivine in ordinary chondrites (MÜLLER et al., 1971). Negative correlations significant at the 0.95 level are observed in the H and L groups. Regression lines drawn through these groups have slopes close to that of the dashed line, an arbitrarily drawn smooth curve connecting the groups. Such concordance is expected if the Fe/Si and oxidation-state fractionations within and among the three groups are part of the same fractionation sequence. Dh = Dhurmsala, Fu = Fukutomi, Gu = Guidder, Kn = Knyahinya, Ot = Ottawa, Ri = Richardton, Zh = Zhovtnevyi

group trends. Heavy solid lines through each group are regression lines; error limits on the slopes of these lines include the smooth curve.

Although there is considerable scatter, the slopes in the H and L groups are those expected from a continuous fractionation sequence; the LL-group data are not correlated. MÜLLER et al. (1971) also show a similar plot for Ir; despite a slightly larger degree of scatter, negative correlations between Ir/Si and Fe in olivine are observed in each group, but the slopes in the H and L groups are distinctly lower than those expected from a continuous fractionation sequence.

These data confirm the observation of TANDON and WASSON (1968), i.e., that the siderophilic-element and oxidation-state fractionations are strongly related. Although these data cannot be considered to confirm the idea that a single fractionation sequence is involved, the only remaining alternative is that each group experienced an independent siderophilic-element oxidation-state fractionation. If, as appears likely, we are dealing with a fractionation which took place in the nebula, it is difficult to envision why individual regions could be isolated so as to evolve separately. As a result, I tentatively conclude that a single fractionation sequence is involved. As discussed in Section C, the hiatus are to be attributed to incomplete sampling of the sequence by the earth.

If the fractionation is of nebular origin, then it must reflect processes which varied as a function of distance from the sun. There are some tentative arguments for believing that the H group formed nearest to the sun and the LL group farthest away. One argument is based on the fact that the systematic decrease in the 1-atm densities of planets (Table I-1) indicates that the Fe/Si decreases with increasing distance from the sun (see UREY, 1966). The second argument is that: a) the nebular pressure probably decreased with increasing distance from the sun; b) material condenses out at higher temperatures where pressure is higher; and c) in a solar mixture of the elements, the higher the condensation temperature, the greater the fraction of reduced Fe present.

LARIMER and ANDERS (1970) proposed a model for the fractionation of ordinary chondrite metal in which metal of constant composition was removed at temperatures of 680—1050 °K. The more recent data of MÜLLER et al. (1971) show that the "lost" metal had a higher Ir/Ni and lower Au/Ni ratio than that remaining, and, since the 90% condensation temperature of Au given by LARIMER and ANDERS is 1220 °K, any fractionation of this type must have occurred at $T > 1220$ °K.

Possible mechanisms for removing appreciable quantities of Fe and at the same time fractionating both refractory (e.g., Ir) and moderately volatile (e.g., Au, Ge) siderophilic elements from one another are: 1. loss of a high-temperature Fe-Ni phase together with refractory elements (see Section K); 2. loss of a high-temperature Fe-Ni phase at a stage following that in which refractory elements were removed; or 3. earlier initiation of Fe-Ni condensation at the H group (as a result of the higher nebular pressure) while the gas phase remained mixed throughout the entire ordinary-chondrite formation location. In the first two models there is no direct relationship between the amount of metal remaining and the oxidation state, whereas in the third both are controlled by the nebular pressure. None of these hypotheses is in a very satisfactory state of development. For some of the arguments pro and con the reader is referred to WASSON (1972a). Earlier models are discussed in LARIMER and ANDERS (1970).

I. Chondrule Formation

At least five different origins for chondrules have been widely discussed during the past decade: 1. they are equilibrium condensates formed in the solar nebula (WOOD, 1963a); 2. they are non-equilibrium condensates formed in the solar nebula (BLANDER and KATZ, 1967; NELSON et al., 1972); 3. they were produced from pre-existing solids by electrical discharges (lightning) (WHIPPLE, 1966); 4. they resulted from impact processes occurring on the surface of the parent body (FREDRIKSSON, 1963; KURAT, 1967; WLOTZKA, 1969; DODD, 1971b); and 5. they were produced by impacts in which the larger body was m-sized or smaller (WASSON, 1972a; CAMERON, 1972; LANGE and LARIMER, 1973; see also DODD, 1971b).

It now seems clear that the first process requires either implausibly high nebular pressures (10^2—10^3 atm) or unrealistically high condensation rates during shock compression and that it can be ruled out. A serious difficulty with the second model is that it demands rather high temperatures, which should lead to chondrules having very low contents of oxidized Fe. In fact, the ferromagnesian minerals in chondrules in unequilibrated chondrites show a wide variety of FeO/(FeO + MgO) ratios ranging from 0.00 to 0.90 (VAN SCHMUS, 1969a). Although there are *ad hoc* mechanisms for producing such a range of compositions by condensation, the following models offer somewhat simpler solutions to this problem.

The third and fifth models involve the formation of chondrules one at a time, either by lightning or impact. As a result, they produce chondrules with essentially the same composition as the starting material. The solid matter in the nebula was probably a mixture of materials ranging from fine-grained low temperature minerals similar to CI chondrites (mean grain size about 1 μm—KERRIDGE, 1964) to high-temperature condensates of unknown size, but surely sometimes as large as mm (the large, high-temperature aggregates in Allende are often larger than 1 cm—see Fig. VI-4). Thus, the fusion of many small grains produces a chondrule with a high FeO/(FeO + MgO) ratio, while the melting of a large high-temperature grain results in low FeO/(FeO + MgO) ratios. It is equally possible that the "high-temperature" chondrules formed by direct metastable liquid condensation.

The formation of chondrules on the surfaces of parent bodies is generally associated with much larger impacts, although this would not have to be the case. The studies of CHRISTOPHE-MICHEL-LÉVY et al. (1972), KING et al. (1972), KURAT et al. (1972) and NELEN et al. (1972) demonstrate that chondrules can be formed on parent-body surfaces and indicate that large impacts are necessary to achieve the devitrification observed in most meteoritic chondrules. However, the kinetic energy of an object impacting the moon is much larger than that of one *accreting* to the surface of an asteroidal sized parent body. In the latter case the relative velocities prior to impact must be small, otherwise more material is removed from the parent body than is added. About 150 cal g^{-1} are needed to melt meteoritic materials. If the entire kinetic energy is converted to heat and this energy is divided equally into a mass equal to 10 times the mass of the projectile, the minimum impact velocity must be 2.5 km sec^{-1}. In contrast, the escape velocity of a 300-km asteroid is only 0.4 km sec^{-1}. Thus, during accretion onto a body of this size or smaller, chondrule formation should be a very inefficient process, and most of those that do form in energetic events would be immediately lost to space (DODD, 1971 b). Once this conclusion is reached, it follows that chondrule production by impact will occur on all bodies in direct proportion to their cross-sectional areas. During the early stages of accretion the largest fractional area will surely reside in the small bodies. Because of fragmentation, the chondrules will be substantially smaller than the larger of the two colliding objects. Thus, these larger chunks must have had minimum sizes of a few cm.

The chief problem with the use of lightning to produce chondrules appears to be inefficient conversion of turbulent energy to electrical energy. This was discussed by CAMERON (1966), who more recently (CAMERON, 1972) has endorsed the small-body impact model.

Chondrules in ordinary chondrites are found to have a narrow range of sizes; the distribution is peaked at a radius of about 0.2 mm (DODD, 1967). Chondrules are often more ellipsoidal than spheroidal; STACEY et al. (1961), WEAVING (1962) and DODD (1965) made use of this fact to search for and find preferred orientations of the long axes of the chondrules. All studied chondrites exhibited foliation, and some also showed lineation in the plane of the foliation. These fabrics are interpreted as sedimentary (accumulation) effects (DODD, 1965). WHIPPLE (1971) proposed that the narrow range of chondrule radii reflects accretion by (low-velocity) impact from a gaseous medium. Particles with radii below a given size would not impact the planetesimal, but follow the gaseous stream lines around it. DODD (1971 b) presents an alternative model involving entrainment of small particles with a transitory atmosphere.

J. Fractionation of Moderately Volatile Elements

The abundances of a number of elements are about a factor of 4 lower in ordinary chondrites than in CI chondrites (ANDERS, 1964). ANDERS (1964, 1971 a; LARIMER and ANDERS, 1967) attributes this to the absence of these elements in chondrules and notes that the matrix content is also about 25% in ordinary chondrites. In some other elements which have abundances between 0.3 and 0.9 the CI values are thought to have been only partially depleted from the chondrules. Ordinary chondrite/CI abundance ratios are plotted in Fig. XVIII-6. Note that, in contrast to the highly volatile elements, these elements show little or no

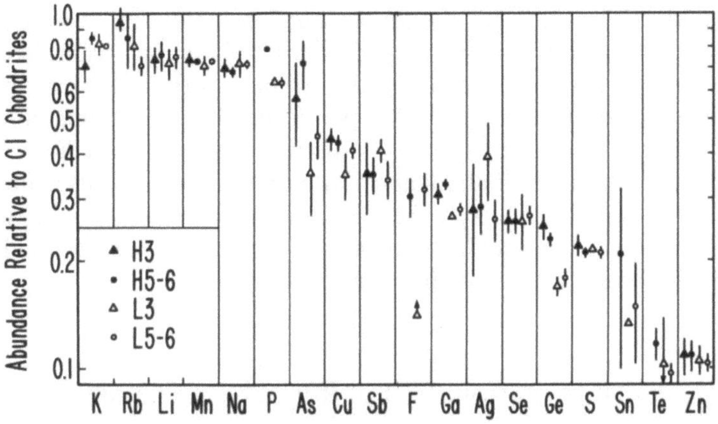

Fig. XVIII-6. Abundance of 18 elements in H and L chondrites relative to those in CI chondrites (WASSON and CHOU, 1974). Note that these elements are not fractionated between petrologic types 3 and 5—6. The difference in abundance between ordinary chondrites and CI chondrites is explained by LARIMER and ANDERS (1967) and KRÄHENBÜHL et al. (1973) in terms of accretion of two distinct nebular components, and by WASSON and CHOU (1974) in terms of continuous gas loss from the nebula during condensation

fractionation between type-3 and type-6 chondrites. According to this two-component model, the volatiles which were removed from (or did not condense with) the chondrules were lost to the system and were not available to recondense during the formation of the low-temperature matrix. It is a consequence of the application of this model to the fractionation of the highly-volatile elements (see Section E) that the chondrules now found in meteorites of higher petrologic types existed prior to the condensation of the highly volatile elements.

An alternative explanation of the abundances of the moderately volatile elements is offered by WASSON and CHOU (1974). They point out that the evidence for a plateau at a ratio of 0.25 is not very strong and that an alternative interpretation is a monotonic decrease. Some of the best determined ratios fall well away from 0.25—the most striking being Zn at 0.11. According to their model, ratios result from the loss of gas from the ordinary chondrite formation region during condensation, and thus, the elements are listed in Fig. XVIII-6 with nebular volatility increasing to the right. When Zn condensed, 89% of the uncondensed gas had been lost.

Two arguments cited in favor of the two-component model are 1. that As and the elements further to the right all have calculated nebular condensation temperatures in a relatively narrow range, 550—950 °K; and 2. that the remarkably constant chondrite/CI abundance ratios in CM, CO, and CV chondrites confirm that the two-component model was operative where these meteorites formed; it seems likely that the same processes were active at the ordinary chondrite formation location (KRÄHENBÜHL et al., 1973). Two of the chief arguments in favor of the gas-loss model are: 1. that, as discussed above, it predicts variable ordinary chondrite/CI abundance ratios, and 2. that because diffusion is a slow process it is unlikely that formation of chondrules by rapid heating and quenching could bring about appreciable loss of moderately volatile elements (WASSON and CHOU, 1974).

K. Fractionation of Refractory Elements

LARIMER and ANDERS (1970) showed that abundances of a large number of refractory elements were distinctly lower in ordinary chondrites than in CI chondrites. The lower abundances of Mg, Al, and Ca are illustrated in Fig. II-2. LARIMER and ANDERS (1970) proposed that material similar to the Ca-Al rich, high-temperature inclusions in Allende has been lost from the formation location of the ordinary chondrites, thus accounting for the lower refractory abundances. CHOU et al. (1973) analyzed separated magnetic and non-magnetic fractions of H-group chondrites and found high Ir/Ni ratios in the non-magnetic fraction, which they interpreted to indicate that some high-temperature material was trapped in the silicate portion of these meteorites; GROSSMAN (1973) showed that Ir is enriched in the Allende inclusions.

There seems to be no reason to doubt that a loss of refractories occurred, but it has proved difficult to find a satisfactory mechanism. As discussed in section VI D, the high-temperature nebular materials may have been either unvaporized residues or early condensates.

Mechanisms for separating these materials include: 1. inhomogeneous accretion, i.e., they accreted first and were buried in the deep interiors of parent bodies (TUREKIAN and CLARK, 1969); and 2. partial separation of the "ordinary-chondrite" gas from unaccreted high-temperature materials as a result of momentum-exchange processes (WASSON, 1972a).

The abundances of highly volatile elements provide strong evidence that the agglomeration of ordinary chondrites occurred at about 450—500 °K (LARIMER and ANDERS, 1967, 1970). A serious difficulty faced by mechanism 1 is that it demands the following implausible scenario: a) 10—20% of the material accretes to parent bodies at high temperatures (≥ 1500 °K); b) accretion stops for a 1000 °K interval; and c) agglomeration and accretion resumes and occurs rapidly at 500 °K.

Mechanism 2 demands that the gas and high-temperature material separate in radial distance from the sun; an early settling of solids into the median plane of the nebula does not prevent mixing with solids condensed later on. Some processes which can result in a momentum exchange between gas and solids are discussed qualitatively in CAMERON (1966) and WASSON (1972a). Because it is partly supported by pressure, cooling of the gas would cause its mean orbital distance to decrease relative to the solid particles. Although mechanism 2 appears slightly more plausible, a quantitative treatment of momentum-exchange processes is needed.

L. Summary: Temperature History of the Nebula

Summaries of models for the formation of ordinary chondrites can be found in ANDERS (1971a) and WASSON (1972a). The sequence of events outlined in the latter paper (with some modifications) is the following:

1. An interstellar gas and dust cloud underwent gravitational collapse and fragmentation. One fragment became the solar nebula.
2. The collapse of the presolar nebula accelerated, and eventually the release rate of gravitational energy became greater than the rate of energy loss by infrared radiation, and heating and vaporization of interstellar grains occurred.
3. Further collapse of the outer portion of the nebula was hindered by the angular momentum, and the material collected into a nebular disk. Temperatures fell.
4. Perhaps as a result of gas-solid momentum exchange, the ordinary-chondrite gas separated from refractory solids.
5. As a result of the falling temperatures, condensation commenced.
6. The Fe/Si ratio became fractionated as a function of solar distance, perhaps because of non-uniform nucleation of Fe-Ni; metal condensates nucleated earlier near the sun, while the gas phase remained mixed throughout the entire formation location.
7. Condensation continued. Gas was lost from the nebula, resulting in increasingly lower abundances for elements of increasing volatility. Condensation occurring nearer the sun produced opacity, and the temperature at the formation location dropped quite low (i.e., lower than that of a particle in a transparent nebula).
8. Chondrule formation occurred during (and possibly after) condensation.
9. The temperature approached a minimum. Condensation ceased. Agglomeration occurred, resulting in m to 100-m sized planetesimals (and possibly some larger proto-parent bodies).

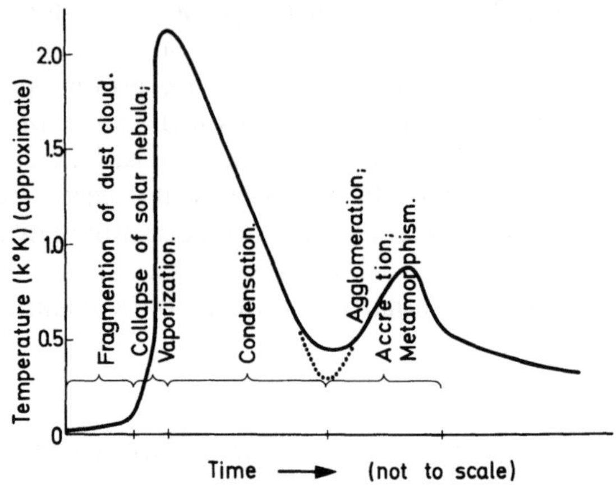

Fig. XVIII-7. Schematic thermal history of ordinary chondrite material and the adjacent nebula. The first peak is attributed to release of gravitational energy during nebular collapse, the minimum to cooling after the collapse was halted by angular momentum, and the second peak to an increase in solar radiation heating as the opacity of the intervening region decreased. An alternative model calls for the second peak to occur only inside the planetesimals

10. The planetesimals were subjected to heating, probably by solar radiation or T-Tauri solar wind-induced electrical currents. Metamorphic recrystallization occurred, and a portion of the highly volatile elements escaped.

11. The heat source declined in intensity and temperatures fell, Planetesimals accreted to parent bodies. In some cases materials which had experienced different degrees of metamorphism were mixed. If the material had an appreciable surface residence time, detectable amounts of solar-type rare gas were implanted by the solar wind.

12. On a much longer time scale, parent bodies were broken up, resulting in the observed distribution of cosmic-ray and gas-retention ages.

13. Debris from three or four ordinary chondrite parent bodies was perturbed into earth-crossing orbits and eventually captured by the earth as meteorites.

As discussed in Section C, the formation location of the ordinary chondrites was probably somewhere in the region extending 1.5 to 3.5 AU from the sun.

The inferred early temperature history of the solar nebula is shown in Fig. XVIII-7. The first maximum results from the release of gravitational energy during nebular collapse. It is drawn at 2000 °K (the limits are probably about ±300 °K), a temperature which will result in either complete or nearly complete vaporization of interstellar grains, depending chiefly on the duration of the heating period. The minimum is shown with two alternatives: the higher value is based on the assumption that the general absence of magnetite in ordinary chondrites sets the minimum at about 450 °K (LARIMER, 1973); the lower minimum is meant to suggest that the nebula reached still lower temperatures, but that slow kinetics prevented the formation of appreciable amounts of reaction products such as magnetite. The height of the second maximum is uncertain, since estimated metamorphic temperatures vary over a wide range. This is a nebular temperature only if the metamorphic reheating was caused by solar radiation.

The time scale in Fig. XVIII-7 shows only the relative sequence, since the rate of change of temperature is not known. Data of GRAY et al. (1973) indicate that Allende inclusions formed about 10 Myr before the eucrites. This may date the interval between the first maximum and the minimum; the interval between the minimum and the second maximum may be about 70 Myr, which is the inferred difference in age between the eucrites and the Guareña L6 chondrite (WASSERBURG et al., 1969b).

References

Ahrens, L. H.: Si-Mg fractionation in chondrites. Geochim. Cosmochim. Acta **28**, 411—423 (1964).

Ahrens, L. H.: Composition of stony meteorites. IX. Abundance trends of the refractory elements in chondrites, basaltic achondrites, and Apollo 11 fines. Earth Planet. Sci. Letters **10**, 1—6 (1970).

Alexander, E. C., Lewis, R. S., Reynolds, J. H., Michel, M. C.: Plutonium-244: Confirmation as an extinct radioactivity. Science **172**, 837—840 (1971).

Alexeyeva, K. N.: Physical properties of stony meteorites and their interpretation in the light of hypotheses regarding the origin of the meteorites (in Russian). Meteoritika **16**, 67—77 (1958).

Alexeyeva, K. N.: New data on the physical properties of stony meteorites (in Russian). Meteoritika **18**, 68—76 (1960).

Alexeyeva, K. N., Guskova, E. G.: Comparative characteristics of electromagnetic parameters of the lunar surface, meteorites and tektites (in Russian). Meteoritika **29**, 128—131 (1969).

Alexeyeva, K. N., Tovarenko, K. A.: Dielectric constants of stony meteorites (in Russian). Meteoritika **20**, 121—123 (1961).

Alfvén, H.: On the cosmogony of the solar system. Stockholms Obs. Ann. **14**, (2), 32 pp. (1942).

Alfvén, H.: On the Origin of the Solar System. 194 pp. Oxford: University Press (1954).

Alfvén, H.: Asteroidal jet streams. Astrophys. Space Sci. **4**, 84—102 (1969).

Alfvén, H., Arrhenius, G.: Structure and evolutionary history of the solar system, I. Astrophys. Space Sci. **8**, 338—421 (1970 a).

Alfvén, H., Arrhenius, G.: Origin and evolution of the solar system, II. Astrophys. Space Sci. **9**, 3—33 (1970 b).

Alfvén, H., Arrhenius, G.: Structure and evolutionary history of the solar system, III. Astrophys. Space Sci. **8**, 338—421 (1973).

Allen, R. O., Mason, B.: Minor and trace elements in some meteoritic minerals. Geochim. Cosmochim. Acta **37**, 1435—1456 (1973).

Anders, E.: Meteorite ages. In: Middlehurst, B. M., Kuiper, G. P. (Eds.): The Solar System — IV. The Moon, Meteorites and Comets, pp. 402—495 Chicago: University Chicago 1963.

Anders, E.: Origin, age and composition of meteorites. Space Sci. Rev. **3**, 583—714 (1964).

Anders, E.: Fragmentation history of asteroids. Icarus **4**, 398—408 (1965).

Anders, E.: Chemical processes in the early solar system, as inferred from meteorites. Accts. Chem. Res. **1**, 289—298 (1968).

Anders, E.: Meteorites and the early solar system. Ann. Rev. Astron. Astrophys. **9**, 1—34 (1971 a).

Anders, E.: How well do we know "cosmic" abundances? Geochim. Cosmochim. Acta **35**, 516—522 (1971 b).

Anders, E.: Interrelations of meteorites, asteroids, and comets. In: Gehrels, T. (Ed.): Physical Studies of Minor Planets, NASA Rep. SP-267, pp. 429—446, (1971 c).

ANDERS, E., HAYATSU, R., STUDIER, M. H.: Organic compounds in meteorites. Science **182**, 781—790 (1973).

ANDERS, E., HEYMANN, D.: Elements 112 to 119: Were they present in meteorites? Science **164**, 821—823 (1969).

ANDERS, E., HEYMANN, D., MAZOR, E.: Isotopic composition of primordial helium in carbonaceous chondrites. Geochim. Cosmochim. Acta **34**, 127—132 (1970).

ANDERS, E., LARIMER, J. W.: Extinct superheavy element in meteorites: Attempted characterization. Science **175**, 981—983 (1971).

ANDERS, E., LARIMER, J. W.: Validity of trace element cosmothermometer. Geochim. Cosmochim. Acta **38**, in press (1974).

ANDERS, E., LIPSCHUTZ, M. E.: Critique of paper by CARTER, N. L., KENNEDY, G. C.: "Origin of diamonds in the Canyon Diablo and Novo Urei meteorites". J. Geophys. Res. **71**, 643—661 (1966).

ANDERS, E., MELLICK, P. J.: Orbital clues to the nature of meteorite parent bodies. In: MILLMAN, P. M. (Ed.): Meteorite Research, pp. 559—572, Dordrecht: Reidel 1969.

ANDERS, E., STEVENS, C. M.: Search for extinct lead-205 in meteorites. J. Geophys. Res. **65**, 3043—3047 (1960).

ARMSTRONG, T. W.: Monte Carlo calculations of residual nuclei production in thick iron targets bombarded by 1- and 3-GeV protons and comparison with experiment. J. Geophys. Res. **74**, 1361—1373 (1969).

ARNOLD, J. R.: The origin of meteorites as small bodies. In: CRAIG, H., MILLER, S. L., WASSERBURG, G. J. (Eds.): Isotopic and Cosmic Chemistry, pp. 347—363, Amsterdam: North-Holland 1964.

ARNOLD, J. R.: The origin of meteorites as small bodies. II. The model. Astrophys. J. **141**, 1536—1547 (1965a).

ARNOLD, J. R.: The origin of meteorites as small bodies. III. General considerations. Astrophys. J. **141**, 1548—1556 (1965b).

ARNOLD, J. R.: Asteroid families and jet streams. Astron. J. **74**, 1235—1242 (1969).

ARNOLD, J. R., SUESS, H. E.: Cosmochemistry. Ann. Rev. Phys. Chem. **20**, 293—314 (1969).

ARONSON, J. R., EMSLIE, A. G., ALLEN, R. V., MCLINDEN, H. G.: Studies of the middle- and far-infrared spectra of mineral surfaces for application in remote compositional mapping of the moon and planets. J. Geophys. Res. **72**, 687—703 (1967).

ARRHENIUS, G.: Chemical effects in plasma condensation. In: ELVIUS, A. (Ed.): From Plasma to Planet, pp. 117—129. Stockholm: Almqvist and Wiksell 1972.

ARRHENIUS, G., ALFVÉN, H.: Fractionation and condensation in space. Earth Planet. Sci. Letters **10**, 253—267 (1971).

ASTOPOVICH, I. S.: Air waves caused by the fall of the meteorite on 30th June, 1908, in Central Siberia. Quart. J. Roy. Meteorol. Soc. **60**, 493—504 (1934).

AXON, H. J.: Metallurgy of meteorites. Prog. Mat. Sci. **13** 183—228 (1968).

AXON, H. J.: Preterrestrial deformation effects in iron meteorites. In: MILLMAN, P. M. (Ed.): Meteorite Research, pp. 796—805. Dordrecht: Reidel 1969.

AXON, H. J., BOUSTEAD, J., YARDLEY, E. D.: The mechanical and thermal alteration of iron meteorite structures. In: FRENCH, B. M., SHORT, N. M. (Eds.): Shock Metamorphism of Natural Materials, pp. 585—599. Mono Book Corp. 1968.

AXON, H. J., FAULKNER, D.: Hot-working effects in the parent γ-phase of iron meteorites. Geochim. Cosmochim. Acta **31**, 1539—1542 (1967).

BAEDECKER, P. A., WASSON, J. T.: Elemental fractionations among enstatite chondrites. Geochim. Cosmochim. Acta 38, in press (1974).

BALDWIN, B., SHAEFFER, Y.: Ablation and breakup of large meteoroids during atmospheric entry. J. Geophys. Res. **76**, 4653—4668 (1971).

BANERJEE, S. K.: Fractionation of iron in the solar system. Nature **216**, 781 (1967).

BANERJEE, S. K., HARGRAVES, R. B.: Natural remanent magnetization of carbonaceous chondrites. Earth Planet. Sci. Letters **10**, 392—396 (1971).

BANERJEE, S. K., HARGRAVES, R. B.: Natural remanent magnetizations of carbonaceous chondrites and the magnetic field in the early solar system. Earth Planet. Sci. Letters **17**, 110—119 (1972).

BARBER, D. J., COWSIK, R., HUTCHEON, I. D., PRICE, P. B., RAJAN, R. S.: Solar flares, the lunar surface, and gas-rich meteorites. Proc. Second Lunar Sci. Conf. (LEVINSON, A. A., Ed.), Geochim. Cosmochim. Acta Suppl. **2**, 2705—2714 (1971).

BEGEMANN, F.: Edelgasmessungen an Eisenmeteoriten und deren Einschlüssen. Z. Naturforsch. **20a**, 950—960 (1965)

BEGEMANN, F., HEINZINGER, K.: Content and isotopic composition of carbon in the light and dark portions of gas-rich chondrites. In: MILLMAN, P. M. (Ed.): Meteorite Research, pp. 125—143. Dordrecht: Reidel 1969.

BEGEMANN, F., VILCSEK, E.: Durch Spallationsreaktionen und Neutroneneinfang erzeugtes ^{36}Cl in Meteoriten und die präatmospherische Größe von Steinmeteoriten. Z. Naturforsch. **20a**, 533—540 (1966).

BEGEMANN, F., VILCSEK, E., NYQUIST, L. E., SIGNER, P.: Exposure history of the Pitts meteorite. Earth Planet. Sci. Letters **9**, 317—321 (1970).

BEGEMANN, F., VILCSEK, E., WÄNKE, H.: Origin of the "excess" argon-39 in stone meteorites. Earth Planet. Sci. Letters **3**, 207—213 (1967).

BEGEMANN, F., WLOTZKA, F.: Shock induced thermal metamorphism and mechanical deformations in Rumsdorf chondrite. Geochim. Cosmochim. Acta **33**, 1351—1370 (1969).

BELSKY, T., KAPLAN, I. R.: Light hydrocarbon gases, C^{13}, and origin of organic matter in carbonaceous chondrites. Geochim. Cosmochim. Acta **34**, 257—278 (1970).

BERNAS, R., GRADSZTAJN, E., REEVES, H., SCHATZMANN, E.: On the nucleosynthesis of lithium, beryllium, and boron. Ann. Phys. (N. Y.) **44**, 426—478 (1967).

BERWERTH, F.: Verzeichnis der Meteoriten im k. k. naturhistorischen Hofmuseum, Ende Oktober, 1902. Ann. K. K. Naturhist. Hofmus. Wien **18**, 1—90 (1903).

BINNS, R. A.: Stony meteorites bearing maskelynite. Nature **213**, 1111—1112 (1967a).

BINNS, R. A.: Olivine in enstatite chondrites. Am. Mineralogist **52**, 1549—1554 (1967b).

BINNS, R. A.: Structure and evolution of noncarbonaceous chondritic meteorites. Earth Planet. Sci. Letters **2**, 23—28 (1967c).

BINNS, R. A.: Cognate xenoliths in chondritic meteorites: Examples in Mezö-Madaras and Ghubara. Geochim. Cosmochim. Acta **32**, 299—317 (1968).

BINNS, R. A.: A chondritic inclusion of unique type in the Cumberland Falls meteorite. In: MILLMAN, P. M. (Ed.): Meteorite Research, pp. 696—704. Dordrecht: Reidel 1969.

BINNS, R. A., DAVIS, R. J., REED, S. J. B.: Ringwoodite, natural $(Mg, Fe)_2SiO_4$ spinel in the Tenham meteorite. Nature **221**, 943—944 (1969).

BIOT, J. B.: Relation d'un voyage fait dans le département de l'Orne, pour constater la réalité d'un météore observé à l'Aigle le 6 floréal an II. Mem. l'Institut France **7**, 224—266 (1803).

BLACK, D. C.: Trapped helium-neon isotopic correlations in gas-rich meteorites and carbonaceous chondrites. Geochim. Cosmochim. Acta **34**, 132—140 (1970).

BLACK, D. C.: Trapped neon-argon isotopic correlations in gas-rich meteorites and carbonaceous chondrites. Geochim. Cosmochim. Acta **35**, 230—235 (1971).

BLACK, D. C.: On the origins of trapped helium, neon and argon isotopic variations in meteorites — I. Gas-rich meteorites, lunar soil and breccia. Geochim. Cosmochim. Acta **36**, 347—375 (1972a).

BLACK, D. C.: On the origins of trapped helium, neon and argon isotopic variations in meteorites — II. Carbonaceous meteorites. Geochim. Cosmochim. Acta **36**, 377—394 (1972b).

BLACK, D. C., PEPIN, R. O.: Trapped neon in meteorites — II. Earth Planet. Sci. Letters **6**, 395—405 (1969).

BLAIR, I. M., EDGINGTON, J. A.: Luminescent properties of rocks, meteorites and natural glasses under proton bombardment. Nature **217**, 157—160 (1968).

BLANDER, M.: Thermodynamic properties of orthopyroxenes and clinopyroxenes based on the ideal two-site model. Geochim. Cosmochim. Acta **36**, 787—799 (1972).

BLANDER, M.: Critical comments on a proposed cosmothermometer. Geochim. Cosmochim. Acta **38**, in press (1974).

BLANDER, M., ABDEL-GAWAD, M.: The origin of meteorites and the constrained equilibrium condensation theory. Geochim. Cosmochim. Acta **33**, 701—716. (1969).

BLANDER, M., KATZ, J. L.: Condensation of primordial dust. Geochim. Cosmochim. Acta **31**, 1025—1034 (1967).

BOATO, G.: The isotopic composition of hydrogen and carbon in the carbonaceous chondrites. Geochim. Cosmochim. Acta **6**, 209—220 (1954).

BOCHSLER, P., EBERHARDT, P., GEISS, J., GRÖGLER, N.: Rare-gas measurements in separate mineral phases of the Otis and Elenovka chondrites. In: MILLMAN, P. M. (Ed.): Meteorite Research, pp. 857—874. Dordrecht: Reidel 1969.

BODENHEIMER, P.: Stellar evolution toward the main sequence. Rep. Prog. Phys. **35**, 1—54 (1972).

BOECKL, R.: Terrestrial age of nineteen stony meteorites derived from their radiocarbon content. Nature **236**, 25—26 (1972).

BOGARD, D. D.: Noble gases in meteorites. Trans. Am. Geophys. Union **52**, IUGG 429—435 (1971).

BOGARD, D. D., BURNETT, D. S., EBERHARDT, P., WASSERBURG, G. J.: ^{87}Rb-^{87}Sr isochron and ^{40}K-^{40}Ar ages of the Norton Country achondrite. Earth Planet. Sci. Letters **3**, 179—189 (1967).

BOGARD, D. D., CLARK, R. S., KEITH, J. E., REYNOLDS, M. A.: Noble gases and radionuclides in Lost City and other recently fallen meteorites. J. Geophys. Res. **76**, 4076—4083 (1971a).

BOGARD, D. D., CRESSY, P. J.: Spallation production of ^{3}He, ^{21}Ne, and ^{38}Ar from target elements in the Bruderheim chondrite. Geochim. Cosmochim. Acta **37**, 527—546 (1973).

BOGARD, D. D., HUNEKE, J. C., BURNETT, D. S., WASSERBURG, G. J.: Xe and Kr analysis of silicate inclusions from iron meteorites. Geochim. Cosmochim. Acta **35**, 1231—2354 (1971b).

BOSTROM, K., FREDRIKSSON, K.: Surface conditions of the Orgueil meteorite parent body as indicated by mineral associations. Smithsonian Misc. Coll. **151**, 39pp. (1966).

BRECHER, A.: I. Vapor Condensation of Ni-Fe Phases and Related Problems. II. The Paleomagnetic Record in Carbonaceous Chondrites. Ph. D. Dissertation, Univ. of California, San Diego, 457 pp. (1972a).

BRECHER, A.: Memory of early magnetic fields in carbonaceous chondrites. In: REEVES, H. (Ed.): On the Origin of the Solar System, pp. 260—272. Paris: CNRS 1972b.

BRENTNALL, W. D., AXON, H. J.: The response of Canyon Diablo meteorite to heat treatment. J. Iron Steel Inst. **200**, 947—955 (1962).

BRETT, R.: Cohenite: its occurrence and a proposed origin. Geochim. Cosmochim. Acta **31**, 143—159 (1967).

BRETT, R., HIGGINS, G. T.: Cliftonite: a proposed origin, and its bearing on the origin of diamonds in meteorites. Geochim. Cosmochim. Acta **33**, 1473—1484 (1969).

BRIDGE, H. S., LAZARUS, A. J., SNYDER, C. W., SMITH, E. J, DAVIS, JR., L., COLEMAN, P. J., JONES, D. E.: Mariner V.: Plasma and magnetic fields observed near Venus. Science **158**, 1669—1673 (1967).

BROUWER, D.: Secular variations of the orbital elements of minor planets Astron. J. **56**, 9—32 (1951).

BROWN, H.: An experimental method for the estimation of the age of the elements. Phys. Rev. **72**, 348 (1947).

BROWN, H., KULLERUD, G., NICHIPORUK, W.: A Bibliography on Meteorites, 686 pp. Chicago: University Press (1953).

BUCHWALD, V. F.: Heat-treated iron meteorites in museum collections. Geochim. Cosmochim. Acta **29**, 603—604 (1965).

BUCHWALD, V. F.: The iron meteorite Föllinge, Sweden. Geochim. Cosmochim. Acta **31**, 1559—1567 (1967).

BUCHWALD, V. F.: Iron meteorites: Tables relating the Widmanstätten angles to the plane of section. Geochim. Cosmochim. Acta **33**, 152—153 (1969a).

BUCHWALD, V. F.: The Gibeon meteorites (Abstract). Meteoritics **4**, 264—265 (1969b).

BUCHWALD, V. F.: A new cutting technique for meteoritic irons. Meteoritics **6**, 27—31 (1971a).

BUCHWALD, V. F.: The Cape York shower, a typical group IIIA iron meteorite, formed by directional solidification in a gravity field (Abstract). Meteoritics **6**, 252—253 (1971b).

BUCHWALD, V. F.: Iron meteorites. In press (1974).

BUCHWALD, V. F., MUNCK, S.: Catalogue of meteorites in the Mineralogical Museum of the University, Copenhagen. Anal. Geol. **1**, 81 pp. (1965).

BUCHWALD, V. F., WASSON, J. T.: The two Colombian iron meteorites, Santa Rosa and Tocavita. Analecta Geologica **3**, 29 pp. (1968).

BUCHWALD, V. F., WASSON, J. T.: On naming meteorites. Meteoritics **7**, 17—21 (1972).

BUDDHUE, J. D.: The compressive strength of meteorites. Contrib. Soc. Res. Meteorites **3**, 39—40 (1942).

BÜHLER, F., EBERHARDT, P., GEISS, J., SCHWARZMÜLLER, J.: Trapped solar wind helium and neon in Surveyor 3 material. Earth Planet. Sci. Letters **10**, 297—306 (1971).

BUFFON, G. L. L.: De la formation des planètes, Paris (1745).

BUNCH, T. E., CASSIDY, W. A.: Impact-induced deformation in the Campo del Cielo meteorite. In: FRENCH, B. M., SHORT, N. M. (Eds.): Shock Metamorphism of Natural Materials, pp. 601—612. Mono Book Corp. 1968.

BUNCH, T. E., KEIL, K.: Chromite and ilmenite in nonchondritic meteorites. Am. Mineralogist **56**, 146—157 (1971).

BUNCH, T. E., KEIL, K., OLSEN, E.: Mineralogy and petrology of silicate inclusions in iron meteorites. Contrib. Mineral. Petrol **25**, 297—340 (1970).

BUNCH, T. E., KEIL, K., SNETSINGER, K. G.: Chromite composition in relation to chemistry and texture of ordinary chondrites. Geochim. Cosmochim. Acta **31**, 1569—1582 (1967).

BUNCH, T. E., OLSEN, E.: Restudy of pyroxene-pyroxene equilibration temperatures for ordinary chondrite meteorites. Contrib. Mineral. Petrol. **43**, 83—90 (1974).

BURBIDGE, E. M., BURBIDGE, G. R., FOWLER, W. A., HOYLE, F.: Synthesis of the elements in stars. Rev. Mod. Phys. **29**, 547—650 (1957).

BURNETT, D. S.: Formation times of meteorites and lunar samples. Trans. Am. Geophys. Union **52**, IUGG435—IUGG440 (1971).

BURNETT, D. S., FOWLER, W. A., HOYLE, F.: Nucleosynthesis in the early history of the solar system. Geochim. Cosmochim. Acta **29**, 1209—1242 (1965).

BURNETT, D. S., LIPPOLT, H. J, WASSERBURG, G. J.: The relative isotopic abundance of ^{40}K in terrestrial and meteoritic samples. J. Geophys. Res. **71**, 1249—1269 (1966).

BURNETT, D. S., WASSERBURG, G. J.: Evidence for the formation of an iron meteorite at 3.8×10^9 years. Earth Planet. Sci. Letters **2**, 137—147 (1967).

BUSECK, P. R., GOLDSTEIN, J. I.: Olivine compositions and cooling rates of pallasitic meteorites. Bull. Geol. Soc. Am. **80**, 2141—2158 (1969).

BUTLER, R. F.: Natural remanent magnetization and thermomagnetic properties of the Allende meteorite. Earth Planet. Sci. Letters **17**, 120—128 (1972).

CAMERON, A. G. W.: The formation of the sun and planets. Icarus **1**, 13—69 (1962).

CAMERON, A. G. W.: Formation of the solar nebula. Icarus **1**, 339—342 (1963a).

CAMERON, A. G. W.: Contraction of the sun toward the main sequence. In: JASTROW, R., CAMERON, A. G. W. (Eds.): Origin of the Solar System, pp. 55—61. New York-London: Academic Press 1963b.

CAMERON, A. G. W.: The accumulation of chondritic material. Earth Planet. Sci. Letters **1**, 93—96 (1966).

CAMERON, A. G. W.: A new table of abundances of the elements in the solar system. In: AHRENS, L. H. (Ed.): Origin and Distribution of the Elements, pp. 125—143. London-New York: Pergamon 1968.

CAMERON, A. G. W.: Models of the primitive solar nebula. In: REEVES, H. (Ed.): On the Origin of the Solar System, pp. 56—67. Paris: CNRS 1972.

CAMERON, A. G. W.: Accumulation processes in the primitive solar nebula. Icarus **18**, 407—450 (1973).

CAMERON, A. G. W., PINE, M. R.: Numerical models of the primitive solar nebula. Icarus **18**, 377—406 (1973).

CANTELAUBE, Y., MAURETTE, M., PELLAS, P.: Traces d'ions lourds dans les mineraux de la chondrite de Saint Séverin. In: Radioactive Dating and Methods of Low Level Counting, pp. 215—229. Vienna: International Atomic Energy Agency 1967.

CANTELAUBE, Y., PELLAS, P., NORDEMANN, D., TOBAILEM, J.: Reconstitution de la météorite Saint-Séverin dans l'espace. In: MILLMAN, P. M., (Ed.): Meteorite Research, pp. 705—713. Dordrecht: Reidel 1969.

CARTER, N. L., KENNEDY, G. C.: Origin of diamonds in the Canyon Diablo and Novo Urei meteorites. J. Geophys. Res. **69**, 2403—2421 (1964).

CARTER, N. L., KENNEDY, G. C.: Origin of diamonds in the Canyon Diablo and Novo Urei meteorites—A reply. J. Geophys. Res. **71**, 666—672 (1966).

CARTER, N. L., RALEIGH, C. B., DECARLI, P. S.: Deformation of olivine in stony meteorites. J. Geophys. Res. **73**, 5439—5461 (1968).

CASE, D. R., LAUL, J. C., PELLY, I. Z., WECHTER, M. A., SCHMIDT-BLEEK, F., LIPSCHUTZ, M. E.: Abundance patterns of thirteen trace elements in primitive carbonaceous and unequilibrated ordinary chondrites. Geochim. Cosmochim. Acta **36**, 19—33 (1972).

CASSIDY, W. A.: Meteorite impact structures at Campo del Cielo, Argentina. In: FRENCH, B. M., SHORT, N. M. (Eds.): Shock Metamorphism of Natural Materials, pp. 117—128. Mono Book Corp. 1968.

CASSIDY, W. A.: A small meteorite crater: structural details. J. Geophys. Res. **76**, 3896—3912 (1971).

CEPLECHA, Z.: Multiple fall of Pribram meteorites photographed. Bull. Astron. Inst. Czech. **12**, 21—47 (1961).

CHAKRABURTTY, A. K., STEVENS, C. M., RUSHING, H. C., ANDERS, E.: Isotopic composition of silver in iron meteorites. J. Geophys. Res. **69**, 505—520 (1964).

CHAMBERLIN, T. C.: On a possible function of disruptive approach in the formation of meteorites, comets, and nebulae. Astrophys. J. **14**, 17—40 (1901).

CHAPMAN, Ç. R., JOHNSON, T. V., MCCORD, T. B.: A review of spectrophotometric studies of asteroids. In: GEHRELS, T. (Ed.): Physical Studies of Minor Planets, pp. 51—65. NASA Rep. SP—267 1971.

CHAPMAN, C. R., MCCORD, T. B., PIETERS, C.: Minor planets and related objects. X. Spectrophotometric study of the composition of (1685) Toro. Astron. J. **78**, 502—505 (1973).

CHAPMAN, C. R., SALISBURY, J. W.: Comparisons of meteorite and asteroid spectral reflectivities. Icarus **19**, 507—522 (1973).

CHEBOTAREV, G. A.: Ephemerides of Minor Planets, 166 pp. Leningrad: Nauka 1971.

CHIRVINSKY, P. N.: The system pyroxene-plagioclase in eucrites and howardites from the physico-chemical viewpoint (in Russian). Meteoritika **2**, 93—102 (1941).

CHIRVINSKY, P. N.: Pallasites, their quantitative chemico-mineralogical composition and quantitative role among other groups of meteorites (in Russian). Meteoritika **6**, 54—63 (1949).

CHLADNI, E. F. F.: Über den Ursprung der von Pallas gefundenen und anderer ihr ähnlicher Eisenmassen und über einige damit in Verbindung stehende Naturerscheinungen. Hartknoch, 63 pp. (1794).

CHLADNI, E. F. F.: Über Feuer-Meteore, und über die mit denselben herabgefallenen Massen, 434 pp. Heubner 1819.

CHOU, C.-L., BAEDECKER, P. A., WASSON, J. T.: Distribution of Ni, Ga, Ge, and Ir between metal and silicate portions of H-group chondrites. Geochim. Cosmochim. Acta **37**, 2159—2171 (1973).

CHOU, C.-L., COHEN, A. J.: Gallium and germanium in the metal and silicates of L- and LL-chondrites. Geochim. Cosmochim. Acta **37**, 315—327 (1973).

CHRISTODOULIDES, C., DURRANI, S. A., ETTINGER, K. V.: Study of thermoluminescence in some stony meteorites. Mod. Geol. **1**, 247—259 (1971).

CHRISTOPHE-MICHEL-LEVY, M.: Un chondre exceptionnel dans la météorite de Vigarano. Bull. Soc. Fr. Mineral. Cristallogr. **91**, 212—214 (1968).

CHRISTOPHE-MICHEL-LÉVY, M.: Etude minéralogique de la chondrite CIII de Lancé. Comparaison avec d'autres météorites du meme groupe. In: MILLMAN, P. M. (Ed.): Meteorite Research, pp. 492—499. Dordrecht: Reidel 1969.

CHRISTOPHE-MICHEL-LÉVY, M., LÉVY, C., PIERROT, R.: Mineralogical aspects of Apollo XIV samples: Lunar chondrules, pink spinel bearing rocks, ilmenites. In: WATKINS, C. (Ed.): Lunar Science III, pp. 136—138. Lunar Sci. Inst. 1972.

CLARK, S. P.: Handbook of Physical Constants, rev. ed., 587 pp. Geol. Soc. Am. 1966.

CLARKE, R. S., JAROSEWICH, E., MASON, B., NELEN, J., GOMEZ, M., HYDE, J. R.: The Allende, Mexico, meteorite shower. Smithson. Contrib. Earth Sci. **5**, 1—53 (1970).

CLAYTON, R. N.: Carbon isotope abundance in meteoritic carbonates. Science **140**, 192—193 (1963).

CLAYTON, R. N., GROSSMAN, L, MAYEDA, T. K.: A component of primitive nuclear composition in carbonaceous meteorites. Science **182**, 485—488 (1973).

CLAYTON, R. N., ONUMA, N. MAYEDA, T. K.: Oxygen isotope fractionation in Apollo 12 rocks and soils. In: LEVINSON, A. A. (Ed.): Proc. Second Lunar Sci. Conf. Geochim. Cosmochim. Acta., Suppl. 2, pp. 1417—1420. New York: M.I.T. Press 1971.

COBB, J. C.: Iron meteorites with low cosmic ray exposure ages. Science **151**, 1524 (1966).

COBB, J. C.: A trace element study of iron meteorites. J. Geophys. Res. **72**, 1329—1341 (1967).

COHEN, E.: Meteoritenkunde. Heft I. Untersuchungsmethoden und Charakteristik der Gemengtheile, 340 pp. Stuttgart: Schweizerbart'sche Verlagsbuchhandlung 1894.

COHEN, E.: Meteoritenkunde. Heft II. Structurformen; Versuche künstlicher Nachbildung von Meteoriten; Rinde und schwarze Adern; Relief der Oberfläche; Gestalt, Zahl und Größe der Meteorite; Nachträge zu Heft I, 301 pp. Stuttgart: Schweizerbart'sche Verlagsbuchhandlung 1903.

COHEN, E.: Meteoritenkunde. Heft III. Classification und Nomenclatur; Körnige bis dichte Eisen; Hexaëdrite; Oktaëdrite mit feinen Lamellen, 419 pp. Stuttgart: Schweizerbart'sche Verlagsbuchhandlung 1905.

COLEMAN, P. J., LICHTENSTEIN, B. R., RUSSELL, C. T., SHARP, L. R., SCHUBERT, G.: Magnetic fields near the moon. Proc. Third Lunar Sci. Conf., Geochim. Cosmochim. Acta, Suppl. **3**, 2271—2286 (1972).

COMPSTON, W., LOVERING, J. F., VERNON, M. J.: The rubidium-strontium age of the Bishopville aubrite and its component enstatite and feldspar. Geochim. Cosmochim. Acta **29**, 1085—1099 (1965).

CORYELL, C. D., CHASE, J. W., WINCHESTER, J. W.: A procedure for geochemical interpretation of terrestrial rare-earth abundance patterns. J. Geophys. Res. **68**, 559—566 (1963).

COWAN, C., ATLURI, C. R, LIBBY, W. F.: Possible antimatter content of the Tunguska Meteor of 1908. Nature **206**, 861—865 (1965).

CRAIG, H.: Isotopic standards for carbon and oxygen and correction factors for mass-spectrometric analysis of carbon dioxide. Geochim. Cosmochim. Acta **12**, 133—149 (1957).

CRAIG, H.: Petrological and compositional relationships in meteorites. In: CRAIG, H., MILLER, S. L., WASSERBURG, G. J. (Eds.): Isotopic and Cosmic Chemistry, pp. 401—451. Amsterdam-London: North-Holland 1964.

CROCKET, J. H.: Some aspects of the geochemistry of Ru, Os, Ir and Pt in iron meteorites. Geochim. Cosmochim. Acta **36**, 517—535 (1972).

CROZAZ, G., DROZD, R., HOHENBERG, C. M., HOYT, H. P., RAGAN, D., WALKER, R. M., YUHAS, D.: Solar flare and galactic cosmic ray studies of Apollo 14 and 15 samples. Proc. Third Lunar Sci. Conf., Geochim. Cosmochim. Acta, Suppl. **3**, 2917—2931 (1972).

DAKOWSKI, M.: The possibility of extinct superheavy elements occuring in meteorites. Earth Planet. Sci. Letters **6**, 152—154 (1969).

DANIELSSON, L.: The profile of a jetstream. In: GEHRELS, T. (Ed.): Physical Studies of Minor Planets, pp. 353—362. NASA Rep. SP—267 1971.

DECARLI, P. S., JAMIESON, J. C.: The formation of diamond by explosive shock, Science **133**, 1821—1822 (1961).

DEINES, P., WICKMAN, F. E.: The isotopic composition of "graphitic" carbon from iron meteorites and some remarks on the troilitic sulfur of iron meteorites. Geochim. Cosmochim. Acta **37**, 1295—1319 (1973).

DELAETER, J. R.: The isotopic composition and elemental abundance of gallium in meteorites and in terrestrial samples. Geochim. Cosmochim. Acta **36**, 735—743 (1972).

DE LAETER, J. R., JEFFERY, P. M.: Tin: its isotopic and elemental abundance. Geochim. Cosmochim. Acta **31**, 969—985 (1967).

DENCE, M. R.: The nature and significance of terrestrial impact structures. Proc. 24th Intern. Geol. Cong. **15**, 77—89 (1972).

DERGE, G., KOMMEL, A. R.: The structures of meteoric irons. Am. J. Sci. **34**, 203—214 (1937).

DEWS, J. R.: The isotopic composition of lithium in chondrules. J. Geophys. Res. **71**, 4011—4020 (1966).

DEWS, J. R., NEWBURY, R. S.: The isotopic composition of silver in the Canyon Diablo meteorite. J. Geophys. Res. **71**, 3069—3081 (1966).

DIETZ, R. S.: Shatter cones (shock fractures) in astroblemes. Proc. 24th Intern. Geol. Cong. **15**, 112—118 (1972).

DOAN, A. S., GOLDSTEIN, J. I.: The formation of phosphides in iron meteorites. In: MILLMAN, P. M. (Ed.): Meteorite Research, pp. 763—779. Dordrecht: Reidel 1969.

DODD, R. T.: Preferred orientation of chondrules in chondrites. Icarus **4**, 308—316 (1965).

DODD, R. T.: Particle sizes and compositions of unequilibrated ordinary chondrites (Abstract). Trans. Am. Geophys. Union **48**, 159 (1967).

DODD, R. T.: Metamorphism of ordinary chondrites. Geochom. Cosmochim. Acta **33**, 161—203 (1969).

DODD, R. T.: Chondrites. Trans. Amer. Geophys. Union **52**, IUGG447-IUGG453 (1971 a).

DODD, R. T.: The petrology of chondrules in the Sharps meteorite. Contrib. Mineral. Petrol. **31**, 201—227 (1971 b).

DODD, R. T.: The metal phase in unequilibrated ordinary chondrites and its implications for calculated accretion temperatures. Geochim. Cosmochim. Acta, in press (1974).

DODD, R. T., VAN SCHMUS, W. R.: Significance of the unequilibrated ordinary chondrites. J. Geophys. Res. **70**, 3801—3811 (1965).

DODD, R. T., VAN SCHMUS, W. R.: Dark-zoned chondrules. Chem. Erde **30**, 59—69 (1971).

DODD, R. T., VAN SCHMUS, W. R., KOFFMAN, D. M.: A survey of the unequilibrated ordinary chondrites. Geochim. Cosmochim. Acta **31**, 921—951 (1967).

DOHNANYI, J. S.: Collisional model of asteroids. J. Geophys. Res. **74**, 2531—2554 (1969).

DOLGINOV, S. S., YEROSHENKO, E. G., ZHUZGOV, L. N.: The magnetic field in the very close neighborhood of Mars according to data from the Mars 2 and Mars 3 spacecraft. J. Geophys. Res. **78**, 4779—4786 (1973).

DUFRESNE, E. R., ANDERS, E.: Chemical evolution of the carbonaceous chondrites. Geochim. Cosmochim. Acta **26**, 1085—1114 (1962).

DUKE, M. B., SILVER, L. T.: Petrology of eucrites, howardites and mesosiderites. Geochim Cosmochim. Acta **31**, 1637—1665 (1967).

DUNLAP, J. L., GEHRELS, T., HOWES, M. L.: Minor planets and related objects. IX. Photometry and polarimetry of (1685) Toro. Astron. J. **78**, 491—501 (1973).

DUNN, J. R., FULLER, M.: On the remanent magnetism of lunar samples with special reference to 10048,55 and 14053,48. Proc. Third Lunar Sci. Conf., Geochim. Cosmochim. Acta, Suppl. **3**, 2363—2386 (1972).

EBERHARDT, P., EUGSTER, O., GEISS, J.: Radiation ages of aubrites. J. Geophys. Res. **70**, 4427—4434 (1965 b).

EBERHARDT, P., EUGSTER, O., GEISS, J., MARTI, K.: Rare gas measurements in 30 stone meteorites. Z. Naturforsch. **21 a**, 416—426 (1966).

EBERHARDT, P., GEISS, J.: Meteorite classes and radiation ages. In: CRAIG, H., MILLER, S. L., WASSERBURG, G. J. F. (Eds.): Isotopic and Cosmic Chemistry, pp. 452—470. Amsterdam-London: North-Holland 1964.

EBERHARDT, P., GEISS, J.: On the mass spectrum of fission xenon in the Pasamonte meteorite. Earth Planet. Sci. Letters **1**, 99—102 (1966).

EBERHARDT, P., GEISS, J., GRAF, H., GRÖGLER, N., MENDIA, M. D., MÖRGELI, M., SCHWALLER, H., STETTLER, A., KRÄHENBÜHL, U., VON GUNTEN, H. R.: Trapped solar wind noble gases in Apollo 12 lunar fines 12001 and Apollo 11 breccia 10046. Proc. Third Lunar Sci. Conf., Geochim. Cosmochim. Acta, Suppl. **3**, 1821—1856 (1972).

EBERHARDT, P., GEISS, J., GRÖGLER, N.: Further evidence on the origin of trapped gases in the meteorite Khor Temiki. J. Geophys. Res. **70**, 4375—4378 (1965a).

EBERHARDT, P., GEISS, J., LUTZ, H.: Neutrons in meteorites. In: GEISS, J., GOLDBERG, E. D. (Eds.): Earth Science and Meteorites, pp. 143—168. Amsterdam-London: North-Holland 1963.

EBERHARDT, P., HESS, D. C.: Helium in stone meteorites. Astrophys. J. **131**, 38—46 (1960).

EGAN, W. G., VEVERKA, J., NOLAND, M., HILGEMAN, T.: Photometric and polarimetric properties of the Bruderheim chondritic meteorite. Icarus **19**, 358—371 (1973).

EHMANN, W. D., BAEDECKER, P. A.: The distribution of gold and iridium in meteoritic and terrestrial materials. In: AHRENS, L. H. (Ed.): Origin and Distribution of the Elements, pp. 301—311. London-New York: Pergamon 1968.

EHMANN, W. D., BAEDECKER, P. A., MCKOWN, D. M.: Gold and iridium in meteorites and some selected rocks. Geochim. Cosmochim. Acta **34**, 493—507 (1970).

EL GORESY, A.: Mineral stability and structures of graphite and sulfide inclusions in iron meteorites. Geochim. Cosmochim. Acta **29**, 1131—1135 (1965).

EL GORESY, A.: Quantitative electron microprobe analyses of coexisting sphalerite, daubreelite, and troilite in the Odessa iron meteorite and their genetic implications. Geochim. Cosmochim. Acta **31**, 1667—1676 (1967).

EL GORESY, A., KULLERUD, G.: Phase relations in the system Cr-Fe-S. In: MILLMAN, P. M. (Ed.): Meteorite Research, pp. 638—656. Dordrecht: Reidel 1969.

ENGELHARDT, W. VON: Der Eukrit von Stannern. Contrib. Mineral. Petrol. **9**, 65—94 (1963).

ENGELHARDT, W. VON: Impact structures in Europe. Proc. 24th Intern. Geol. Cong. **15**, 90—111 (1972).

EPSTEIN, S., TAYLOR, H. P.: O^{18}/O^{16}, Si^{30}/Si^{28}, D/H and C^{13}/C^{12} ratios in lunar samples. In: LEVINSON, A. A. (Ed.): Proc. Second Lunar Sci. Conf., Geochim. Cosmochim. Acta, Supp. **2**, pp. 1421—1441. M. I. T. Press 1971.

EUGSTER, O., EBERHARDT, P., GEISS, J.: Krypton and xenon isotopic composition in three carbonaceous chondrites. Earth Planet Sci. Letters **3**, 249—257 (1967a).

EUGSTER, O., EBERHARDT, P., GEISS, J.: The isotopic composition of krypton in unequilibrated and gas-rich chondrites. Earth Planet. Sci. Letters **2**, 385—393 (1967b).

EUGSTER, O., EBERHARDT, P., GEISS, J.: ^{81}Kr in meteorites and ^{81}Kr radiation ages. Earth Planet. Sci. Letters **2**, 77—82 (1967c).

EUGSTER, O., EBERHARDT, P., GEISS, J.: Isotopic analyses of krypton and xenon in 14 stone meteorites. J. Geophys. Res. **74**, 3874—3896 (1969).

EUGSTER, O., TERA, F., BURNETT, D. S., WASSERBURG, G. J.: Neutron capture effects in Gd from the Norton County achondrite. Earth Planet. Sci. Letters **7**, 436—440 (1970).

EUGSTER, O., TERA, F., WASSERBURG, G. J.: Isotopic analyses of barium in meteorites and in terrestrial samples. J. Geophys. Res. **74**, 3897—3908 (1969).
EZER, D., CAMERON, A. G. W.: The early evolution of the Sun. Icarus **1**, 422—441 (1963).
EZER, D., CAMERON, A. G. W.: A study of solar evolution. Can. J. Phys. **43**, 1497—1517 (1965).
FANALE, F. P., CANNON, W. A.: Origin of planetary primordial rare gas: the possible role of adsorption. Geochim. Cosmochim. Acta **36**, 319—328 (1972).
FARRINGTON, O. C.: Meteorites: Their Structure, Composition, and Terrestrial Relations. Self-published, 233 pp. (1915a).
FARRINGTON, O. C.: Catalogue of the meteorites of North America, to January 1, 1909. Mem. Nat. Acad. Sci. **13**, 1—513 (1915b).
FECHTIG, H., GENTNER, W., LÄMMERZAHL: Argonbestimmungen an Kaliummineralien — XII. Edelgasdiffusionsmessungen an Stein- und Eisenmeteoriten. Geochim. Cosmochim. Acta **27**, 1149—1169 (1963).
FECHTIG, H., KALBITZER, S.: The diffusion of argon in potassium-bearing solids. In: SCHAEFFER, O. A., ZÄHRINGER, J. (Eds.): Potassium Argon Dating, pp. 68—107. Berlin-Heidelberg-New York: Springer 1966.
FENSLER, W. E., KNOTT, E. F., OLTE, A., SIEGEL, K. The electromagnetic parameters of selected terrestrial and extraterrestrial rocks and glasses. In: KOPAL, Z., MIKHAILOV, Z. (Eds.): The Moon, pp. 545—565. London-New York: Academic Press 1962.
FESENKOV, V.: Can comets consist of antimatter? J. Brit. Astron. Ass. **78**, 126—128 (1968).
FIREMAN, E. L.: Distribution of helium-3 in the Carbo meteorite. Nature **181**, 1725 (1958).
FIREMAN, E. L.: Neutron exposure ages of meteorites. Z. Naturforsch. **21a**, 1138—1146 (1966).
FIREMAN, E. L., DEFELICE, J.: Argon-37, argon-39 and tritium in meteorites and the spatial constancy of cosmic rays. J. Geophys. Res. **65**, 3035—3041 (1960).
FIREMAN, E. L., GOEBEL, R.: Argon-37 and argon-39 in recently fallen meteorites and cosmic-ray variations. J. Geophys. Res. **75**, 2115—2124 (1970).
FISH, R. A., GOLES, G. G., ANDERS, E.: The record in the meteorites. III. On the development of meteorites in asteroidal bodies. Astrophys. J. **132**, 243—258 (1960).
FISHER, D. C.: The origin of meteorites: Space erosion and cosmic radiation ages. J. Geophys. Res. **71**, 3251—3259 (1966).
FISHER, D. E.: Uranium content and radiogenic ages of hypersthene, bronzite, amphoterite and carbonaceous chondrites. Geochim. Cosmochim. Acta **36**, 15—33 (1972).
FLEISCHER, R. L., HART, H. R., COMSTOCK, G. M.: Very heavy solar cosmic rays: Energy spectrum and implications for lunar erosion. Science **171**, 1240—1242 (1971).
FLEISCHER, R. L., NAESER, C. W., PRICE, P. B., WALKER, R. M., MARVIN, U. B.: Fossil particle tracks and uranium distributions in minerals of the Vaca Muerta meteorite. Science **148**, 629—632 (1965).
FLEISCHER, R. L., PRICE, P. B., WALKER, R. M.: Identification of Pu^{244} fission tracks and the cooling of the parent body of the Toluca meteorite. Geochim. Cosmochim. Acta **32**, 21—31 (1968).
FLORENSKI, K. P., IVANOV, A. V., ILYIN, N. P., PETRIKOVA, M. N., LOSEVA, L. E.: Chemical composition of cosmic spherules from the region of the Tunguska holocaust and some problems related to the differentiation of cosmic body material. Geokhimiya **1968**, 1163—1173 (1968a).

FLORENSKI, K. P., IVANOV, A. V., KIROVA, O. A., ZASLAVSKAYA, N. I.: Phase composition of finely dispersed extraterrestrial material from the region of the Tunguska holocaust. Geokhimiya **1968**, 1174—1182 (1968b).

FOLINSBEE, R. E., DOUGLAS, J. A. V., MAXWELL, J. A.: Revelstoke, a new type I carbonaceous chondrite. Geochim. Cosmochim. Acta **31**, 1625—1635 (1967).

FOUCHÉ, K. F., SMALES, A. A.: The distribution of gold and rhenium in iron meteorites. Chem. Geol. **1**, 329—339 (1966).

FOWLER, W. A., GREENSTEIN, J. L., HOYLE, F.: Nucleosynthesis during the early history of the solar system. Geophys. J. Roy. Astron. Soc. **6**, 148—220 (1962).

FREDRIKSSON, K.: Chondrules and the meteorite parent bodies. Trans. N. Y. Acad. Sci. **25**, 756—769 (1963).

FREDRIKSSON, K., DECARLI, P., AARAMÄE, A.: Shock-induced veins in chondrites. Space Res. **3**, 974—983 (1963).

FREDRIKSSON, K., JAROSEWICH, E., NELEN, J.: The Sharps chondrite—New evidence on the origin of chondrules and chondrites. In: MILLMAN, P. M. (Ed.): Meteorite Research, pp. 155—165. Dordrecht: Reidel 1969.

FREDERIKSSON, K., KEIL, K.: The light-dark structure in the Pantar and Kapoeta stone meteorites. Geochim. Cosmochim. Acta **27**, 717—739 (1963).

FREDRIKSSON, K., KRAUT, F.: Impact glass in the Cachari eucrite. Geochim. Cosmochim. Acta **31**, 1701—1704 (1967).

FREDRIKSSON, K., NELEN, J., FREDRIKSSON, B. J.: The LL group chondrites. In: AHRENS, L. H. (Ed.): Origin and Distribution of the Elements, pp. 457—466. London-New York: Pergamon 1968.

FREDERIKSSON, K., REID, A. M.: Meteorite investigations by electron microprobe techniques. In: ABELSON, P. H. (Ed.): Researches in Geochemistry, Vol. 2, pp. 143—169. New York: Wiley 1967.

FRICKER, P. E., GOLDSTEIN, J. I., SUMMERS, A. L.: Cooling rates and thermal histories of iron and stony-iron meteorites. Geochim. Cosmochim. Acta **34**, 475—491 (1970).

FRIEDEL, R. A., SHARKEY, A. G.: Similar compositions of alkanes from coal, petroleum, natural gas and Fischer-Tropsch product; calculation of isomers. U.S. Bur. Mines Rep. **RI 7122.** (1968)

FUCHS, L. H.: The phosphate mineralogy of meteorites. In: MILLMAN, P. M. (Ed.): Meteorite Research, pp. 683—695. Dordrecht: Reidel 1969.

FUCHS, L. H., BLANDER, M.: Calcium-aluminium rich inclusions in the Allende meteorite: Textural and mineralogical evidence for a liquid origin (Abstract). Trans. Am. Geophys. Union **54**, 345 (1973).

FUNK, H., PODOSEK, F., ROWE, M. W.: Spallation yields of krypton and xenon from irradiation of strontium and barium with 730 MeV protons. Earth Planet. Sci. Letters **3**, 193—196 (1967).

FUNK, H., ROWE, M. W.: Spallation yield of xenon from 730 MeV proton irradiation of barium. Earth Planet. Sci. Letters **2**, 215—219 (1967).

FUSE, K., ANDERS, E.: Aluminium-26 in meteorites--VI. Achondrites. Geochim. Cosmochim. Acta **33**, 653—670 (1969).

GAFFEY, M. J.: A systematic study of the spectral reflectivity characteristics of the meteorite classes with applications to the interpretation of asteroid spectra for mineralogical and petrological information. Ph. D. Thesis, M. I. T., 355 pp. (1974).

GANAPATHY, R., ANDERS, E.: Ages of calcium-rich achondrites-II. Howardites, nakhlites, and the Angra dos Reis angrite. Geochim. Cosmochim. Acta **33**, 775—787 (1969).

GAST, P. W.: The chemical composition of the earth, the moon, and chondritic meteorites. In: ROBERTSON, E. C. (Ed.): The Nature of the Solid Earth, pp. 19—40. New York: McGraw-Hill 1971.

GAST, P. W., MCCONNELL, R. K.: Evidence for initial chemical layering of the moon. In: WATKINS, C. (Ed.): Lunar Science III, **88**, pp. 289—290. Lunar Sci. Inst. 1972.

GAULT, D. E., HÖRZ, F., HARTUNG, J. B.: Effects of microcratering on the lunar surface. Proc. Third Lunar. Sci. Conf., Geochim. Cosmochim. Acta, Suppl. **3**, 2713—2734 (1972).

GAULT, D. E., WEDEKIND, J. A.: The destruction of tektites by micrometeoroid impact. J. Geophys. Res. **74**, 6780—6794 (1969).

GEAKE, J. E., WALKER, G.: Laboratory investigations of meteorite luminescence. Proc. Roy. Soc. London **296**, 337—346 (1967).

GEISS, J., BÜHLER, F., CERUTTI, H., EBERHARDT, P.: The solar-wind composition experiment. Apollo 14 Preliminary Science Report, NASA Document SP-272, 221—226 (1971).

GEISS, J., EBERHARDT, P., BÜHLER, F., MEISTER, J., SIGNER, P.: Apollo 11 and 12 solar wind composition experiments: fluxes of He and Ne isotopes. J. Geophys. Res. **75**, 5972—5979 (1970).

GELPI, E., HAN, J., NOONER, D. W., ORÓ, J.: Organic compounds in meteorites—III. Distribution and identification of aliphatic hydrocarbons produced by open flow Fischer-Tropsch processes. Geochim. Cosmochim. Acta **34**, 965—979 (1970).

GELPI, E., ORÓ, J.: Organic compounds in meteorites—IV. Gas chromatographic-mass spectrometric studies on the isoprenoids and other isomeric alkanes in carbonaceous chondrites. Geochim. Cosmochim. Acta **34**, 981—994 (1970).

GENTRY, R. V.: Anti-matter content of the Tunguska meteor. Nature **211**, 1071—1072 (1966).

GERLING, E. K., LEVSKII, L. K.: On the origin of inert gases in stony meteorites (in Russian). Dokl. Akad. Nauk. S. S. S. R. **110**, 750—753 (1956).

GOEL, P. S., KOHMAN, T. P.: Cosmogenic carbon-14 in meteorites and terrestrial ages of "finds" and craters. Science **136**, 875—876 (1962).

GOLDBERG, E., UCHIYAMA, A., BROWN, H.: The distribution of nickel, cobalt, gallium, palladium and gold in iron meteorites. Geochim. Cosmochim. Acta **2**, 1—25 (1951).

GOLDSCHMIDT, V. M.: Geochemische Verteilungsgesetze der Elemente. IX. Die Mengenverhältnisse der Elemente und der Atom-Arten. Skr. Nors. Vidensk. Akad. Oslo, Mat.-Naturv. Kl., No. **4**, 148 pp. (1937).

GOLDSTEIN, J. I.: The formation of the kamacite phase in metallic meteorites. J. Geophys. Res. **70**, 6223—6232 (1965).

GOLDSTEIN, J. I.: Distribution of germanium in the metallic phases of some iron meteorites J. Geophys. Res. **72**, 4689—4696 (1967).

GOLDSTEIN, J. I.: The classification of iron meteorites. In: MILLMAN, P. M. (Ed.): Meteorite Research, pp. 721—737. Dordrecht: Reidel 1969.

GOLDSTEIN, J. I., OGILVIE, R. E.: Electron microanalysis of metallic meteorites. Part I — phosphides and sulphides. Geochim. Cosmochim. Acta **27**, 623—637 (1963).

GOLDSTEIN, J. I., OGILVIE, R. E.: The growth of the Widmanstätten pattern in metallic meteorites. Geochim. Cosmochim. Acta **29**, 893—920 (1965).

GOLDSTEIN, J. I., SHORT, J. M.: Cooling rates of 27 iron and stony-iron meteorites. Geochim. Cosmochim. Acta **31**, 1001—1023 (1967a).

GOLDSTEIN, J. I., SHORT, J. M.: The iron meteorites, their thermal history and parent bodies. Geochim. Cosmochim. Acta **31**, 1733—1770 (1967b).

GOLES, G. G., FISH, R. A., ANDERS, E.: The record in the meteorites—I. The former environment of stone meteorites as deduced from K^{40}—Ar^{40} ages. Geochim. Cosmochim. Acta **19**, 177—195 (1960).

GOPALAN, K., WETHERILL, G. W.: Rubidium-strontium age of amphoterite (LL) chondrites. J. Geophys. Res. **74**, 4349—4358 (1969).

GOPALAN, K., WETHERILL, G. W.: Rubidium-strontium studies on enstatite chondrites: whole meteorite and mineral isochrons. J. Geophys. Res. **75**, 3457—3467 (1970)

GOPALAN, K., WETHERILL, G. W.: Rubidium-strontium studies on black hypersthene chondrites: Effects of shock and reheating. J. Geophys. Res. **76**, 8484—8492 (1971).

GORDON, R. B.: Mechanical properties of iron meteorites and the structure of their parent planets. J. Geophys. Res. **75**, 439—447 (1970).

GORSHKOV, E. S., GUSKOVA, E. G., POCHTAREV, V. I.: Physical properties of meteorites. Mod. Geol. **3**, 105—106 (1972).

GRADSZTAJN, E., GUEZ, M.: Mass discrimination introduced by the optical properties of a sputtering ion source during the isotope analysis of lithium in stony meteorites. Int. J. Mass Spectrom. Ion Phys. **4**, 305—313 (1970).

GRAY, C. M., PAPANASTASSIOU, D. A., WASSERBURG, G. J.: The identification of early condensates from the solar nebula. Icarus **20**, 213—239 (1973).

GREENLAND, L., LOVERING, J. F.: Minor and trace element abundances in chondritic meteorites. Geochim. Cosmochim. Acta **29**, 821—858 (1965).

GREENMAN, N. N., GROSS, H. G.: Luminescence of Apollo 14 and Apollo 15 lunar samples. Proc. Third Lunar. Sci. Conf., Geochim. Cosmochim. Acta, Suppl. **3**, 2981—2995 (1972).

GROSSMAN, L.: Condensation in the primitive solar nebula. Geochim. Cosmochim. Acta **36**, 597—619 (1972).

GROSSMAN, L.: Refractory trace elements in Ca-Al-rich inclusions in the Allende chondrite. Geochim. Cosmochim. Acta **37**, 1119—1140 (1973).

GROSSMAN, L., CLARK, S. P.: High-temperature condensates in chondrites and the environment in which they formed. Geochim. Cosmochim. Acta **37**, 635—649 (1973).

GROSSMAN, L., LARIMER, J. W.: Early chemical history of the solar system. Rev. Geophys. Space Phys. **12**, 71—101 (1974).

GUSKOVA, E. G.: Magnetic Properties of Meteorites (in Russian), 108 pp. Leningrad: Nauka (1972).

GUSKOVA, E. G., POCHTAREV, V. I.: Magnetic fields in space according to a study of the magnetic properties of meteorites. Geomag. Aeron. **7**, 245—250 (1967).

GUSKOVA, E. G., POCHTAREV, V. I.: Magnetic properties of meteorites in the Soviet collection (in Russian). In: MILLMAN, P. M. (Ed.): Meteorite Research, pp. 633—637. Dordrecht: Reidel 1969.

HANNEMAN, R. E., STRONG, H. M., BUNDY, F. P.: Hexagonal diamonds in meteorites: implications. Science **155**, 995—997 (1967).

HARRIS, P. G., TOZER, D. C.: Fractionation of iron in the solar system. Nature **215**, 1449—1451 (1967).

HASKIN, L. A., ALLEN, R. O., HELMKE, P. A., PASTER, T. P., ANDERSON, M. R., KOROTEV, R. L., ZWEIFEL, K. A.: Rare earths and other trace elements in Apollo 11 lunar samples. Proc. Apollo 11 Lunar Sci. Conf., Geochim. Cosmochim. Acta Suppl. **1**, 1213—1231 (1970).

HAYASHI, C.: Stellar evolution in early phases of gravitational contraction. Publ. Astron. Soc. Japan **13**, 140—452 (1961).

HAYES, J. M.: Organic constituents of meteorites—a review. Geochim. Cosmochim. Acta **31**, 1395—1440 (1967).

HEIDE, F.: Kleine Meteoritenkunde, 2nd edition. 142 pp. Springer 1957. Engl. trans. Meteorites, 144 pp. Chicago: University Press 1964.

HELLYER, B.: Statistics of meteor falls. Earth Planet. Sci. Letters **7**, 148—150 (1969).
HENDERSON, E. P., PERRY, S. H.: A discussion of the densities of iron meteorites. Geochim. Cosmochim. Acta **6**, 221—240 (1954).
HENDERSON, E. P., PERRY, S. H.: Studies of seven siderites. Proc. U. S. Nat. Mus. **107**, 339—403 (1958).
HERBIG, G. H.: The properties and problems of T Tauri stars and related objects. Advan. Astron. Astrophys. **1**, 47—103 (1962).
HERBIG, G. H.: Early stellar evolution at intermediate masses. In: HERBIG, G. H. (Ed.): Spectroscopic Astrophysics, pp. 237—248. San Francisco: Univ. Calif. Press 1970.
HERBIG, G. H., PEIMBERT, M.: The distribution of emission-line stars in the Taurus dark nebulae. Trans. Intern. Astron. Union **12B**, 412—416 (1966).
HERNDON, J. M., ROWE, M. W., LARSON, E. E., WATSON, D. E.: Magnetism of meteorites: a review of Russian studies. Meteoritics **7**, 263—284 (1972).
HERR, W., HOFFMEISTER, W., HIRT, B., GEISS, J., HOUTERMANS, F. G.: Versuch zur Datierung von Eisenmeteoriten nach der Rhenium-Osmium-Methode. Z. Naturforsch. **16a**, 1053—1058 (1961).
HERZOG, G. F., ANDERS, E.: Absolute scale for radiation ages of stony meteorites. Geochim. Cosmochim. Acta **35**, 605—611 (1971a).
HERZOG, G. F., ANDERS, E.: Radiation age of the Norton Country meteorite. Geochim. Cosmochim. Acta **35**, 239—244 (1971b).
HESS, H. H., HENDERSON, E. P. The Moore Country meteorite: a further study with comment on its primordial environment. Am. Mineralogist **34**, 494—507 (1949).
HEY, M. H.: Catalogue of Meteorites, 3rd. Ed., 637 pp. London: British Museum 1966.
HEYMANN, D.: The origin of hypersthene chondrites: ages and shock effects of black chondrites. Icarus **6**, 189—221 (1967).
HEYMANN, D., LIPSCHUTZ, M. E., NIELSEN, B., ANDERS, E.: Canyon Diablo meteorite. Metallographic and mass-spectrometric study of 56 fragments. J. Geophys. Res. **71**, 619—641 (1966).
HEYMANN, D., MAZOR, E.: St. Mesmin, a gas-rich amphoteric chondrite. J. Geophys. Res. **71**, 4695—4697 (1966).
HEYMANN, D., MAZOR, E.: Primordial rare gases in unequilibrated ordinary chondrites. Science **155**, 701—702 (1967).
HEYMANN, D., MAZOR, E.: Noble gases in unequilibrated ordinary chondrites. Geochim. Cosmochim. Acta **32**, 1—19 (1968).
HEYMANN, D. YANIV, A.: Inert gases in the fines from the Sea of Tranquillity. Geochim. Cosmochim. Acta, Suppl. **1**, 1269—1282 (1970).
HINTENBERGER, H., KÖNIG, H., SCHULTZ, L., WÄNKE, H., WLOTZKA, F.: Die relativen Produktionsquerschnitte für ^{3}He und ^{21}Ne aus Mg, Si, S und Fe in Steinmeteoriten. Z. Naturforsch. **19a**, 88—92 (1964).
HINTENBERGER, H., SCHULTZ, L., WEBER, H.: Rare gases in the iron and in the inclusions of the Campo del Cielo meteorite, El Taco. In: MILLMAN, P. M. (Ed.): Meteorite Research, pp. 895—900. Dordrecht: Reidel 1969.
HINTENBERGER, H., VILCSEK, E., WÄNKE, H.: Über die Isotopenzusammensetzung und über den Sitz der leichten Uredelgase in Steinmeteoriten. Z. Naturforsch. **20a**, 939—945 (1965b).
HINTENBERGER, H., VOSHAGE, H., SARKAR, H.: Durch die kosmische Strahlung produziertes Lithium und Calcium in Eisenmeteoriten. Z. Naturforsch. **20a**, 965—967 (1965a).
HIRAYAMA, K.: Groups of asteroids probably of common origin. Astron. J. **31**, 185—188 (1918).

HOFFMAN, J. H., NIER, A. O.: Production of helium in iron meteorites by the action of cosmic rays. Phys. Rev. **112**, 2112—2117 (1958).

HOFFMAN, J. H., NIER, A. O.: The cosmogenic He^3 and He^4 distribution in the meteorite Carbo. Geochim. Cosmochim. Acta **17**, 32—36 (1959).

HOHENBERG, C. M.: Radioisotopes and the history of nucleosynthesis in the galaxy. Science **166**, 212—215 (1969).

HOHENBERG, C. M., MUNK, M. N., REYNOLDS, J. H.: Spallation and fissiogenic xenon and krypton from stepwise heating of the Pasamonte achondrite; The case for extinct plutonium-244 in meteorites; Relative ages of chondrites and achondrites. J. Geophys. Res. **72**, 3139—3177 (1967b).

HOHENBERG, C. M., PODOSEK, F. A., REYNOLDS, J. H.: Xenon-iodine dating: Sharp isochronism in chondrites. Science **156**, 202—206 (1967a).

HOLMES, A.: Principles of Physical Geology, 2nd Ed., 1288 pp. London: Nelson 1965.

HONDA, M., ARNOLD, J. R.: Effects of cosmic rays on meteorites. In: FLÜGGE, S. (Ed.): Handbuch der Physik, pp. 613—632. Berlin-Heidelberg-New York: Springer 1967.

HORBACK, H., OLSEN, E.: Catalog of the Collection of Meteorites in Chicago Natural History Museum. Fieldiana Geol. **15**, 173—319 (1965).

HOYLE, F.: On the origin of the solar nebula. Quart. J. Roy. Astron. Soc. **1**, 28—55 (1960).

HOYLE, F.: Formation of the planets. In: JASTROW, R., CAMERON, A. G. W. (Eds.): Origin of the Solar System, pp. 63—71. New York—London: Academic Press 1963.

HOYLE, F., WICKRAMASINGHE, N. C.: Condensation of the planets. Nature **217**, 415—418 (1968).

HUEY, J. M., KOHMAN, T. P.: Search for extinct natural radioactivity of ^{205}Pb via thallium-isotope anomalies in chondrites and lunar soil. Earth Planet. Sci. Letters **16**, 401—412 (1972).

HUEY, J. M., KOHMAN, T. P.: ^{207}Pb-^{206}Pb isochron and age of chondrites. J. Geophys. Res. **78**, 3227—3244 (1973).

HULSTON, J. R., THODE, H. G.: Variations in the ^{33}S, ^{34}S and ^{36}S contents of meteorites and their relation to chemical and nuclear effects. J. Geophys. Res. **70**, 3475—3484 (1965).

HUNEKE, J. C., NYQUIST, L. E., FUNK, H., KÖPPEL, V., SIGNER, P.: The thermal release of rare gases from separated minerals of the Mocs meteorite. In: MILLMAN, P. M. (Ed.): Meteorite Research, pp. 901—921. Dordrecht: Reidel 1969.

HUNT, J. N., PALMER, R., PENNEY, W.: Atmospheric waves caused by large explosions. Phil. Trans. Roy. Soc. London **252 A**, 275—315 (1960).

IVANOVA, I. A., LEBEDINETS, V. N., MAKSAKOV, B. I., PORTNYAGIN, Y. I.: Rate of vaporization of stone meteorites as determined by comparison with the rate of vaporization of iron. Geokhimiya **1968**, 239—242. Engl. trans. Geochem. Intern. **1968**, 190—193 (1968).

JAFFEY, A. H., FLYNN, K. F., GLENDENIN, L. E., BENTLEY, W. C., ESSLING, A. M.: Precision measurement of half-lives and specific activities of U^{235} and U^{238}. Phys. Rev. **C4**, 1889—1906 (1971).

JAIN, A. V., LIPSCHUTZ, M. E.: Shock history of iron meteorites and their parent bodies: a review, 1967—1971. Chem. Erde **30**, 199—215 (1971).

JAROSEWICH, E.: Chemical analyses of ten stony meteorites. Geochim. Cosmochim. Acta **30**, 1261—1265 (1966).

JAROSEWICH, E.: Chemical analyses of seven stony meteorites and one iron with silicate inclusions. Geochim. Cosmochim. Acta **31**, 1103—1106 (1967).

JEANS, J.: The origin of the solar system. Nature **128**, 432—435 (1931).

JEFFREYS, H.: Collision and the origin of rotation in the solar system. Mon. Not. Roy. Astron. Soc. **89**, 636—641 (1929).

JÉROME, D. Y., GOLES, G. G.: A re-examination of relationships among pyroxene-plagioclase achondrites. In: BRUNFELT, A. O., STEINNES, E. (Eds.): Activation Analysis in Geochemistry and Cosmochemistry, pp. 261—266. Universitetsforlaget 1971.

JOHNSON, T. V., FANALE, F. P.: Optical properties of carbonaceous chondrites and their relationship to asteroids. Geochim. Cosmochim. Acta, in press (1974).

KANT, I.: Allgemeine Naturgeschichte und Theorie des Himmels (1755).

KAPLAN, I. R., HULSTON, J. R.: The isotopic abundance and content of sulfur in meteorites. Geochim. Cosmochim. Acta **30**, 479—496 (1966).

KARR, M. L., LEWIS, C. F., MOORE, C. B.: Catalog of Meteorites in the Collections of Arizona State University including the Nininger Meteorite Collection, 257 pp. Arizona State University 1970.

KAUFMAN, L., COHEN, M.: The martensitic transformation in the iron-nickel system. Trans. AIME **206**, 1393—1401 (1956).

KAULA, W. M.: An Introduction to Planetary Physics: The Terrestrial Planets, 490 pp. New York: Wiley 1968.

KAULA, W. M.: Selenodesy and planetary geodesy. Trans. Am. Geophys. Union **52**, IUGG 1—IUGG 4 (1971).

KAUSHAL, S. K., WETHERILL, G. W.: Rb^{87}—Sr^{87} age of bronzite (H group) chondrites. J. Geophys. Res. **74**, 2717—2726 (1969).

KAUSHAL, S. K., WETHERILL, G. W.: Rubidium-87—strontium-87 age of carbonaceous chondrites. J. Geophys. Res. **75**, 463—468 (1970).

KEAYS, R. R., GANAPATHY, R., ANDERS, E.: Chemical fractionations in meteorites—IV. Abundances of 14 trace elements in L chondrites; implications for cosmothermometry. Geochim. Cosmochim. Acta **35**, 337—363 (1971).

KEIDEL, W.: Untersuchungen am Meteoriten von Borkut und anderen Chondriten über Form, Aufbau und Entstehung der Chondren. Contrib. Mineral. Petrol. **11**, 487—506 (1965).

KEIL, K.: Quantitativ-erzmikroskopische Integrationsanalyse der Chondrite (Zur Frage des mittleren Verhältnisses von Nickeleisen-: Troilit-: Chromit-: Silikatanteil in den Chondriten). Chem. Erde **22**, 281—348 (1962).

KEIL, K.: The electron microprobe X-ray microanalyser and its applications to mineralogy. Fortschr. Mineral. **44**, 4—66 (1967).

KEIL, K.: Mineralogical and chemical relationships among enstatite chondrites. J. Geophys. Res. **73**, 6945—6976 (1968).

KEIL, K.: Meteorite composition. In: WEDEPOHL, K. H. (Ed.): Handbook of Geochemistry, Vol. I. pp. 78—115, Berlin—Heidelberg—New York: Springer 1969a.

KEIL, K.: Titanium distribution in enstatite chondrites and achondrites, and its bearing on their origin. Earth Planet. Sci. Letters **7**, 243—248 (1969b).

KEIL, K., FREDRIKSSON, K.: The iron, magnesium, and calcium distribution in coexisting olivines and rhombic pyroxenes of chondrites. J. Geophys. Res. **69**, 3487—3515 (1964).

KEIL, K., HUSS, G. I., WIIK, H. B.: The Leoville, Kansas, meteorite: A polymict breccia of carbonaceous chondrites and achondrites (abstract). In: MILLMAN, P. M. (Ed.): Meteorite Research, p. 217. Dordrecht: Reidel 1969.

KEIL, K., MASON, B., WIIK, H. B., FREDRIKSSON, K.: The Chainpur meteorite. Amer. Mus. Novitates **2173**, 1—28 (1964).

KEMPE, W., MÜLLER, O.: The stony meteorite Krähenberg: Its chemical composition and the Rb-Sr age of the light and dark portions. In: MILLMAN, P. M. (Ed.): Meteorite Research, pp. 418—428. Dordrecht: Reidel 1969.

KERRIDGE, J. F.: Low-temperature minerals from the fine-grained matrix of some carbonaceous meteorites. Ann. Rev. N.Y. Acad. Sci. **119**, 41—53 (1964).

KIELBASINSKI, J., WANAT, L.: Isotopic and elemental composition of meteorites. A bibliography: 1947—1967. Nuclear Energy Information Center (Warsaw) Rept. NEIC-RR-32, 133 pp. (1968).

KING, E. A., BUTLER, J. C., CARMAN, M. F.: Chondrules in Apollo 14 samples and size analyses of Apollo 14 and 15 fines. Proc. Third. Lunar Sci. Conf., Geochim. Cosmochim. Acta, Suppl. **3**, 673—686 (1972).

KIRSTEN, T., KRANKOWSKY, D., ZÄHRINGER, J.: Edelgas- und Kalium-Bestimmungen an einer größeren Zahl von Steinmeteoriten. Geochim. Cosmochim. Acta **27**, 13—42 (1963).

KIRSTEN, T. A., SCHAEFFER, O. A.: High energy interactions in space. In: YUAN, L. C. L. (Ed.): Elementary Particles: Science, Technology and Society, pp. 76—157. New York-London: Academic Press 1971.

KNOX, R.: The yield strength of meteoritic iron. Meteoritics **5**, 63—74 (1970).

KOHMAN, T. P., BENDER, M. L.: Nuclide production by cosmic rays in meteorites and on the moon. In: SHEN, B. S. P. (Ed.): High Energy Nuclear Reactions in Astrophysics, pp. 169—245. New York: W. A. Benjamin 1967.

KOLESNIKOV, E. M., LAVRUKHINA, A. K., FISENKO, A. V., LEVSKY, L. K.: Radiation ages of different fragments of the Sikhote-Alin meteorite fall. Geochim. Cosmochim. Acta **36**, 573—576 (1972).

KOPAL, Z., RACKHAM, T. W.: Excitation of lunar luminescence by solar activity. Icarus **2**, 5—6, 481—500 (1963).

KRÄHENBÜHL, U., MORGAN, J. W., GANAPATHY, R., ANDERS, E.: Abundance of 17 trace elements in carbonaceous chondrites. Geochim. Cosmochim. Acta **37**, 1353—1370 (1973).

KRANKOWSKY, D., MÜLLER, O.: Isotopic composition and abundance of lithium in meteoritic matter. Geochim. Cosmochim. Acta **31**, 1833—1844 (1967).

KRETZ, R.: Distribution of magnesium and iron between orthopyroxene and calcic pyroxene in natural mineral assemblages. J. Geol. **71**, 773—785 (1963).

KRINOV, E. L.: Principles of Meteoritics, 535 pp. London—New York: Pergamon 1960 (Russian original published in 1955).

KROUSE, H. R., MODZELESKI, V. E.: Carbon-13/carbon-12 abundances in components of carbonaceous chondrites and terrestrial samples. Geochim. Cosmochim. Acta **34**, 459—474 (1970).

KRUGER, S. T., HEYMANN, D.: Cosmic-ray-produced hydrogen-3 and helium-3 in stony meteorites. J. Geophys. Res. **73**, 4784—4787 (1968).

KRUMMENACHER, D., MERRIHUE, C. M., PEPIN, R. O., REYNOLDS, J. H.: Meteoritic krypton and barium versus the general isotopic anomalies in meteoritic xenon. Geochim. Cosmochim. Acta **26**, 231—249 (1962).

KUHI, L. V.: Mass loss from T Tauri stars. Astrophys. J. **140**, 1409—1433 (1964).

KUHI, L. V.: Mass loss from T Tauri stars. II. Astrophys. J. **143**, 991—992 (1966).

KURAT, G.: Formation of chondrules. Geochim. Cosmochim. Acta **31**, 491—502 (1967).

KURAT, G.: Zur Genese der Ca-Al-reichen Einschlüsse im Chondriten von Lancé. Earth Planet. Sci. Letters **9**, 225—231 (1970).

KURAT, G., FREDRIKSSON, K., NELEN, J.: Der Meteorit von Siena. Geochim. Cosmochim. Acta **33**, 765—773 (1969).

KURAT, G., KEIL, K., PRINZ, M., NEHRU, C. E.: Chondrules of lunar origin. Proc. Third Lunar Sci. Conf., Geochim. Cosmochim. Acta, Suppl. **3**, 707—722 (1972).

KURODA, P. K.: Nuclear fission in the early history of the earth. Nature **187**, 36—38 (1960).

KURODA, P. K., MANUEL, O. K.: Mass fractionation and isotope anomalies in neon and xenon. Nature **227**, 1113—1116 (1970).

KVASHA, L. G.: Catalog of the Meteorite Collection of the Committee on Meteorites, Academy of Sciences, USSR, as of 1 July 1961 (in Russian). Meteoritika **22**, 127—156 (1962).

KVENHOLDEN, K. A., LAWLESS, J., PERING, K., PETERSON, E., FLORES, J., PONNAMPERUMA, C., KAPLAN, I. R., MOORE, C. B.: Evidence for extraterrestrial amino-acids and hydrocarbons in the Murchison meteorite. Nature **228**, 923—926 (1970).

LACROIX, A.: Les météorites tombées en France et dans les colonies et conservé au Muséum National d'Histoire Naturelle, avec remarques sur la classification des meteorites. Bull. Mus. Nat. Hist. Natur. Paris **33**, 411—455 (1928).

LAL, D.: Recent advances in the study of fossil tracks in meteorites due to heavy nuclei of the cosmic radiation. Space Sci. Rev. **9**, 623—650 (1969).

LAL, D., RAJAN, R. S.: Observations on space irradiation of individual crystals of gas-rich meteorites. Nature **223**, 269—271 (1969).

LALOU, C., NORDEMANN, D., LABEYRIE, J.: Etude préliminaire de la thermoluminescence de la météorite Saint-Séverin, C. R. Acad. Sci. Paris **270**, 2401—2404 (1970).

LANCET, M. S., ANDERS, E.: Carbon isotope fractionation in the Fischer-Tropsch synthesis and in meteorites. Science **170**, 980—982 (1970).

LANCET, M. S., ANDERS, E.: Solubilities of noble gases in magnetite: Implications for planetary gases in meteorites. Geochim. Cosmochim. Acta **37**, 1371—1388 (1973).

LANGE, D. E., LARIMER, J. W.: Chondrules: An origin by impacts between dust grains. Science **82**, 920—922 (1973).

LA PAZ, L.: Meteorite craters and the hypothesis of the existence of contraterrene meteorites. Pop. Astron. **49**, 99—105; Contrib. Soc. Res. Meteorites **2**, 244—247 (1941).

LAPLACE, P. S, DE: Exposition du Système du Monde, Paris. (1796).

LARIMER, J. W.: Chemical fractionations in meteorites—I. Condensation of the elements Geochim. Cosmochim. Acta **31**, 1215—1238 (1967).

LARIMER, J. W.: Experimental studies on the system Fe-MgO-SiO_2-O_2 and their bearing on the petrology of chondritic meteorites. Geochim. Cosmochim. Acta **32**, 1187—1207 (1968 a).

LARIMER, J. W.: An experimental investigation of oldhamite, CaS; and the petrologic significance of oldhamite in meteorites. Geochim. Cosmochim. Acta **32**, 965—982 (1968 b).

LARIMER, J. W.: Chemical fractionations in meteorites—VII. Cosmothermometry and cosmobarometry. Geochim. Cosmochim. Acta **37**, 1603—1623 (1973)

LARIMER, J. W., ANDERS, E.: Chemical fractionations in meteorites. II. Abundance patterns and their interpretation. Geochim. Cosmochim. Acta **31**, 1239—1270 (1967).

LARIMER, J. W., ANDERS, E.: Chemical fractionations in meteorites—III. Major element fractionations in chondrites. Geochim. Cosmochim. Acta **34**, 367—387 (1970).

LARSON, R. B.: Numerical calculations of the dynamics of a collapsing proto-star. Monthly Notices. Roy. Astron. Soc. **145**, 271—295 (1969).
LARSON, R. B.: Collapse calculations and their implications for the formation of the solar system. In: REEVES, H. (Ed.): On the Origin of the Solar System, pp. 142—150. Paris: CNRS 1972.
LAUL, J. C., CASE, D. R., SCHMIDT-BLEEK, F., LIPSCHUTZ, M. E.: Bismuth contents of chondrites. Geochim. Cosmochim. Acta **34**, 89—103 (1970).
LAUL, J. C., GANAPATHY, R., ANDERS, E., MORGAN, J. W.: Chemical fractionations in meteorites—VI. Accretion temperatures of H-, LL- and E-chondrites, from abundance of volatile elements. Geochim. Cosmochim. Acta **37**, 329—357 (1973).
LAUL, J. C., KEAYS, R. R., GANAPATHY, R., ANDERS, E., MORGAN, J. W.: Chemical fractionations in meteorites—V. Volatile and siderophile elements in achondrites and ocean ridge basalts. Geochim. Cosmochim. Acta **36**, 329—345 (1972).
LEQUEX, J.: Observations of interstellar matter and protostars. In: REEVES, H. (Ed.): On the Origin of the Solar System, pp. 118—131. Paris: CNRS 1972.
LEVIN, B. Y.: Origin of meteorites. Planet. Space Sci. **13**, 243—259 (1965).
LEVIN, B. Y.: Origin of meteorites and planetary cosmogony. In: MILLMAN, P. M. (Ed.): Meteorite Research, pp. 16—30. Dordrecht: Reidel 1969.
LEVIN, B. Y., SIMONENKO, A. N.: Meteorite radiants and orbits. In: MILLMAN, P. M. (Ed.): Meteorite Research, pp. 552—558. Dordrecht: Reidel 1969.
LEWIS, C. F., MOORE, C. B.: Chemical analyses of thirty-eight iron meteorites. Meteoritics **6**, 195—205 (1971).
LEWIS, J. S., KROUSE, H. R.: Isotopic composition of sulfur and sulfate produced by oxidation of FeS. Earth Planet. Sci. Letters **5**, 425—428 (1969).
LIENER, A., GEISS, J.: Thermoluminescence measurements on chondritic meteorites. In: MCDOUGALL, D. J. (Ed.): Thermoluminescence of Geological Materials, pp. 559—568 1968.
LIGHTNER, B. D., MARTI, K.: Lunar trapped xenon. Lunar Science V, 447—449 (1974).
LINDBLAD, B. A., SOUTHWORTH, R. B.: A study of asteroid families and streams by computer techniques. In: GEHRELS, T. (Ed.): Physical Studies of Minor Planets. NASA Rep. SP-267, pp. 337—352 1971.
LINDSLEY, D. H., ANDREASEN, G. E., BALSLEY, J. R.: Magnetic properties of rocks and minerals. In: CLARK, S. P. (Ed.): Handbook of Physical Constants, pp. 543—552. Geol. Soc. Am. 1966.
LIPSCHUTZ, M. E.: Origin of diamonds in the ureilites. Science **143**, 1431—1434 (1964).
LIPSCHUTZ, M. E.: X-ray diffraction analysis of cohenite from iron meteorites. Geochim. Cosmochim. Acta **31**, 621—633 (1967).
LIPSCHUTZ, M. E.: Shock effects in meteorites. In: FRENCH, B. M., SHORT, N. M. (Eds.): Shock Metamorphism of Natural Materials, pp. 571—583. Mono Book Corp. 1968.
LIPSCHUTZ, M. E., ANDERS, E.: Origin of diamonds in iron meteorites. Geochim. Cosmochim. Acta **24**, 83—105 (1961).
LIPSCHUTZ, M. E., ANDERS, E.: Cohenite as a pressure indicator in iron meteorites? Geochim. Cosmochim. Acta **28**, 699—711 (1964).
LORD, H. C.: Molecular equilibria and condensation in a solar nebula and cool stellar atmospheres. Icarus **4**, 279—288 (1965).
LOVERING, J. F.: Pressures and temperatures within a typical parent meteorite body. Geochim. Cosmochim. Acta **12**, 253—261 (1957).
LOVERING, J. F.: The magnetic field in a primary meteorite body. Am. J. Sci. **257**, 271—275 (1959).

LOVERING, J. F.: The evolution of the meteorites — evidence for the co-existence of chondritic, achondritic and iron meteorites in a typical parent meteorite body. In: MOORE, C. B. (Ed.): Researches on Meteorites, pp. 179—197. New York: Wiley 1962.

LOVERING, J. F., NICHIPORUK, W., CHODOS, A., BROWN, H.: The distribution of gallium, germanium, cobalt, chromium, and copper in iron and stony-iron meteorites in relation to nickel content and structure. Geochim. Cosmochim. Acta **11**, 263—278 (1957).

LOVERING, J. F., PARRY, L. G.: Thermomagnetic analysis of co-existing nickel-iron metal phases in iron meteorites and the thermal histories of the meteorites. Geochim. Cosmochim. Acta **26**, 361—382 (1962).

LOWREY, B. E.: Orbital evolution of Lost City Meteorite. J. Geophys. Res. **76**, 4084—4089 (1971).

MAGNOLIA, L. R.: Interplanetary matter: A bibliography. ASTIA Doc. No. AD 276064. 591 pp. (1962).

MAGNOLIA, L. R.: Interplanetary matter: A bibliography. STL Res. Biblio. No. 46, Space Tech. Lab., Redondo Beach, Calif., 294 pp. (1963).

MANUEL, O. K.: Noble gases in the Fayetteville meteorite. Geochim. Cosmochim. Acta **31**, 2413—2431 (1967).

MANUEL, O. K., HENNECKE, E. W., SABU, D. D.: Xenon in carbonaceous chondrites. Nature Phys. Sci. **240**, 99—101 (1972).

MANUEL, O. K., KURODA, P. K.: Isotopic composition of the rare gases in the Fayetteville meteorite. J. Geophys. Res. **69**, 1413—1419 (1964).

MARINGER, R. E., MANNING, G. K.: Some observations on deformation and thermal alterations in meteoritic iron. In: MOORE, C. B. (Ed.): Researches on Meteorites, pp. 123—144. New York: Wiley 1962.

MARSDEN, B. G.: Catalogue of Cometary Orbits. Smithsonian Astrophys. Obs., 70 pp. (1972).

MARSHALL, L.: Non-anti-matter nature of the Tunguska meteor. Nature **212**, 1226 (1966).

MARTI, K.: Trapped xenon and the classification of chondrites. Earth Planet. Sci. Letters **2**, 193—196 (1967a).

MARTI, K.: Isotopic composition of trapped krypton and xenon in chondrites. Earth. Planet. Sci. Letters **2**, 243—248 (1967b).

MARTI, K.: Mass-spectrometric detection of cosmic-ray-produced ^{81}Kr in meteorites and the possibility of Kr-Kr dating. Phys. Rev. Letters **18**, 264—266 (1967c).

MARTI, K.: Solar type xenon: a new isotopic composition of xenon in the Pesyanoe meteorite. Science **166**, 1263—1265 (1969).

MARTI, K., EBERHARDT, P., GEISS, J.: Spallation, fission, and neutron capture anomalies in meteoritic Kr and Xe. Z. Naturforsch. **21a**, 398—413 (1966).

MARTI, K., SHEDLOVSKY, J. P., LINDSTROM, R. M., ARNOLD, J. R., BHANDARI, N. G.: Cosmic-ray produced radionuclides and rare gases near the surface of Saint-Séverin meteorite. In: MILLMAN, P. M. (Ed.): Meteorite Research, pp. 246—265. Dordrecht: Reidel 1969.

MARVIN, U. B., WOOD, J. A.: The Haverö ureilite: petrographic notes. Meteoritics **7**, 601—610 (1972).

MARVIN, U. B., WOOD, J. A., DICKEY, J. S.: Calcium-aluminum rich phases in the Allende meteorite. Earth Planet. Sci. Letters **7**, 346—350 (1970).

MASON, B.: Meteorites. 274 pp. New York: Wiley 1962a.

MASON, B.: Meteorite lists and catalogues, 1949—62. Mineral. Mag. **33**, 265—269 (1962b).

MASON, B.: Olivine composition in chondrites. Geochim. Cosmochim. Acta **27**, 1011—1023 (1963a).
MASON, B.: The hypersthene achondrites. Am. Mus. Novitates **2155**, 13 pp. (1963b).
MASON, B.: The carbonaceous chondrites. Space Sci. Rev. **1**, 621—646 (1963c).
MASON, B.: The meteorite and tektite collection of the American Museum of Natural History. Am. Mus. Novitates **2190**, 40 pp. (1964).
MASON, B.: The chemical composition of olivine-bronzite and olivine-hypersthene chondrites. Am. Mus. Novitates **2223**, 1—38 (1965a).
MASON, B.: Feldspar in chondrites. Science **148**, 943 (1965b).
MASON, B.: The enstatite chondrites. Geochim. Cosmochim. Acta **30**, 23—39 (1966).
MASON, B.: Meteorites. Am. Sci. **55**, 429—455 (1967a).
MASON, B.: Olivine composition in chondrites—a supplement. Geochim. Cosmochim. Acta **31**, 1100—1103 (1967b).
MASON, B.: The Bununu meteorite, and a discussion of the pyroxene-plagioclase achondrites. Geochim. Cosmochim. Acta **31**, 107—115 (1967c).
MASON, B.: Extraterrestrial mineralogy. Am. Mineralogist **52**, 307—325 (1967d).
MASON, B.: The Woodbine meteorite, with notes on silicates in iron meteorites. Mineralog. Mag. **36**, 120—126 (1967e).
MASON, B.: Pyroxenes in meteorites. Lithos **1**, 1—11 (1968).
MASON, B.: The carbonaceous chondrites—A selective review. Meteoritics **6**, 59—70 (1971a).
MASON, B.: Handbook of Elemental Abundances in Meteorites (edited volume). 555 pp. New York: Gordon and Breach 1971b.
MASON, B.: The mineralogy of meteorites. Meteoritics **7**, 309—326 (1972).
MASON, B.: Hammond Downs, a new chondrite from the Tenham area, Queensland, Australia. Meteoritics **8**, 1—7 (1973).
MASON, B.: List of meteorites in the National Collection. Smithson. Contrib. Earth Sci., in press (1974).
MASON, B., GRAHAM, A. L.: Minor and trace elements in meteoritic minerals. Smithson. Contrib. Earth Sci. **3**, 1—17 (1970).
MASON, B., JAROSEWICH, E.: The Barea, Dyarrl Island, and Emery meteorites, and a review of the mesosiderites. Mineral. Mag. **39**, 204—215 (1973).
MASON, B., WIIK, H. B.: The composition of the Ottawa, Chateau Renard, Mocs, and New Concord meteorites. Am. Mus. Novitates **2069**, 25 pp. (1961).
MASON, B., WIIK, H. B.: The amphoterites and meteorites of similar composition. Geochim. Cosmochim Acta **28**, 533—538 (1964).
MASON, B., WIIK, H. B.: The composition of the Barratta, Carraweena, Kapoeta, Mooresfort, and Ngawi meteorites. Am. Mus. Novitates **2273**, 1—25 (1966a).
MASON, B., WIIK, H. B.: The composition of the Bath, Frankfort, Kakangari, Rose City, and Tadjera meteorites. Am. Mus. Novitates **2272**, 1—24 (1966b).
MASSALSKAYA, K. P.: Bibliographic guide to the literature on meteorites (in Russian). Issues No. 9—10 (1959—1960). Meteoritika **23**, 101—134 (1963).
MASSALSKI, T. B., PARK, F. R., VASSAMILLET, L. F.: Speculations about plessite. Geochim. Cosmochim. Acta **30**, 649—662 (1966).
MASUDA, A.: Regularities in variation of relative abundances of lanthanide elements and an attempt to analyse separation-index patterns of some minerals. J. Earth Sci. Nagoya Univ. **10**, 173—187 (1962).
MASUDA, A.: Lanthanide concentrations in the olivine phase of the Brenham pallasite. Earth Planet. Sci. Letters **5**, 59—62 (1968).
MATSON, D. L.: Infrared observations of asteroids. In: GEHRELS, T. (Ed.): Physical Studies of Minor Planets, pp. 45—50. NASA Rep. SP-267, 1971.

MATSUDA, H., UMEMOTO, S., HONDA, M.: Manganese-53 produced by 730 MeV proton bombardment of iron. Radiochim. Acta **15**, 51—53 (1971).
MAURETTE, M., THRO, P., WALKER, R., WEBBINK, R.: Fossil tracks in meteorites and the chemical abundance and energy spectrum of extremely heavy cosmic rays. In: MILLMAN, P.M. (Ed.): Meteorite Research, pp. 286—315. Dordrecht: Reidel 1969.
MAZOR, E., ANDERS, E.: Primordial gases in the Jodzie howardite and the origin of gas-rich meteorites. Geochim. Cosmochim. Acta **31**, 1441—1456 (1967).
MAZOR, E., HEYMANN, D., ANDERS, E.: Noble gases in carbonaceous chondrites. Geochim. Cosmochim. Acta **34**, 781—824 (1970).
MCCALL, G.J.H.: Meteorites and Their Origins, 352 pp. New York: Wiley 1973.
MCCALL, G.J.H., CLEVERLY, W.H.: New stony meteorite finds including two ureilites from the Nullarbor Plain, Western Australia. Mineral. Mag. **36**, 691—716 (1968).
MCCARTHY, T.S., AHRENS, L.H.: The composition of stony meteorites. X. The Ca/Al ratio in mesosiderites. Earth Planet. Sci. Letters **11**, 35—36 (1971).
MCCARTHY, T.S., AHRENS, L.H.: Chemical sub-groups amongst HL chondrites. Earth Planet Sci. Letters **14**, 97—102 (1972).
MCCARTHY, T.S., AHRENS, L.H., ERLANK, A.J.: Further evidence in support of the mixing model for howardite origin. Earth Planet. Sci. Letters **15**, 86—93 (1972).
MCCORD, T.B., ADAMS, J.B., JOHNSON, T.V.: Asteroid Vesta: spectral reflectivity and compositional implications. Science **168**, 1445—1447 (1970).
MCCREA, W.H.: The origin of the solar system. Proc. Roy. Soc. London **256 A**, 245—266 (1960).
MCCROSKY, R.E.: Orbits of photographic meteors. In: KRESÁK, L., MILLMAN, P.M. (Eds.): Physics and Dynamics of Meteors, pp. 265—279. Dordrecht: Reidel 1968.
MCCROSKY, R.E.: Structure of meteoroids (abstract). Trans. Am. Geophys. Union **53**, 724 (1972).
MCCROSKY, R.E., CEPLECHA, Z.: Photographic networks for fireballs. In: MILLMAN, P.M. (Ed.): Meteorite Research, pp. 600—612. Dordrecht: Reidel 1969.
MCCROSKY, R.E., POSEN, A., SCHWARTZ, G., SHAO, G.-Y.: Lost City meteorite—its recovery and a comparison with other fireballs. J. Geophys. Res. **76**, 4090—4108 (1971).
MEADOWS, A.I.: Remanent magnetization in meteorites. Nature **237**, 274 (1972).
MEDARIS, L.G.: Partitioning of Fe^{++} and Mg^{++} between coexisting synthetic olivine and orthopyroxene. Am. J. Sci. **267**, 945—968 (1969).
MEGRUE, G.H.: Distribution and origin of primordial helium, neon, and argon in the Fayetteville and Kapoeta meteorites. In: MILLMAN, P.M. (Ed.): Meteorite Research, pp. 992—930. Dordrecht: Reidel 1969.
MEHL, R.F.: On the Widmanstätten structure. In: SMITH, C.S. (Ed.): Sorby Centennial Symposium on History of Metallurgy, pp. 245—269. New York: Gordon and Breach 1965.
MERRIHUE, C.M.: Xenon and krypton in the Bruderheim meteorite. J. Geophys. Res. **71**, 263—313 (1966).
MERRIHUE, C.M., PEPIN, R.O., REYNOLDS, J.H.: Rare gases in the chondrite Pantar. J. Geophys. Res. **67**, 2017—2021 (1962).
MERRIHUE, C.M., TURNER, G.: Potassium-argon dating by activation with fast neutrons. J. Geophys. Res. **71**, 2852—2857 (1966).
MERRILL, G.P.: Handbook and descriptive catalogue of the meteorite collections in the United States National Museum. Bull. U.S. Nat. Mus. No. 94, 207 pp. (1916).
MERRILL, G.P.: Composition and structure of meteorites. Bull. U. S. Nat. Mus. No. 149, 62 pp. (1930).

MEUNIER, S.: Meteorites, 532 pp. Paris: Dunod 1884.

MICHAELIS, H., VON, AHRENS, L.H., WILLIS, J.P.: The composition of stony meteorites. II. The analytical data and an assessment of their quality. Earth Planet. Sci. Letters **5**, 387—394 (1969).

MIDDLEHURST, B.M.: An analysis of lunar events. Rev. Geophys. **5**, 173—189 (1967).

MILLER, S.L., UREY, H.C.: Organic compound synthesis on the primitive earth. Science **130**, 245—251 (1959).

MONSTER, J., ANDERS, E., THODE, H.G.: $^{34}S/^{32}S$ ratios for the different forms of sulfur in the Orgueil meteorite and their mode of formation. Geochim. Cosmochim. Acta **29**, 773—779 (1965).

MOORE, C.B., BIRRELL, P.J., LEWIS, C.F.: Variations in the chemical and mineralogical composition of rim and plains specimens of the Canyon Diablo meteorite. Geochim. Cosmochim. Acta **31**, 1885—1892 (1967).

MOORE, C.B., LEWIS, C.F.: Total carbon content of ordinary chondrites. J. Geophys. Res. **72**, 6289—6292 (1967).

MOORE, C.B., LEWIS, C.F., NAVA, D.: Superior analyses of iron meteorites. In: MILLMAN, P.M. (Ed.): Meteorite Research, pp. 738—748. Dordrecht: Reidel 1969.

MORGAN, J.W., LOVERING, J.F.: Rhenium and osmium abundances in chondritic meteorites. Geochim. Cosmochim. Acta **31**, 1893—1909 (1967).

MORRISON, D.: New techniques for determining sizes of satellites and asteroids. Comments Astrophys. Space Phys., **4**, 51, 56 (1973).

MOULTON, F.R.: Evolution of the solar system. Astrophys. J. **22**, 165—181 (1905).

MÜLLER, H.W., ZÄHRINGER, J.: Rare gases in stony meteorites. In: MILLMAN, P.M. (Ed.): Meteorite Research, pp. 845—856. Dordrecht: Reidel 1969.

MÜLLER, O., BAEDECKER, P.A., WASSON, J.T.: Relationship between siderophilic-element content and oxidation state of ordinary chondrites. Geochim. Cosmochim. Acta **35**, 1121—1137 (1971).

MÜLLER, O., ZÄHRINGER, J.: Chemische Unterschiede bei uredelgashaltigen Steinmeteoriten. Earth Planet. Sci. Letters **1**, 25—29 (1966).

MUELLER, R.F.: Phase equilibration and the crystallization of chondritic meteorites. Geochim. Cosmochim. Acta **28**, 189—207 (1964).

MUELLER, R.F.: System Fe-MgO-SiO_2-O_2 with applications to terrestrial rocks and meteorites. Geochim. Cosmochim. Acta **29**, 967—976 (1965).

MUELLER, R.F., OLSEN, E.J.: Olivine, pyroxene, and metal content of chondritic meteorites as a consequence of Prior's rule. Mineral. Mag. **36**, 311—318 (1967).

MÜLLER, W.F., HORNEMANN, U.: Shock-induced planar deformation structure in experimentally shock-loaded olivines and in olivines from chondritic meteorites. Earth Planet. Sci. Letters **7**, 251—264 (1969).

MURTHY, V.R.: The isotopic composition of silver in iron meteorites. Geochim. Cosmochim. Acta **26**, 481—488 (1962).

MURTHY, V.R., SANDOVAL, P.: Chromium isotopes in meteorites. J. Geophys. Res. **70**, 4379—4382 (1965).

MURTHY, V.R., SCHMITT, R.A.: Isotope abundances of rare-earth elements in meteorites. I. Implications of samarium, europium, and gadolinium to the early history of the solar system. J. Geophys. Res. **68**, 911—917 (1963).

NAGATA, T.: Rock Magnetism, 350 pp. Tokyo: Maruzen 1961.

NAGATA, T., FISHER, R.M., SCHWERER, F.C., FULLER, M.D., DUNN, J.R.: Rock magnetism of Apollo 14 and 15 materials. Proc. Third Lunar Sci. Conf., Geochim. Cosmochim. Acta, Suppl. 3, 2423—2447 (1972).

NELEN, J., NOONAN, A., FREDRIKSSON, K.: Lunar glasses, breccias, and chondrules. Proc. Third Lunar Sci. Conf., Geochim. Cosmochim. Acta, Suppl. 3, 723—738 (1972).

NELSON, L. S., BLANDER, M., SKAGGS, S. R., KEIL, K.: Use of a CO_2 laser to prepare chondrule like spherules from supercooled molten oxide and silicate droplets. Earth Planet. Sci. Letters **14**, 338—344 (1972).

NESS, N. F.: The interplanetary medium. In: HESS, W. N., MEAD, G. D. (Eds.): Introduction to Space Science, rev. edition, pp. 345—371. New York: Gordon and Breach 1968.

NEUGEBAUER, G., BECKLIN, E.: The brightest infrared sources. Sci. Am. **228**, (4) 28—40 (1973).

NEUGEBAUER, G., BECKLIN, E., HYLAND, A. R.: Infrared sources of radiation. Ann. Rev. Astron. Astrophys. **9**, 67—102 (1971).

NEWBURN, R. L., GULKIS, S.: A survey of the outer planets Jupiter, Saturn, Uranus, Neptune, Pluto, and their satellites. Space Sci. Rev. **3**, 179—271 (1973).

NICHIPORUK, W.: Former location of the Admire and Springwater pallasites. Earth Planet. Sci. Letters **4**, 433—435 (1968).

NININGER, H. H.: Out of the Sky, 336 pp. Dover 1952.

NININGER, H. H.: Find a Falling Star, 254 pp. New York: Eriksson 1972.

NÖLKE, F.: Der Entwicklungsgang unseres Planetensystems, 359 pp. Bonn: Dümmler 1930.

NOONER, D. W., ORÓ, J.: Organic compounds in meteorites—I. Aliphatic hydrocarbons. Geochim. Cosmochim. Acta **31**, 1359—1394 (1967).

NORDENSKJÖLD, A. E.: On the composition and common origin of certain meteorites. Nature **18**, 510—511 (1878).

NYQUIST, L. E., HUNEKE, J. C., SIGNER, P.: Spallogenic rare gases in the El Taco meteorite. Earth Planet. Sci. Letters **2**, 241—248 (1967).

ÖPIK, E. J.: Collision probabilities with the planets and distribution of interplanetary matter. Proc. Roy. Irish Acad. **54 A**, 165—199 (1951).

ÖPIK, E. J.: Photometry, dimensions, and ablation rate of comets. Irish. Astron. J. **6**, 93—112 (1963).

ÖPIK, E. J.: The stray bodies in the solar system. Part II. The cometary origin of meteorites. Advan. Astron. Astrophys. **4**, 301—336 (1966).

ÖPIK, E. J.: The cometary origin of meteorites. Irish Astron. J. **8**, 185—208 (1968).

OLSEN, E., FREDRIKSSON, K.: Phosphates in iron and pallasite meteorites. Geochim. Cosmochim. Acta **30**, 459—470 (1966).

OLSEN, E., FUCHS, L. H.: The state of oxidation of some iron meteorites. Icarus **6**, 242—253 (1967).

ONUMA, N., CLAYTON, R. N., MAYEDA, T. K.: Oxygen isotope temperatures of "equilibrated" ordinary chondrites. Geochim. Cosmochim. Acta **36**, 157—168 (1972a).

ONUMA, N., CLAYTON, R. N., MAYEDA, T. K.: Oxygen isotope cosmothermometer. Geochim. Cosmochim. Acta **36**, 169—188 (1972b).

ONUMA, N., CLAYTON, R. N., MAYEDA, T. K.: Oxygen isotope cosmothermometer revisited. Geochim. Cosmochim. Acta **38**, 189—191 (1974).

OROWAN, E.: Density of the moon and nucleation of planets. Nature **222**, 867 (1969).

ORSINI, P. G., CENTO, L.: Inclusioni di solfuro nelle leghe meteoriche: segregazione cromo. Ric. Sci. **37**, 725—730 (1967).

OSBORN, T. W., SCHMITT, R. A.: Sodium and manganese homogeneity in chondritic meteorites. Icarus **13**, 207—214 (1970).

OSTIC, R. G.: Physical conditions in gaseous spheres. Monthly Notices Roy. Astron. Soc. **131**, 191—197 (1965).

OSTIC, R. G., EL-BADRY, H. M., KOHMAN, T. P.: Isotopic composition of meteoritic thallium. Earth Planet. Sci. Letters **7**, 12—16 (1969).

OTTING, W., ZÄHRINGER, J.: Total carbon content and primordial rare gases in chondrites. Geochim. Cosmochim. Acta **31**, 1949—1960 (1967).

OVERSBY, V. M.: The isotopic composition of lead in iron meteorites. Geochim. Cosmochim. Acta **34**, 65—75 (1970).

PALACHE, C.: Notes on new or incompletely described meteorites in the Mineralogical Museum of Harvard University. Am. J. Sci. **212**, 136—150 (1926).

PANETH, F. A.: The discovery and earliest reproductions of the Widmanstätten figures. Geochim. Cosmochim. Acta **18**, 176—182 (1960).

PAPANASTASSIOU, D. A., RAJAN, R. S., HUNEKE, J. C., WASSERBURG, G. J.: Rb-Sr ages and lunar analogs in a basaltic achondrite; implications for early solar system chronologies. Lunar Science V, 583—585 (1974).

PAPANASTASSIOU, D. A., WASSERBURG, G. J.: Intitial strontium isotopic abundances and the resolution of small time differences in the formation of planetary objects. Earth Planet. Sci. Letters **5**, 361—376 (1969).

PATTERSON, C. C.: Age of meteorites and the earth. Geochim. Cosmochim. Acta **10**, 230—237 (1956).

PEARCE, G. W., STRANGWAY, D. W., GOSE, W. A.: Remanent magnetization of the lunar surface. Proc. Third Lunar Sci. Conf., Geochim. Cosmochim. Acta, Suppl. 3, 2449—2464 (1972).

PELLAS, P.: Irradiation history of grain aggregates in ordinary chondrites. Possible clues to the advanced stages of accretion. In: ELVIUS, A. (Ed.): From Plasma to Planet, pp. 65—90. Stockholm: Almqvist and Wiksell 1972.

PELLAS, P., POUPEAU, G., LORIN, J. C., REEVES, H., AUDOUZE, J.: Primitive low-energy particle irradiation of meteoritic crystals. Nature **223**, 272—274 (1969).

PELLY, I., LIPSCHUTZ, M. E., BALSIGER, H.: Vanadium isotopic composition and contents in chondrites. Geochim. Cosmochim. Acta **34**, 1033—1036 (1970).

PEPIN, R. O.: Trapped neon in meteorites. Earth Planet. Sci. Letters **2**, 13—18 (1967a).

PEPIN, R. O.: Neon and xenon in carbonaceous chondrites. In: AHRENS, L. H. (Ed.): Origin and Distribution of the Elements, pp. 379—386. London-New York: Pergamon 1967b.

PEPIN, R. O., NYQUIST, L. E., PHINNEY, D., BLACK, D. C.: Rare gases in Apollo 11 lunar material. Geochim. Cosmochim. Acta, Suppl. 1, 1435 1454 (1970).

PEPIN, R. O., SIGNER, P.: Primordial rare gases in meteorites. Science **149**, 253—265 (1965).

PERRY, S. H.: The metallography of meteoric iron. Bull. U. S. Nat. Mus. **184**, 206 pp. (1944).

PODOSEK, F. A.: Dating of meteorites by the high-temperature release of iodine-correlated Xe^{129}. Geochim. Cosmochim. Acta **34**, 341—365 (1970a).

PODOSEK, F. A.: The abundance of ^{244}Pu in the early solar system. Earth Planet. Sci. Letters **8**, 183—187 (1970b).

PODOSEK, F. A.: Neutron-activation potassium-argon dating in meteorites. Geochim. Cosmochim. Acta **35**, 157—173 (1971).

PODOSEK, F. A.: Gas retention chronology of Petersburg and other meteorites. Geochim. Cosmochim. Acta **36**, 755—772 (1972).

PODOSEK, F. A.: Thermal history of the nakhlites by the ^{40}Ar—^{39}Ar method. Earth Planet. Sci. Letters **19**, 135—144 (1973).

PODOSEK, F. A., HOHENBERG, C. M.: I-Xe-dating: Evidence for cold assembly of an unequilibrated chondrite. Earth Planet. Sci. Letters **8**, 443—447 (1970).

PODOSEK, F. A., HUNEKE, J. C.: Isotopic composition of ^{244}Pu fission xenon in meteorites: reevaluation using lunar spallation xenon systematics. Earth Planet. Sci. Letters **12**, 73—82 (1971).

PODOSEK, F. A., HUNEKE, J. C.: Argon-40—argon-39 chronology of four calcium-rich achondrites. Geochim. Cosmochim. Acta **37**, 667—684 (1973).

PODOSEK, F. A., HUNEKE, J. C., BURNETT, D. S., WASSERBURG, G. J.: Isotopic composition of xenon and krypton in the lunar soil and in the solar wind. Earth Planet. Sci. Letters **10**, 199—216 (1971).

POLLACK, S. S.: Disordered pyroxene in chondrites. Geochim. Cosmochim. Acta **32**, 1209—1217 (1968).

POWELL, B. N.: Petrology and chemistry of mesosiderites—I. Textures and composition of nickel-iron. Geochim. Cosmochim. Acta **33**, 789—810 (1969).

POWELL, B. N.: Petrology and chemistry of mesosiderites—II. Silicate textures and compositions and metal-silicate relationships. Geochim. Cosmochim. Acta **35**, 5—34 (1971).

PRICE, P. B., RAJAN, R. S., TAMHANE, A. S.: On the pre-atmospheric size and maximum space erosion rate of the Patwar stony-iron meteorite. J. Geophys. Res. **72**, 1377—1388 (1967).

PRIOR, G. R.: On the genetic relationship and classification of meteorites. Mineral. Mag. **18**, 26—44 (1916).

PRIOR, G. T.: The classification of meteorites. Mineral. Mag. **19**, 51—63 (1920).

QUIJANO-RICO, M., WÄNKE, H.: Determination of boron, lithium, and chlorine in meteorites. In: MILLMAN, P. M. (Ed.): Meteorite Research, pp. 132—145. Dordrecht: Reidel 1969.

RAMDOHR, P.: The opaque minerals in stony meteorites. J. Geophys. Res. **68**, 2011—2036 (1963a).

RAMDOHR, P.: Opaque ores in some meteorites, mainly from New South Wales. Chem. Erde **23**, 119—145 (1963b).

RAMDOHR, P.: Einiges über die Opakerze in Achondriten und Enstatitchondriten. Sitzber. Deut. Akad. Wiss. Berlin, Kl. Chem. Geol. Biol. **5**, 1—40 (1964).

RAMDOHR, P.: Chromite and chromite chondrules in meteorites. I. Geochim. Cosmochim. Acta **31**, 1961—1967 (1967).

RAMDOHR, P.: The Opaque Minerals in Stony Meteorites, 245 pp. Berlin: Akademie-Verlag 1973.

REED, G. W.: Heavy elements in the Pantar meteorite. J. Geophys. Res. **68**, 3531—3535 (1963).

REED, G. W., JOVANOVIC, S.: Mercury in chondrites. J. Geophys. Res. **72**, 2219—2228 (1967).

REED, G. W., JOVANOVIC, S.: Mercury-196 and mercury-202 isotopic ratios in chondrites. J. Inorg. Nucl. Chem. **31**, 3783—3788 (1969).

REED, S. J. B.: Electron-probe microanalysis of the metallic phases in iron meteorites. Geochim. Cosmochim. Acta **29**, 535—549 (1965a).

REED, S. J. B.: Electron-probe microanalysis of schreibersite and rhabdite in iron meteorites. Geochim. Cosmochim. Acta **29**, 513—534 (1965b).

REED, S. J. B.: Perryite in the Kota-Kota and South Oman enstatite chondrites. Mineral. Mag. **36**, 850—854 (1968).

REED, S. J. B.: Phosphorus in meteoritic nickel-iron. In: MILLMAN, P. M. (Ed.): Meteorite Research, pp. 749—762. Dordrecht: Reidel 1969.

REED, S. J. B.: Determination of Ni, Ga and Ge in iron meteorites by X-ray fluorescence analysis. Meteoritics **7**, 257—262 (1972).

REEDY, R. C., ARNOLD, J. R.: Interaction of solar and galactic cosmic-ray particles with the moon. J. Geophys. Res. **77**, 537—555 (1972).

REEVES, H.: On the Origin of the Solar System (edited volume), 383 pp. Paris: CNRS 1972a.

REEVES, H.: Spatial inhomogeneities of nucleosynthesis. Astron. Astrophys. **19**, 215—223 (1972b).

REEVES, H., AUDOUZE, J.: Early heat generation in meteorites. Earth Planet. Sci. Letters **4**, 135—141 (1968).

REID, A. M., COHEN, A. J.: Some characteristics of enstatite from enstatite achondrites. Geochim. Cosmochim. Acta **31**, 661—672 (1967).

REID, A. M., FREDRIKSSON, K.: Chondrules and chondrites. In: ABELSON, P. H. (Ed.): Researches in Geochemistry, Vol. 2, pp. 170—203. New York: Wiley 1967.

REUTER, J. H., EPSTEIN, S., TAYLOR, H. P., JR.: O^{18}/O^{16} ratios of some chondritic meteorites and terrestrial ultramafic rocks. Geochim. Cosmochim. Acta **29**, 481—488 (1965).

REYNOLDS, J. H.: Determination of the age of the elements. Phys. Rev. Letters **4**, 8—10 (1960).

REYNOLDS, J. H.: Isotopic abundance anomalies in the solar system. Ann. Rev. Nucl. Sci. **17**, 253—316 (1967).

REYNOLDS, J. H.: Plutonium-244 in the early solar system. Nature **218**, 1024—1028 (1968).

REYNOLDS, M. A., BOGARD, D. D., POLO, C. M.: Dwalene-a new gas-rich chondrite. (abstract). Trans. Amer. Geophys. Union **52**, 269 (1971).

RIEDER, R., WÄNKE, H.: Study of trace element abundance in meteorites by neutron activation. In: MILLMAN, P. M. (Ed.): Meteorite Research, pp. 75—86. Dordrecht: Reidel 1969.

RINGWOOD, A. E.: Chemical and genetic relationships among meteorites. Geochim. Cosmochim. Acta **24**, 159—197 (1961).

RINGWOOD, A. E.: Cohenite as a pressure indicator in iron meteorites—III. Comments on a paper by Lipschutz and Anders. Geochim. Cosmochim. Acta **29**, 573—579 (1965).

RINGWOOD, A. E.: Genesis of chondritic meteorites. Rev. Geophys. **4**, 113—175 (1966).

RINGWOOD, A. E., ESSENE, E.: Petrogenesis of Apollo-11 basalts, internal constitution and origin of the moon. Proc. Apollo 11 Lunar Sci. Conf., Geochim. Cosmochim. Acta, Suppl. 1, Pergamon, 769—799 (1970).

RINGWOOD, A. E., KAUFMAN, L.: The influence of high pressure on transformation equilibria in iron meteorites. Geochim. Cosmochim. Acta **26**, 999—1009 (1962).

ROEMER, E.: The dimensions of cometary nuclei. In: Nature et Origine des Cometes, Mem. Soc. Roy. Sci. Liège, Ser. 15, **12**, 23—28 (1966).

ROWE, M. W., BOGARD, D. D.: Anomalies in the Pasamonte meteorite. J. Geophys. Res. **71**, 686—687 (1966).

ROWE, M. W., KURODA, P. K.: Fissiogenic xenon from the Pasamonte meteorite. J. Geophys. Res. **70**, 709—714 (1965).

RUNCORN, S. K.: Fossil magnetic fields. In: ELVIUS, A. (Ed.): From Plasma to Planet, Nobel Symp. 21, pp. 373—376. New York: Wiley 1972.

RUSKOL, E. L.: On the possible differences in the bulk chemical composition of the earth and the moon forming in the circumterrestrial swarm. In: UREY, H. C., RUNCORN, S. K. (Eds.): The Moon, pp. 426—428. IAU 1972.

RUSSELL, H. N.: The Solar System and its Origin, 144 pp. London: Macmillan 1935.

SABLENA, N. D., YAVNEL, A. A.: (1965) Bibliographic guide to the literature on meteorites (in Russian, but with citations in original language, and with indices in English). Issues No. 11—13 (1961—1963), Academy of Sciences USSR, 305 pp. (in 2 parts).

SABLENA, N. D., YAVNEL, A. A.: (1968) Bibliographic guide to the literature on meteorites (in Russian, but with citations in original language, and with indices in English). Issues No. 14—16 (1964—1966), Academy of Sciences USSR, 407 pp.

SABLENA, N. D., YAVNEL, A. A.: (1971) Bibliographic guide to the literature on meteorites (in Russian, but with citations in original language, and with indices in English). Issues No. 17—18 (1967—1968), Academy of Sciences USSR, 344 pp.

SABU, D. D., KURODA, P. K.: Plutonium-244 in the early solar system and concordant plutonium/xenon and iodine/xenon decay intervals of achondrites. Nature **216**, 442—446 (1967).

SABU, D. D., MANUEL, O. K.: Superheavy elements: Were they present in meteorites? Trans. Missouri Acad. Sci. **5**, 16—21 (1971).

SAFRONOV, V. S.: Evolution of the Protoplanetary Cloud and the Formation of the Earth and Planets. (in Russian) Nauka Moscow (1969). Engl. translation as NASA TT F-677. NTIS, Springfield, Va., 1972.

SANZ, H. G., WASSERBURG, G. J.: Determination of an internal ^{87}Rb-^{87}Sr isochron for the Olivenza chondrite. Earth. Planet. Sci. Letters **6**, 335—345 (1969).

SCHATZMAN, E.: A theory of the role of magnetic activity during star formation. Ann. Astrophys. **25 B**, 18—29 (1962).

SCHATZMAN, E.: Cosmogony of the solar system and origin of the deuterium. Ann. Astrophys. **30**, 963—973 (1967).

SCHAUDY, R., WASSON, J. T., BUCHWALD, V. F.: The chemical classification of iron meteorites. VI. A reinvestigation of irons with Ge concentrations lower than 1 ppm. Icarus **17**, 174—192 (1972).

SCHMIDT, O. Y.: A meteoric theory of the origin of the earth and planets (in Russian). Dokl. Akad. Nauk. SSSR **45**, 229—233 (1944).

SCHMITT, R. A., SMITH, R. H.: Indium abundances in chondritic and achondritic meteorites and terrestrial. In: AHRENS, L. H. (Ed.): Origin and Distribution of the Elements, pp. 283—300. London-New York: Pergamon 1968.

SCHMITT, R. A., SMITH, R. H., LASCH, J. E., MOSEN, A. W., OLEHY, D. A., VASILEVSKIS, J.: Abundances of the fourteen rare-earth elements, scandium and yttrium in meteoritic and terrestrial matter. Geochim. Cosmochim. Acta **27**, 577—622 (1963).

SCHNETZLER, C. C., BOTTINO, M. L.: Some alkali, alkaline earth, and rare earth element concentrations and the Rb-Sr age of the Lost City meteorite and separated phases. J. Geophys. Res. **76**, 4061—4066 (1971).

SCHNETZLER, C. C., PHILPOTTS, J. A.: Genesis of the calcium-rich achondrites in light of rare-earth and barium concentrations. In: MILLMAN, P. M. (Ed.): Meteorite Research, pp. 206—216. Dordrecht: Reidel 1969.

SCHRAMM, D. N., TERA, F., WASSERBURG, G. J.: The isotopic abundance of ^{26}Mg and limits on ^{26}Al in the early solar system. Earth. Planet. Sci. Letters **10**, 44—59 (1970).

SCHULTZ, L.: Die Bestrahlungsgeschichte Xenolithischer Chondrite. Habilitationsschrift ETH Zürich, 108 pp (1973).

SCHULTZ, L., FUNK, H., NYQUIST, L., SIGNER, P.: Helium, neon and argon in separated phases of iron meteorites. Geochim. Cosmochim. Acta **35**, 77—88 (1971 b).

SCHULTZ, L., HINTENBERGER, H.: Edelgasmessungen an Eisenmeteoriten. Z. Naturforsch. **22 a**, 773—779 (1967).

SCHULTZ, L., SIGNER, P., LORIN, J. C., PELLAS, P.: Complex irradiation history of the Weston chondrite. Earth Planet. Sci. Letters **15**, 403—410 (1972).

SCHULTZ, L., SIGNER, P., PELLAS, P., POUPEAU, G.: Assam: A new gas-rich hypersthene chondrite. Earth Planet. Sci. Letters **12**, 119—123 (1971a).

SCOTT, E. R. D.: Chemical fractionation in iron meteorites and its interpretation. Geochim. Cosmochim. Acta **36**, 1205—1236 (1972).

SCOTT, E. R. D.: The nature of dark-etching rims in meteoritic taenite. Geochim. Cosmochim. Acta, **37**, 2283—2294 (1973).

SCOTT, E. R. D., BILD, R. W.: Structure and formation of the San Cristobal iron meteorite, other group IB members and group IIICD. Geochim. Cosmochim. Acta **38**, in press (1974).

SCOTT, E. R. D., WASSON, J. T., BUCHWALD, V. F.: The chemical classification of iron meteorites—VII. A reinvestigation of irons with Ge concentrations between 25 and 80 ppm. Geochim. Cosmochim. Acta **37**, 1957—1983 (1973).

SHEDLOVSKY, J. P., RAYUDU, G. V. S.: Radionuclide productions in thick iron targets bombarded with 1- and 3-Gev protons. J. Geophys. Res. **69**, 2231—2242 (1964).

SHIRCK, J., HOPPE, M., MAURETTE, M., WALKER, R.: Recent fossil track studies bearing on extinct Pu^{244} in meteorites. In: MILLMAN, P. M. (Ed.): Meteorite Research, pp. 41—50. Dordrecht: Reidel 1969.

SHOEMAKER, E. M., LOWERY, C. J.: Airwaves associated with large fireballs and the frequency distribution of energy of large meteoroids. (Abstract). Meteoritics **3**, 123—124 (1967).

SHORT, J. M., ANDERSEN, A.: Electron microprobe analyses of the Widmanstätten structure of nine iron meteorites. J. Geophys. Res. **70**, 3745—3749 (1965).

SHORT, J. M., GOLDSTEIN, J. L.: Rapid methods of determining cooling rates of iron and stony iron meteorites. Science **156**, 59—61 (1967).

SHORT, N. M., BUNCH, T. E.: A worldwide inventory of features characteristic of rocks associated with presumed meteorite impact structures. In: FRENCH, B. M., SHORT, N. M. (Eds.): Shock Metamorphism of Natural Materials, pp. 255—266. Mono 1968.

SHUR, A. S., ELKINA, N. T., YUDIN, I. A.: Ultra- and microporosity of iron meteorites. (in Russian). Tr. Inst. Geol. Geokhim., Akad. Nauk SSSR, Ural. Filial **86**, 151—156 (1970).

SIGNER, P.: Primordial rare gases in meteorites. In: BRANCAZIO, P. J., CAMERON, A. G. W. (Eds.): Origin and Evolution of Atmospheres and Oceans, pp. 183—190. New York: Wiley 1964.

SIGNER, P., SUESS, H. E.: Rare gases in the sun, in the atmosphere, and in meteorites. In: GEISS, J., GOLDBERG, E. D. (Eds.): Earth Science and Meteoritics, pp. 241—272. Amsterdam: North-Holland 1963.

SILL, C. W., WILLIS, C. P.: Beryllium content of some meteorites. Geochim. Cosmochim. Acta **20**, 1209—1214 (1962).

SMALES, A. A., MAPPER, D., FOUCHE, K. F.: The distribution of trace elements in iron meteorites, as determined by neutron activation. Geochim. Cosmochim. Acta **31**, 673—720 (1967).

SMITH, C. S.: Note on the history of the Widmannstätten structure. Geochim. Cosmochim. Acta **26**, 971—972 (1962).

SMITH, J. V., MASON, B.: Pyroxene-garnet transformation in Coorara meteorite. Science **168**, 832—833 (1970).

SMITH, J. W., KAPLAN, I. R.: Endogenous carbon in carbonaceous meteorites. Science **167**, 1367—1370 (1970).

SNETSINGER, K. G., KEIL, K.: Ilmenite in ordinary chondrites. Am. Mineralogist **54**, 780—786 (1969).
SONETT, C. P.: The relationship of meteoritic parent body thermal histories and electromagnetic heating by a pre-main sequence T Tauri sun. In: GEHRELS, T. (Ed.): Physical Studies of Minor Planets, pp. 239—245. NASA Rept. SP-267 1971.
SONETT, C. P., COLBURN, D. S., SCHWARTZ, K.: Electrical heating of meteorite parent bodies and planets by dynamo induction from a pre-main sequence T Tauri "solar wind". Nature **219**, 924—926 (1968).
SONETT, C. P., COLBURN, D. S., SCHWARTZ, K., KEIL, K.: The melting of asteroidal-sized bodies by unipolar dynamo induction from a primordial T Tauri sun. Astrophys. Space. Sci. **7**, 446—488 (1970).
SPENCER, L. J.: A list of catalogues of meteorite collections. Mineral. Abs. **28**, 471—478 (1949).
SPITZER, L.: Dissipation of planetary filaments. Astrophys. J. **90**, 675—688 (1939).
SPITZER, L.: Star formation. In: JASTROW, R., CAMERON, A. G. W. (Eds.): Origin of the Solar System, pp. 39—53. New York: Academic Press 1963.
STACEY, F. D.: Physics of the Earth, 324 pp. New York: Wiley 1969.
STACEY, F. D., LOVERING, J. F., PARRY, C. G.: Thermo-magnetic anisotropies of some chondritic meteorites. J. Geophys. Res. **60**, 1523—1534 (1961).
STAUB, R. E., MCCALL, J. L.: Metallographic studies of an iron-nickel meteorite following long-time heat treatments. Proc. Int. Metallogr. Soc. **2**, 337—344 (1970).
STAUFFER, H.: Primordial argon and neon in carbonaceous chondrites and ureilites. Geochim. Cosmochim. Acta **24**, 70—82 (1961).
STAUFFER, H., HONDA, M.: Cosmic-ray-produced stable isotopes in iron meteorites. J. Geophys. Res. **67**, 3503—3512 (1962).
STÖFFLER, D.: Deformation and transformation of rock-forming minerals by natural and experimental shock processes. I. Behavior of minerals under shock compression. Fortschr. Miner. **49**, 50—113 (1972).
STUDIER, M. H., HAYATSU, R., ANDERS, E.: Origin of organic matter in early solar system—I. Hydrocarbons. Geochim. Cosmochim. Acta **32**, 151—173 (1968).
STUDIER, M. H., HAYATSU, R., ANDERS, E.: Origin of organic matter in early solar system—V. Further studies of meteoritic hydrocarbons and a discussion of their origin. Geochim. Cosmochim. Acta **36**, 189—215 (1972).
SUESS, H. E.: The Urey-Craig groups of chondrites and their states of oxidation. In: CRAIG, H., MILLER, S. L., WASSERBURG, G. J. (Eds.): Isotopic and Cosmic Chemistry, pp. 385—400. Amsterdam: North Holland 1964.
SUESS, H. E.: Chemical evidence bearing on the origin of the solar system. Ann. Rev. Astron. Astrophys. **3**, 217—234 (1965).
SUESS, H. E., UREY, H. C.: Abundance of the elements. Rev. Mod. Phys. **28**, 53—74 (1956).
SUESS, H. E., WÄNKE, H.: Radiocarbon content and terrestrial age of twelve stony meteorites and one iron meteorite. Geochim. Cosmochim. Acta **26**, 475—480 (1962).
SUESS, H. E., WÄNKE, H., WLOTZKA, F.: On the origin of gas-rich meteorites. Geochim. Cosmochim. Acta **28**, 595—607 (1964).
TANDON, S. N., WASSON, J. T.: Indium variations in a petrologic suite of L-group chondrites. Science **158**, 259—261 (1967).
TANDON, S. N., WASSON, J. T.: Gallium, germanium, indium and iridium variations in a suite of L-group chondrites. Geochim. Cosmochim. Acta **32**, 1087—1109 (1968).
TANENBAUM, A. S.: Clustering of the cosmic ray ages of stone meteorites. Earth Planet. Sci. Letters **2**, 33—35 (1967).

TATSUMOTO, M., KNIGHT, R. J., ALLEGRE, C. J.: Time differences in the formation of meteorites as determined from the ratio of lead-207 to lead-206. Science **180**, 1279—1283 (1973).

TAYLOR, G. J., HEYMANN, D.: Shock, reheating, and the gas retention ages of chondrites. Earth Planet. Sci. Letters **7**, 151—161 (1969).

TAYLOR, H. P. JR., DUKE, M. B., SILVER, L. T., EPSTEIN, S.: Oxygen isotope studies of minerals in stony meteorites. Geochim. Cosmochim. Acta **29**, 489—512 (1965).

TER HAAR, D., CAMERON, A. G. W.: Historical review of theories of the origin of the solar system. In: JASTROW, R., CAMERON, A. G. W. (Eds.): Origin of the Solar System, pp. 1—37. New York-London: Academic Press 1963.

TRIVEDI, B. M. P., GOEL, P. S.: Production of ^{22}Na and ^{3}H in a thick silicate target and its application to meteorites. J. Geophys. Res. **74**, 3909—3917 (1969).

TSCHERMAK, G.: Beitrag zur Classification der Meteoriten. Sitzber. Akad. Wiss. Wien, Math.-Naturw. Cl. **85**, Abt. 1, 347—371 (1883).

TUREKIAN, K. K., CLARK, S. P.: Inhomogeneous accumulation of the earth from the primitive solar nebula. Earth Planet. Sci. Letters **6**, 346—348 (1969).

TURNER, G.: Thermal histories of meteorites by the ^{39}Ar—^{40}Ar method. In: MILLMAN, P. M. (Ed.): Meteorite Research, pp. 407—417. Dordrecht: Reidel 1969.

TURNER, G.: ^{40}Ar—^{39}Ar ages from the lunar maria. Earth Planet. Sci. Letters **11**, 169—191 (1971).

UHLIG, H. H.: Contribution of metallurgy to the origin of meteorites. Part II—The significance of Neumann bands in meteorites. Geochim. Cosmochim. Acta **7**, 34—42 (1955).

UNESCO: Directory of Meteorite Collections and Meteorite Research, 50 pp. Paris: UNESCO 1968.

UREY, H. C.: The thermodynamic properties of isotopic substances. J. Chem. Soc. **1947**, 562—581 (1947).

UREY, H. C.: The origin and development of the earth and other terrestrial planets. Geochim. Cosmochim. Acta **1**, 209—277 (1951).

UREY, H. C.: The Planets: Their Origin and Development, 236 pp. Yale Univ. 1952.

UREY, H. C.: On the dissipation of gas and volatilized elements from protoplanets. Astrophys. J. Suppl. **1** 147—173 (1954).

UREY, H. C.: Comments on two papers by John F. Lovering concerning a typical meteorite parent body. Geochim. Cosmochim. Acta **13**, 335—338 (1958).

UREY, H. C.: Criticism of Dr. B. Mason's paper on "The origin of meteorites". J. Geophys. Res. **66**, 1988—1991 (1961).

UREY, H. C.: The origin of the moon and its relationship to the origin of the solar system. In: KOPAL, Z., MIKHAILOV, Z. K. (Eds.): The Moon, IAU Symposium No. **14**, pp. 133—148. New York-London: Academic Press 1962.

UREY, H. C.: A review of atomic abundances in chondrites and the origin of meteorites. Rev. Geophys. **2**, 1—34 (1964).

UREY, H. C.: Chemical evidence relative to the origin of the solar system. Mon. Not. Roy. Astron. Soc. **131**, 199—223 (1966).

UREY, H. C., CRAIG, H.: The composition of the stone meteorites and the origin of the meteorites. Geochim. Cosmochim. Acta **4**, 36—82 (1953).

UREY, H. C., LEWIS, J. S.: Organic matter in carbonaceous chondrites. Science **152**, 102—104 (1966).

VAN SCHMUS, W. R.: Polymict structure of the Mezö-Madaras chondrite. Geochim. Cosmochim. Acta **31**, 2027—2042 (1967).

VAN SCHMUS, W. R.: Mineralogy and petrology of chondritic meteorites. Earth Sci. Rev. **5**, 145—184 (1969a).

Van Schmus, W. R.: Mineralogy, petrology, and classification of types 3 and 4 carbonaceous chondrites. In: Millman, P. (Ed.): Meteorite Research, pp. 480—491. Dordrecht: Reidel 1969b.

Van Schmus, W. R., Hayes, J. M.: Chemical and petrographic correlations among carbonaceous chondrites. Geochim. Cosmochim. Acta **38**, 47—64 (1974).

Van Schmus, W. R., Koffman, D. M.: Equilibration temperatures of iron and magnesium in chondritic meteorites. Science **155**, 1009—1011 (1967).

Van Schmus, W. R., Ribbe, P. H.: The composition and structural state of feldspar from chondritic meteorites. Geochim. Cosmochim. Acta **32**, 1327—1342 (1968).

Van Schmus, W. R., Ribbe, P. H.: Composition of phosphate minerals in ordinary chondrites. Geochim. Cosmochim. Acta **33**, 637—640 (1969).

Van Schmus, W. R., Wood, J. A.: A chemical-petrologic classification for the chondritic meteorites. Geochim. Cosmochim. Acta **31**, 747—765 (1967).

Vdovykin, G. P.: Carbon polytypism in meteorites. In: Schenck, P. A., Havenaar, I. (Eds.): Advances in Organic Geochemistry, **1968**, pp. 593—604. London-New York: Pergamon 1969.

Vdovykin, G. P.: Ureilites. Space Sci. Rev. **10**, 483—510 (1970).

Veverka, J.: The polarization curve and the absolute diameter of Vesta. Icarus **15**, 11—17 (1971a).

Veverka, J.: Photopolarimetric observations of the minor planet Flora. Icarus **15**, 454—458 (1971b).

Veverka, J., Noland, M.: Asteroid reflectivities from polarization curves: Calibration of the "slope-albedo" relationship. Icarus **19**, 230—239 (1973).

Vilcsek, E., Wänke, H.: Cosmic-ray exposure ages and terrestrial ages of stone and iron meteorites derived from Cl^{36} and Ar^{39} measurements. In: Radioactive Dating, pp. 381—393. Vienna: IAEA 1963.

Vinogradov, A. P., Kropotova, O. I., Vdovykin, G. P., Grinenko, V. A.: Isotopic composition of different phases of carbon in carbonaceous meteorites. Geokhimiya **1967**, 267—273; Engl. translation Geochim. Int. **4**, 229—235 (1967).

Vinogradov, A. P., Zadorozhnyi, I. K.: Inert gases in stony meteorites (in Russian). Geokhimiya **1964**, 587—600 (1964).

Vogel, R.: Über das Verhalten des Troilits in einem Chondriten (Pultusk) beim Erhitzen. Chem. Erde **24**, 244—253 (1965).

Volarovich, M. P., Leontieva, L. L.: A study of the viscosity of meteorites and tektites (in Russian). Meteoritika **1**, 33—42 (1941).

Voshage, H.: Bestrahlungsalter und Herkunft der Eisenmeteorite. Z. Naturforsch. **22a**, 477—506 (1967).

Voshage, H.: Massenspektrometrische Element- und Isotopenhäufigkeitsanalysen zur Erforschung der Geschichte der Meteorite und des Planetensystems. Int. J. Mass. Spectrom. Ion Phys. **1**, 157—190 (1968a).

Voshage, H.: Massenspektrometrische Element- und Isotopenhäufigkeitsanalysen zur Erforschung der Geschichte der Meteorite und des Planetensystems. II. Int. J. Mass. Spectrom. Ion. Phys. **1**, 237—281 (1968b).

Wänke, H.: Der Sonnenwind als Quelle der Uredelgase in Steinmeteoriten. Z. Naturforsch. **20a**, 946—949 (1965).

Wänke, H.: Meteoritenalter und verwandte Probleme der Kosmochemie. Fortschr. Chem. Forsch. **7**, 322—408 (1966).

Wänke, H., Baddenhausen, H., Balacescu, A., Teschke, F., Spettel, B., Dreibus, G., Quijano-Rico, M., Kruse, H., Wlotzka, F., Begemann, F.: Multielement analyses of lunar samples (abstract). In: Watkins, C. (Ed.): Lunar Science-III, pp. 779—781. Lunar Science Institute Contr. No. 88 1972.

WÄNKE, H., KÖNIG, H.: Eine neue Methode zur Kalium-Argon-Altersbestimmung und ihre Anwendung auf Steinmeteorite. Z. Naturforsch. **14a**, 860—866 (1959).

WÄNKE, H. and coworkers: On the chemistry of the Allende inclusions and their origin as high-temperature condensates. In preparation (1973).

WAHL, W.: The brecciated stony meteorites and meteorites containing foreign fragments. Geochim. Cosmochim. Acta **2**, 91—117 (1952).

WAI, C. M., WETHERILL, G. W., WASSON, J. T.: The distribution of trace quantities of germanium between iron, silicate and sulfide phases. Geochim. Cosmochim. Acta **32**, 1269—1278 (1968).

WASSERBURG, G. J., HUNEKE, J. C., BURNETT, D. S.: Correlation between fission tracks and fission type xenon in meteoritic whitlockite. J. Geophys. Res. **74**, 4221—4232 (1969a).

WASSERBURG, G. J., PAPANASTASSIOU, D. A., BURNETT, D. S.: Rb-Sr ages of lunar rocks from the Sea of Tranquillity. Earth Planet. Sci. Letters **8**, 1—19 (1970).

WASSERBURG, G. J., PAPANASTASSIOU, D. A., SANZ, H. G.: Initial strontium for a chondrite and the determination of a metamorphism or formation interval. Earth Planet. Sci. Letters **7**, 33—43 (1969b).

WASSERBURG, G. J., SANZ, H. G., BENCE, A. E.: Potassium-feldspar phenocrysts in the surface of Colomera, an iron meteorite. Science **161**, 684—687 (1968).

WASSERBURG, G. J., SCHRAMM, D. N., HUNEKE, J. C.: Nuclear chronologies for the galaxy. Astrophys. J. **157**, L91—L96 (1969c).

WASSON, J. T.: The chemical classification of iron meteorites: I. A study of iron meteorites with low concentrations of gallium and germanium. Geochim. Cosmochim. Acta **31**, 161—180 (1967).

WASSON, J. T.: Concentrations of nickel, gallium, germanium, and iridium in Canyon Diablo and other Arizona octahedrites. J. Geophys. Res. **73**, 3207—3211 (1968).

WASSON, J. T.: The chemical classification of iron meteorites—III. Hexahedrites and other irons with germanium concentrations between 80 and 200 ppm. Geochim. Cosmochim. Acta **33**, 859—876 (1969).

WASSON, J. T.: The chemical classification of iron meteorites—IV. Irons with Ge concentrations greater than 190 ppm and other meteorites associated with group I. Icarus **12**, 407—423 (1970).

WASSON, J. T.: Differentiated meteorites. Trans. Am. Geophys. Union **52**, IUGG441—IUGG447 (1971a).

WASSON, J. T.: An equation for the determination of iron-meteorite cooling rates. Meteoritics **6**, 139—147 (1971b).

WASSON, J. T.: Volatile elements on the Earth and Moon. Earth Planet. Sci. Letters **11**, 219—225 (1971c).

WASSON, J. T.: Formation of ordinary chondrites. Rev. Geophys. Space Phys. **10**, 711—759 (1972a).

WASSON, J. T.: Parent-body models for the formation of iron meteorites. Proc. 24th Intern. Geol. Cong. **15**, 161—168 (1972b).

WASSON, J. T., CHOU, C.-L.: Fractionation of moderately volatile elements in ordinary chondrites. Meteoritics **9**, in press (1974).

WASSON, J. T., KIMBERLIN, J.: The chemical classification of iron meteorites—II. Irons and pallasites with germanium concentrations between 8 and 100 ppm. Geochim. Cosmochim. Acta **31**, 2065—2093 (1967).

WASSON, J. T., SCHAUDY, R.: The chemical classification of iron meteorites—V. Groups IIIC and IIID and other irons with germanium concentrations between 1 and 25 ppm. Icarus **14**, 59—70 (1971).

WASSON, J. T., SEDWICK, S. P.: Meteoritic material from Hopewell Indian burial mounds: Chemical data regarding possible sources. Nature **222**, 22—24 (1969).
WASSON, J. T., WAI, C. M.: Composition of the metal, schreibersite and perryite of enstatite achondrites and the origin of enstatite chondrites and achondrites. Geochim. Cosmochim. Acta **34**, 169—184 (1970).
WATSON, F. G.: Reflectivity and color of meteorites. Proc. Nat. Acad. Sci. (Wash.). **24**, 532—537 (1938).
WATSON, F. G.: Between the Planets, 224 pp. Doubleday 1962.
WEAVING, B.: Magnetic anisotropy in chondritic meteorites. Geochim. Cosmochim. Acta **26**, 451—455 (1962).
WETHERILL, G. W.: Isotopic composition and concentration of molybdenum in iron meteorites. J. Geophys. Res. **69**, 4403—4408 (1964).
WETHERILL, G. W.: Collisions in the asteroid belt. J. Geophys. Res. **72**, 2429—2444 (1967).
WETHERILL, G. W.: Stone meteorites: Time of fall and origin. Science **159**, 79—82 (1968 a).
WETHERILL, G. W.: Dynamical study of asteroidal and cometary orbits and their relation to the origin of meteorites. In: AHRENS, L. H. (Ed.): Origin and Distribution of the Elements, pp. 423—443. London-New York: Pergamon 1968 b.
WETHERILL, G. W.: Mutual gravitational and collisional scattering of bodies in the asteroid belt. Z. Naturforsch. **23a**, 791—795 (1968 c).
WETHERILL, G. W.: Relationships between orbits and sources of chondritic meteorites. In: MILLMAN, P. M. (Ed.): Meteorite Research, pp. 573—589. Dordrecht: Reidel 1969.
WETHERILL, G. W.: Cometary versus asteroidal origin of chondritic meteorites. In: GEHRELS, T. (Ed.): Physical Studies of Minor Planets, pp. 447—460. NASA Rep. SP-267 1971.
WETHERILL, G. W., MARK, R., LEE-HU, C.: Chondrites: Initial strontium-87/strontium-86 ratios and the early history of the solar system. Science **182**, 281—283 (1973).
WETHERILL, G. W., WILLIAMS, J. G.: Evaluation of the Apollo asteroids as sources of stone meteorites. J. Geophys. Res. **73**, 635—648 (1968).
WHIPPLE, F. J. W.: The great Siberian meteor and the waves, seismic and aerial, which it produced. Quart. J. Roy. Meteorol. Soc. **56**, 287—304 (1930).
WHIPPLE, F. J. W.: On phenomena related to the great Siberian meteor. Quart. J. Roy. Meteorol. Soc. **60**, 505—512 (1934).
WHIPPLE, F. L.: A comet model. I. The acceleration of Comet Encke. Astrophys. J. **111**, 375—394 (1950).
WHIPPLE, F. L.: Chondrules: suggestions concerning their origin. Science **153**, 54—56 (1966).
WHIPPLE, F. L.: Accumulation of chondrules on asteroids. In: GEHRELS, T. (Ed.): Physical Studies of Minor Planets, pp. 251—256. NASA Rep. SP-267, 1971.
WHIPPLE, F. L., FIREMAN, E. L.: Calculation of erosion in space from the cosmic-ray exposure ages of meteorites. Nature **183**, 1315 (1959).
WIIK, H. B.: The chemical composition of some stony meteorites. Geochim. Cosmochim. Acta **9**, 279—289 (1956).
WIIK, H. B.: On regular discontinuities in the composition of meteorites. Commun. Phys.-Math. (Helsinki) **34**, 135—145 (1969).
WILKENING, L., LAL, D., REID, A. M.: The evolution of the Kapoeta howardite based on fossil track studies. Earth Planet. Sci. Letters **10**, 334—340 (1971).
WILLIAMS, I. P., CREMIN, A. W.: A survey of theories relating to the origin of the solar system. Quart. J. Roy. Astron. Soc. **9**, 40—62 (1968).

WILLIAMS, J. G.: Proper elements, families, and belt boundaries. In: GEHRELS, T. (Ed.): Physical Studies of Minor Planets, pp. 177—180. NASA Rep. SP-267, 1971.

WILLIAMS, R.: Equilibrium temperatures, pressures, and oxygen fugacities of the equilibrated chondrites. Geochim. Cosmochim. Acta **35**, 407—411 (1971).

WLOTZKA, F.: Über die Hell-Dunkel-Struktur der urgashaltigen Chondrite Breitscheid und Pantar. Geochim. Cosmochim. Acta **27**, 419—429 (1963).

WLOTZKA, F.: On the formation of chondrules and metal particles by "shock melting". In: MILLMAN, P. M. (Ed.): Meteorite Research, pp. 174—184. Dordrecht: Reidel 1969.

WLOTZKA, F.: Haverö ureilite: Evidence for recrystallization and partial reduction. Meteoritics **7**, 591—600 (1972).

WOOD, J. A.: Origin of chondrules and chondrites. Icarus **2**, 152—180 (1963a).

WOOD, J. A.: Physics and chemistry of meteorites. In: MIDDLEHURST, B. M., KUIPER, G. P. (Eds.): The Moon, Meteorites, and Comets. III. The Solar System, Vol. IV, pp. 337—401. Univ. Chicago Press 1963b.

WOOD, J. A.: The cooling rates and parent planets of several iron meteorites. Icarus **3**, 429—459 (1964).

WOOD, J. A.: Olivine and pyroxene compositions in type II carbonaceous chondrites. Geochim. Cosmochim. Acta **31**, 2095—2108 (1967a).

WOOD, J. A.: Chondrites: Their metallic minerals, thermal histories, and parent planets. Icarus **6**, 1—49 (1967b).

WOOD, J. A.: Criticism of paper by H. E. Suess and H. Wänke, "Metamorphosis and equilibration in chondrites." J. Geophys. Res. **72**, 6379—6383 (1967c).

WOOD, J. A.: Meteorites and the Origin of Planets, 117 pp. New York: McGraw-Hill 1968.

WOOLFSON, M. M.: A capture theory of the origin of the solar system. Proc. Roy. Soc. London **282 A**, 485—507 (1964).

WOOLFSON, M. M.: The evolution of the solar system. Rep. Prog. Phys. **32**, 135—185 (1969).

YAVNEL, A. A.: Genetic relationships in the chemical composition of chondrites (in Russian). Meteoritika **23**, 36—41 (1963).

YAVNEL, A. A.: On the degree of equilibration reflected in the ferromagnesian silicates in ordinary chondrites (in Russian). Meteoritika **28**, 19—29. Engl. trans., Meteoritics **5**, 153—168 (1970).

YOUNG, J.: The crystal structure of meteoric iron as determined by X-rays. Proc. Roy. Soc. (London) **A 112**, 630—641 (1926).

ZÄHRINGER, J.: Isotopie-Effekt und Häufigkeiten der Edelgase in Steinmeteoriten und auf der Erde. Z. Naturforsch. **17a**, 460—471 (1962).

ZÄHRINGER, J.: Primordial argon and the metamorphism of chondrites. Earth Planet. Sci. Letters **1**, 379—382 (1966a).

ZÄHRINGER, J.: Primordial helium detection by microprobe technique. Earth Planet. Sci. Letters **1**, 20—22 (1966b).

ZÄHRINGER, J.: Die Chronologie der Chondriten auf Grund von Edelgasanalysen. Meteoritika **27**, 25—40 (1966c).

ZÄHRINGER, J.: Rare gases in stony meteorites Geochim. Cosmochim. Acta **32**, 209—237 (1968).

ZIMMERMANN, P. D., WETHERILL, G. W.: Asteroidal source of meteorites. Science **182**, 51—53 (1973).

ZUKAS, E. G.: Metallurgical results from shock-loaded iron alloys applied to a meteorite. J. Geophys. Res. **74**, 1993—2001 (1969).

Appendix I: Glossary

Some words which are unique to the meteorite literature, or for which the usages are slightly different from those in other fields are defined below. The names and symbols of meteorite groups are defined in Tables II-5, II-7 and AII-1. Mineral names are either defined in Table V-1 or when used. Italicized words are defined elsewhere in the glossary.

Ablation. Loss of material from the surface of a *meteorite* as a result of deceleration processes in the earth's atmosphere.

Abundance. The atomic ratio of an element to another "normalizing" element (generally Si or Mg; stellar abundances are usually relative to H). Abundances are more suitable than concentrations (units of mass per mass of sample) for comparing materials with substantially different elemental or phase compositions.

Accretion. Gravitational accumulation of material onto the surface of a body. See also *agglomeration.*

Achondrite. A silicate-rich meteorite with a nonvolatile element distribution which is fractionated relative to that in the sun. Not a good generic term, since some meteorites designated achondrites are closely related to the *chondrites.*

Agglomeration. Growth in space of small (mm — m-sized) objects as a result of short-range (chemical, magnetic) forces following low-velocity collisions. See also *accretion.*

Ataxite. An iron meteorite in which nearly no macroscopic structure is observed, although considerable petrographic information is recorded in the microscopic structure. See Section II D and Chapter XIV.

Breccia. A fragmental rock type including components (the larger pieces called xenoliths or clasts) which were previously part of another rock. In a **monomict** breccia all components originated in the same rock; in a **genomict** breccia the components originated in distinct but genetically closely related rocks; in **polymict** breccias the components originated in two or more unrelated rocks.

Chondrite. A silicate-rich meteorite with a roughly solar distribution of nonvolatile elements. Most but not all chondrites contain *chondrules.* See Chapter II.

Chondrule. Although used by some authors to designate any mm-sized spheroidal body in a meteorite, usage should be restricted to those bodies which formed as molten droplets by the same process or processes which produced the chondrules observed in *chondrites*. Spheroidal bodies which have not been molten, or which may have formed by processes other than those which formed the chondrite chondrules, should be called chondrule-like objects. See Chapter VI.

Fall. A meteorite recovered immediately after falling to the earth; in some cases also applied to meteorites recovered months or years after observations of fall phenomena.

Find. A meteorite which cannot be associated with an observed fall; find should also be applied to weathered meteorites recovered months or years after an observed fall.

Friable. Easily crushed, as between thumb and forefinger.

Genomict. See *breccia.*

Hexahedrite. An iron meteorite in which the Ni-Fe is present as the body-centered cubic α alloy. Called hexahedrite because cubic cleavage is observed. See Section II D.

Metal-Rich Meteorite and **Silicate-Rich Meteorite** are descriptions rather than generic terms. Metal-rich meteorites are those with enough metal (> ca. 10—20 wt %) to allow straightforward separation of a representative sample for analysis. Silicate-rich meteorites contain enough silicates (> ca. 1 wt %) to allow petrographic characterization on the basis of the silicates. These definitions allow a number of meteorites to be characterized as both metal- and silicate-rich. See Chapter II.

Metamorphism. Solid-state *recrystallization* and replacement of less stable by more stable phases as a result of the application of heat. Metamorphism in chondritic meteorites is much less severe than that observed in most terrestrial metamorphic rocks.

Meteorite. Extraterrestrial material which survives passage through the earth's atmosphere and reaches the earth's surface as a recoverable object (or objects).

Meteoroid. An object in a heliocentric orbit such that it can be captured by the earth and, after undergoing ablation and fragmentation in the atmosphere, fall to the earth's surface as a *meteorite.*

Monomict. See *breccia.*

Octahedrite. An iron meteorite in which the low-Ni, α alloy has precipitated as octahedrally oriented lamellae on the 111 planes of the parental γ alloy. See Sections II D, V E.

Paired. Said of meteorites which are cataloged separately, but are in fact portions of the same fall (see Chapter XIII).

Parent Body. Planet- or comet-like solar-system bodies in which meteorites were stored and, in some cases, formed.

Planetary-Type Rare Gas. A rare-gas distribution in which elements with higher atomic number are more abundant than those with lower atomic number. This type of rare gas is found in all *chondrites*. See *solar-type rare gas* and Chapter IX.

Planetesimal. Bodies of intermediate (perhaps m—100-m) size, most of which finally accreted to larger *parent bodies*.

Polymict. See *breccia*.

Recrystallization. Either a coarsening of the texture of a rock as a result of the solid-state growth of larger crystals at the expense of smaller ones or the solid-state replacement of mechanically strained phases by unstrained ones.

Silicate-Rich Meteorite. See *metal-rich meteorite*.

Solar Nebula. The fragment of an interstellar gas and dust cloud which collapsed to form the solar system; also applied to the gas and dust extrasolar portion of this material from which the planets, asteroids, comets and meteorite parent bodies formed. See Chapter I.

Solar-Type Rare Gas. A rare-gas distribution in which interelement rare-gas ratios are roughly solar (i.e., unfractionated). See also *planetary-type rare gas* and Chapter IX.

Widmanstätten Structure. The octahedral structure of an *octahedrite*.

Appendix II: Lists of Classified Meteorites

A. Introduction: Alphabetical List of Well-Classified Meteorites

Classificational information for meteorites which have been studied by modern analytical techniques during the past two decades is given in this Appendix. The lists were compiled by computer from a data bank at the Max-Planck-Institut für Chemie, Mainz. With few exceptions, entries for silicate-rich meteorites (Appendices II B and II C) include compositional data for the ferromagnesian minerals, pyroxene and olivine; with the exception of some meteorites which BUCHWALD (1974) pairs with previously analyzed irons, all entries for metal-rich meteorites (Appendix II D) include Ni, Ga, Ge, and (with rare exceptions) Ir data. The compiled data were obtained by a reasonably thorough survey of the literature published through June 1973.

In Appendix II A well-classified meteorites are listed alphabetically together with the group symbol and a six-space abbreviation of the meteorite's name. The names are as given in Hey (1966) or in the Meteoritical Bulletin (see Chapter III); long names are truncated after 19 spaces. In a few cases revised names proposed by BUCHWALD (1974) are used; names of some new meteorites are taken from published descriptions. In Appendices II B—II D meteorites are listed by class, and only the abbreviation of the name is given. Persons familiar with the meteorite literature will find that they are able to identify most meteorites from the abbreviation alone. Exceptions exist where there are several names beginning with the same 3 or 4 letters. Note that the abbreviations have been chosen in order to maintain the same alphabetical order as the full name. The classification of meteorites is discussed in Chapter II. Group symbols are listed in Table A II-1 as well as in Tables II-5 and II-7. Arabic numeral suffixes indicate petrologic types (of chondrites). The suffix −AN indicates that a meteorite possesses most of the properties considered typical for that group, but differs in some important respects and is therefore designated an "anomalous member" of that group. Chondrites which are not group members are designated CHANOM (anomalous chondrite). Anomalous differentiated silicate-rich meteorites and metal-rich meteorites are designated ACANOM (anomalous achon-

Table AII-1. Symbols and names for groups of 5 or more genetically related meteorites

Symbol	Group
AUB	aubrites
CI	Ivuna-group chondrites
CM	Mighei-group chondrites
CO	Ornans-group chondrites
CV	Vigarano-group chondrites
DIO	diogenites
E	enstatite chondrites
EUC	eucrites
H	H-group chondrites
HOW	howardites
L	L-group chondrites
LL	LL-group chondrites
MES	mesosiderites
PAL	pallasites
URE	ureilites
IA	group-IA iron meteorites
IB	group-IB iron meteorites
IIA, etc.	— other iron-meteorite groups

drite) and IRANOM (anomalous iron), respectively. If a non-chondritic meteorite contains both metal and silicate in amounts sufficient for classificational investigation, it is designated MEANOM (anomalous meteorite). Genetic relationships among members of these anomalous categories are indicated in the "Remarks" columns of Appendices IIB—IID.

Appendix II A

CLASS	ABBREV	N A M E
H	AARHUS	AARHUS
IIIA	ABAKAN	ABAKAN
IIIF	ABANCA	ABANCAY
H	ABBOTT	ABBOTT
E4	ABEE++	ABEE
L6	ABERNA	ABERNATHY
H	ABO+++	ABO
L	ACCALA	ACCALANA
H	ACHILL	ACHILLES
L6	ACHIRA	ACHIRAS
H	ACME++	ACME
EUC	ADALIA	ADALIA
H5	ADAMSC	ADAMS COUNTY
IIIB	ADARGA	ADARGAS
H	ADDAHB	AD-DAHBUBAH
L5	ADELIE	ADELIE LAND
E4	ADHIKO	ADHI KOT
PAL	ADMIRE	ADMIRE
H4	ADRIAN	ADRIAN
H5	AGEN++	AGEN
IIIA	AGGIEC	AGGIE CREEK
L6	AGUADA	AGUADA
PAL	AHUMAD	AHUMADA
IIB	AINSWO	AINSWORTH
L	AIR+++	AIR
L6	AKABA+	AKABA
H4	AKBARP	AKBARPUR
IIIA	AKPOHO	AKPOHON

CLASS	ABBREV	N A M E
H	AKR040	AKRON (1940)
H	AKR054	AKRON (1954)
L	AKR061	AKRON (1961)
H	AKWANG	AKWANGA
CI1	ALAIS+	ALAIS
H5	ALAMOG	ALAMOGORDO
L	ALAMOS	ALAMOSA
L4	ALBARE	ALBARETO
PAL	ALBIPA	ALBIN (PALLASITE)
L	ALBIST	ALBIN (STONE)
LL5	ALDSWO	ALDSWORTH
L6	ALEPPO	ALEPPO
H	ALESSA	ALESSANDRIA
H	ALEXKY	ALEXANDROVSKY
L6	ALFIAN	ALFIANELLO
H	ALGHST	AL-GHANIM (STONE)
IRANOM	ALGOMA	ALGOMA
IRANOM	ALIKAT	ALIKATNIMA
H5	ALLEGA	ALLEGAN
CV3	ALLEND	ALLENDE
CV2-AN	ALRAIS	AL RAIS
IVA	ALTONA	ALTONAH
IA	AMATES	AMATES
H5	AMBAPU	AMBAPUR NAGLA
L	AMBER+	AMBER
L	AMHERS	AMHERST
PAL	ANDERS	ANDERSON
L6	ANDOVE	ANDOVER
L	ANDRYU	ANDRYUSHKI
H	ANDURA	ANDURA
IIIA	ANGELI	ANGELICA
L6	ANGERS	ANGERS
IIA	ANGRIR	ANGRA DOS REIS (IRO
ACANOM	ANGRST	ANGRA DOS REIS (STO
H4	ANKOBE	ANKOBER
IA-AN	ANNAHE	ANNAHEIM
IIIC	ANOKA+	ANOKA
H5	ANTHON	ANTHONY
PAL	ANTOFA	ANTOFAGASTA
IIIB	APOALA	APOALA
IIIA	APOAPS	APOALA, PSEUDO-
LL6	APPLEY	APPLEY BRIDGE
L6	APT+++	APT
IIA	ARAGON	ARAGON
L5	ARAPAH	ARAPAHOE
LL6	ARCADI	ARCADIA
H	ARCHIE	ARCHIE
PAL	ARGONI	ARGONIA
IRANOM	ARISPE	ARISPE
IIE	ARLING	ARLINGTON
IID-AN	ARLTUN	ARLTUNGA
L	ARMEL+	ARMEL
L5	ARRIBA	ARRIBA
H	ARROYO	ARROYO AGUIAR
L6	ARTRAC	ARTRACOONA
IIIA	ASARCO	ASARCO MEXICANA
L	ASCO++	ASCO
L6	ASHDON	ASHDON
IA	ASHFOR	ASHFORK
H5	ASHMOR	ASHMORE
H	ASHSHA	ASH-SHALFAH
L5	ASSAM+	ASSAM
H5	ASSISI	ASSISI
IIIA	ASWAN+	ASWAN
L4	ATARRA	ATARRA
L	ATEMAJ	ATEMAJAC
LL6	ATHENS	ATHENS
E6	ATLANT	ATLANTA
L	ATOKA+	ATOKA
L6	ATWOOD	ATWOOD
AUB	AUBRES	AUBRES
IRANOM	AUBURN	AUBURN
IIIA-AN	AUGUCO	AUGUSTA COUNTY
IIIB	AUGUST	AUGUSTINOVKA
L6	AUMALE	AUMALE
L6	AUMIER	AUMIERES
H4	AURORA	AURORA
L5	AUSSON	AUSSON
H	AVANHA	AVANHANDAVA
IIA	AVCE++	AVCE
H	AVILEZ	AVILEZ
IIIA	AVOCWE	AVOCA (WESTERN AUST
L4	AWERE+	AWERE
L6	AZTEC+	AZTEC
IRANOM	BABBBL	BABB≠S MILL (BLAKE≠
IRANOM	BABBTR	BABB≠S MILL (TROOST
L6	BACHMU	BACHMUT
IRANOM	BACUBI	BACUBIRITO
IIIA	BAGDAD	BAGDAD
IA-AN	BAHJOI	BAHJOI
IIIB	BALDEA	BALD EAGLE
L4	BALDMO	BALD MOUNTAIN
L6	BALDWY	BALDWYN
IA	BALFOU	BALFOUR DOWNS
CV3	BALI++	BALI
IA-AN	BALLER	BALLINGER
IIC	BALLOO	BALLINOO
LL6	BANDON	BANDONG
L	BANSWA	BANSWALL
IIIA	BAQUED	BAQUEDANO
H5	BARBOT	BARBOTAN
MES	BAREA+	BAREA
L6	BAROTI	BAROTI
IIA	BARRAB	BARRABA
IIE	BARRAN	BARRANCA BLANCA
L4	BARRAT	BARRATTA
IIIA	BARTLE	BARTLETT
L	BARWEL	BARWELL

CLASS	ABBREV	N A M E
H	BARWIS	BARWISE
IIIA	BASEDO	BASEDOW RANGE
H4	BATH++	BATH
L6	BATHFU	BATH FURNACE
L	BAXTER	BAXTER
IIIB	BEARCR	BEAR CREEK
H5	BEARDS	BEARDSLEY
H4	BEAVER	BEAVER CREEK
H5	BEDDGE	BEDDGELERT
LL	BEELER	BEELER
L5	BEENHA	BEENHAM
IIIB	BELLAR	BELLA ROCA
L	BELLEP	BELLE PLAINE
CM2	BELLS+	BELLS
IRANOM	BELLSB	BELLSBANK
H	BELLYR	BELLY RIVER
H	BELMON	BELMONT
LL6	BENARE	BENARES
MEANOM	BENCUB	BENCUBBIN
IRANOM	BENDEG	BENDEGO
H6	BENLD+	BENLD
IIA	BENNET	BENNETT COUNTY
H	BENONI	BENONI
LL6	BENTON	BENTON
LL	BERDYA	BERDYANSK
EUC	BEREBA	BEREBA
L6	BERLAN	BERLANGUILLAS
H	BETHLE	BETHLEHEM
E4,5	BETHUN	BETHUNE
L5	BEUSTE	BEUSTE
L	BEYROU	BEYROUT
L6	BHAGUR	BHAGUR
L6	BHERAI	BHERAI
LL	BHOLA+	BHOLA
HOW	BHOLGH	BHOLGHATI
EUC	BIALYS	BIALYSTOK
H4	BIELOK	BIELOKRYNITSCHIE
IIIA	BILLIN	BILLINGS
L	BILLYG	BILLYGOAT DONGA
HOW	BINDA+	BINDA
IIA	BINGER	BINGERA
H	BIRHAD	BIR HADI
H	BIRNIN	BIRNI N≠KONNI
IA	BISCHT	BISCHTUEBE
IVA	BISHOC	BISHOP CANYON
AUB	BISHOV	BISHOPVILLE
L3	BISHUN	BISHUNPUR
IB	BITBUR	BITBURG
H6	BJELAJ	BJELAJA ZERKOV
L4	BJURBO	BJURBOELE
L	BLACMP	BLACK MOSHANNAN PAR
IA	BLACMT	BLACK MOUNTAIN
L5	BLACWE	BLACKWELL
L6	BLANKE	BLANKET
H6	BLANSK	BLANSKO
E6	BLITHF	BLITHFIELD
IA	BLOODY	BLOODY BASIN
L5	BLUFF+	BLUFF
L6	BOCAS+	BOCAS
IVA	BODAIB	BODAIBO
LL	BOELUS	BOELUS
H6	BOERNE	BOERNE
H	BOGOSL	BOGOSLOVKA
IA	BOGOU+	BOGOU
IIA	BOGUSL	BOGUSLAVKA
IA	BOHUMI	BOHUMILITZ
IA	BOLIVI	BOLIVIA
H	BOLSHA	BOLSHAYA KORTA
MES	BONDOC	BONDOC PENINSULA
H5	BONITA	BONITA SPRINGS
IVA	BOOGAL	BOOGALDI
LL6	BORGOS	BORGO SAN DONINO
L6	BORI++	BORI
CM2	BORISK	BORISKINO
L5	BORKUT	BORKUT
H	BORODI	BORODINO
L	BOTSCH	BOTSCHETSCHKI
L	BOVEDY	BOVEDY
H	BOWDEN	BOWDEN
L	BOWESM	BOWESMONT
IIIA	BOXHOL	BOXHOLE
L	BRADY+	BRADY
PAL	BRAHIN	BRAHIN
IIA	BRAUNA	BRAUNAU
IIIB	BREECE	BREECE
H	BREITS	BREITSCHEID
H3	BREMER	BREMERVOERDE
PAL	BRENHA	BRENHAM
L6	BREWST	BREWSTER
IID	BRIDGE	BRIDGEWATER
HOW	BRIENT	BRIENT
IIIA	BRIGGS	BRIGGSDALE
L	BRISCO	BRISCOE
IVA	BRISTO	BRISTOL
H4	BROKEN	BROKEN BOW
H3	BROW37	BROWNFIELD (1937)
H	BROW64	BROWNFIELD (1964)
IID	BROW66	BROWNFIELD (1966)
L6	BRUDER	BRUDERHEIM
IIA	BRUNO+	BRUNO
MES	BUDULA	BUDULAN
HOW	BUNUNU	BUNUNU
H5	BURDET	BURDETT
IA	BURGAV	BURGAVLI
H5	BURGHE	BUR-GHELUAI
IA	BURKET	BURKETT
IIIEA	BURLIN	BURLINGTON
H	BURNAB	BURNABBIE

CLASS	ABBREV	N A M E
L	BURRIK	BURRIKA
L	BURSA+	BURSA
L6	BUSCHH	BUSCHHOF
IVA	BUSHMA	BUSHMAN LAND
H4	BUSHNE	BUSHNELL
AUB	BUSTEE	BUSTEE
IRANOM	BUTLER	BUTLER
H6	BUTSUR	BUTSURA
L6	CABEZO	CABEZO DE MAYO
IIIA	CACARI	CACARIA
EUC	CACHAR	CACHARI
IRANOM	CACHIO	CACHIYUYAL (OM)
H	CACILA	CACILANDIA
L	CADELL	CADELL
IIA	CALICO	CALICO ROCK
IA	CALIFO	CALIFORNIA
L	CALLIH	CALLIHAM
IRANOM	CAMBRI	CAMBRIA
IIIB	CAMPBE	CAMPBELLSVILLE
IA	CAMPOD	CAMPO DEL CIELO
IA	CAMPVE	CAMP VERDE
L	CANAKK	CANAKKALE
H	CANELL	CANELLAS
H5	CANGAS	CANGAS DE ONIS
IIIA	CANTON	CANTON
H	CANY57	CANYON
IIIA	CANYCI	CANYON CITY
IA	CANYD1	CANYON DIABLO (1891
IA-AN	CANYD6	CANYON DIABLO (1936
IA-AN	CANYD9	CANYON DIABLO (1949
H6	CAPEGI	CAPE GIRARDEAU
IVB	CAPEOF	CAPE OF GOOD HOPE
IIIA	CAPERR	CAPERR
IIIA	CAPEYO	CAPE YORK
H	CAPILL	CAPILLA DEL MONTE
LL6	CARATA	CARATASH
IID	CARBO+	CARBO
H5	CARCOT	CARCOTE
L	CARDAN	CARDANUMBI
IIIC	CARLTO	CARLTON
H	CAROLI	CAROLINE
L3	CARRAW	CARRAWEENA
IIIA	CARTHA	CARTHAGE
L	CARTOO	CARTOONKANA
IIIA	CASASG	CASAS GRANDES
IA	CASEYC	CASEY COUNTY
H	CASHIO	CASHION
IIIA	CASIMI	CASIMIRO DE ABREU
H5	CASTAL	CASTALIA
L6	CASTIN	CASTINE
H6	CAVOUR	CAVOUR
H6	CEDAKA	CEDAR (KANSAS)
H4	CEDATE	CEDAR (TEXAS)
IIA	CEDATO	CEDARTOWN
H	CENTER	CENTERVILLE
IIB	CENTRA	CENTRAL MISSOURI
H5	CERESE	CERESETO
H	CHAIL+	CHAIL
LL3	CHAINP	CHAINPUR
H	CHAMBE	CHAMBERLIN
IIIA	CHAMBO	CHAMBORD
IIIA	CHANAR	CHANARAL
L5	CHANDA	CHANDAKAPUR
L6	CHANDP	CHANDPUR
H5	CHANNI	CHANNING
L6	CHANTO	CHANTONNAY
IIIA	CHARCA	CHARCAS
IVA	CHARLO	CHARLOTTE
H6	CHARSO	CHARSONVILLE
H6	CHARWA	CHARWALLAS
ACANOM	CHASSI	CHASSIGNY
L6	CHATEA	CHATEAU-RENARD
HOW	CHAVES	CHAVES
IRANOM	CHEBAN	CHEBANKOL
LL5	CHEROK	CHEROKEE SPRINGS
L5	CHERVE	CHERVETTAZ
EUC	CHERVO	CHERVONY KUT
IIA	CHESTE	CHESTERVILLE
LL	CHICO+	CHICO
H	CHICOH	CHICO HILLS
IIA	CHICOM	CHICO MOUNTAINS
LL6	CHICOR	CHICORA
IRANOM	CHIHUA	CHIHUAHUA CITY
IIIA	CHILKO	CHILKOOT
IVA-AN	CHINAU	CHINAUTLA
IVB-AN	CHINGA	CHINGA
MES	CHINGU	CHINGUETTI
IIIA	CHULAF	CHULAFINEE
IIIB	CHUPAD	CHUPADEROS
L	CLARET	CLARETON
IIIF	CLARKC	CLARK COUNTY
L	CLAYTO	CLAYTONVILLE
IIIB	CLEVEL	CLEVELAND
L	CLOHAR	CLOHARS
MES	CLOVER	CLOVER SPRINGS
H3	CLOVI1	CLOVIS (NO.1)
L	CLOVI2	CLOVIS (NO.2)
IIA	COAHUI	COAHUILA
H6	COBIJA	COBIJA
L	COCKBU	COCKBURN
H	COCKLE	COCKLEBIDDY
L	COCUND	COCUNDA
H5	COLBKA	COLBY (KANSAS)
L6	COLBWI	COLBY (WISCONSIN)
PAL	COLDBA	COLD BAY
CM2	COLDBO	COLD BOKKEVELD
H5	COLDST	COLDWATER (STONE)
IB	COLFAX	COLFAX

CLASS	ABBREV	N A M E
H5	COLLES	COLLESCIPOLI
IIE	COLOME	COLOMERA
IA	COMAIR	COMANCHE (IRON)
H	COMAST	COMANCHE (STONE)
L6	CONCHO	CONCHO
IA	COOKEV	COOKEVILLE
IA	COOLAC	COOLAC
L	COOLAM	COOLAMON
CV4	COOLID	COOLIDGE
H	COOMAN	COOMANDOOK
H	COONAN	COONANA
L6	COONBU	COON BUTTE
IIIE	COOPER	COOPERTOWN
L5	COORAR	COORARA
H	COPE++	COPE
IA	COPIAP	COPIAPO
IRANOM	COROWA	COROWA
IA	CORRIZ	CORRIZATILLO
H	CORTEZ	CORTEZ
IA	COSBYS	COSBY≠S CREEK
H5	COSINA	COSINA
IIIA	COSTIL	COSTILLA PEAK
L	COTESF	COTESFIELD
H	COTTON	COTTONWOOD
H5	COVERT	COVERT
IIIA	COWELL	COWELL
IRANOM	COWRA+	COWRA
IIA	COYANO	COYA NORTE
MES	CRABOR	CRAB ORCHARD
IA	CRANBO	CRANBOURNE
H	CRANFI	CRANFILLS GAP
L6	CRANGA	CRANGANORE
IVA	CRAT31	CRATHEUS (1931)
IIC	CRAT50	CRATHEUS (1950)
L	CREDO+	CREDO
CM2	CRESCE	CRESCENT
H5	CRONST	CRONSTAD
H	CROSBY	CROSBYTON
H5	CROSSR	CROSS ROADS
L5	CRUMLI	CRUMLIN
IRANOM	CRUZDE	CRUZ DEL AIRE
IIIB	CUERNA	CUERNAVACA
H5	CUERO+	CUERO
H	CULBER	CULBERTSON
H4	CULLIS	CULLISON
AUB	CUMBEA	CUMBERLAND FALLS (A
CHANOM	CUMBEC	CUMBERLAND FALLS (C
IIIA	CUMPAS	CUMPAS
H4	CUSHIN	CUSHING
L4	CYNTHI	CYNTHIANA
L	DALEDR	DALE DRY LAKE
MES	DALGAR	DALGARANGA
L4	DALGET	DALGETY DOWNS
IIIA	DALTON	DALTON
L6	DANDAP	DANDAPUR
E6	DANIEL	DANIEL≠S KUIL
L6	DANVIL	DANVILLE
L	DAOURA	DAOURA
H	DARMST	DARMSTADT
IIIA	DAVISM	DAVIS MOUNTAINS
L	DAVY++	DAVY
IIID	DAYTON	DAYTON
L	DEAL++	DEAL
H	DEATHV	DEATH VALLEY
H	DECEWS	DE CEWSVILLE
IA	DEELFO	DEELFONTEIN
H	DEKALB	DE KALB
IRANOM	DEEPSP	DEEP SPRINGS
IIIB-AN	DELEGA	DELEGATE
L	DELHI+	DELHI
IRANOM	DELRIO	DEL RIO
L	DEMINA	DEMINA
L	DENOVA	DE NOVA
L6	DENS79	DENSMORE (1879)
IIIA	DENTON	DENTON COUNTY
L6	DENVER	DENVER
IA	DEPORT	DEPORT
IRANOM	DERMBA	DERMBACH
IIIA	DESCUB	DESCUBRIDORA
H	DESURI	DESURI
IIIA	DEXTER	DEXTER
LL6	DHURMS	DHURMSALA
L	DIEPRI	DIEP RIVER
H	DIMBOO	DIMBOOLA
IIIA	DIMITR	DIMITROVGRAD
H3,4	DIMMIT	DIMMITT
URE	DINGOP	DINGO PUP DONGA
H	DISPAT	DISPATCH
H	DISTRI	DISTRITO QUEBRACHO
L	DIX+++	DIX
H6	DJATIP	DJATI-PENGILON
H	DJERMA	DJERMAIA
H5	DOKACH	DOKACHI
L	DOLGOV	DOLGOVOLI
L	DOMANI	DOMANITCH
H	DONGAK	DONGA KOHROD
IRANOM	DOROFE	DOROFEEVKA
H6	DORONI	DORONINSK
L	DOSSO+	DOSSO
LL6	DOUARM	DOUAR MGHILA
H	DOYLEV	DOYLEVILLE
L6	DRAKEC	DRAKE CREEK
H	DRESKA	DRESDEN (KANSAS)
H	DRESON	DRESDEN (ONTARIO)
IIIA	DRUMMO	DRUM MOUNTAINS
IVA	DUCHES	DUCHESNE
IVA-AN	DUEL54	DUEL HILL (1854)
IIIA	DUKETO	DUKETON

CLASS	ABBREV	N A M E
H	DUMAS+	DUMAS
H	DUNCAN	DUNCANVILLE
H5	DUNDRU	DUNDRUM
IA	DUNGAN	DUNGANNON
L6	DURALA	DURALA
IIIA	DURANG	DURANGO
L6	DURUMA	DURUMA
H	DWALEN	DWALENI
L6	DWIGHT	DWIGHT
URE	DYALPU	DYALPUR
MES	DYARRL	DYARRL ISLAND
PAL-AN	EAGLES	EAGLE STATION
IIA	EDMOTC	EDMONTON (CANADA)
IIIC	EDMOTK	EDMONTON (KENTUCKY)
CV3	EFREMO	EFREMOVKA
H	EHOLE+	EHOLE
IA	EHRENB	EHRENBERG
H5	EICHST	EICHSTAEDT
H	EKEBY+	EKEBY
H	EKHKHE	EKH KHERA
IIA	ELBERT	ELBERTON
IID	ELBOGE	ELBOGEN
IIB	ELBURR	EL BURRO
IIIB	ELCAPI	EL CAPITAN
L	ELENOV	ELENOVKA
IIE	ELGA++	ELGA
L6	ELIELW	ELI ELWAH
H	ELKHAR	ELKHART
DIO	ELLEME	ELLEMEET
L	ELLERS	ELLERSLIE
H4	ELMCRE	ELM CREEK
H	ELPERD	EL PERDIDO
IRANOM	ELQOSE	EL QOSEIR
H	ELSINO	ELSINORA
IRANOM	ELTON+	ELTON
MES	EMERY+	EMERY
EUC	EMMAVI	EMMAVILLE
IRANOM	EMSLAN	EMSLAND
H4	ENIGMA	ENIGMA
MEANOM	ENON++	ENON
LL6	ENSISH	ENSISHEIM
H	EPINAL	EPINAL
CM2	ERAKOT	ERAKOT
L5	ERGHEO	ERGHEO
L	ERIE++	ERIE
H	EROFEE	EROFEEVKA
H6	ERXLEB	ERXLEBEN
H	ESNAND	ESNANDES
PAL	ESQUEL	ESQUEL
CM2	ESSEBI	ESSEBI
H6	ESTACA	ESTACADO
MES	ESTHER	ESTHERVILLE
H	ESU+++	ESU
H4	EUSTIS	EUSTIS

CLASS	ABBREV	N A M E
IA	FAIROA	FAIR OAKS
H5	FARLEY	FARLEY
L5	FARMIN	FARMINGTON
H4	FARMVI	FARMVILLE
L5	FARNUM	FARNUM
H	FAUCET	FAUCETT
H5	FAVARS	FAVARS
H	FAYETT	FAYETTEVILLE
H	FEIDCH	FEID CHAIR
CO3	FELIX+	FELIX
H	FENBAR	FENBARK
H	FENGHS	FENGHSIEN-KU
H	FERGSW	FERGUSON SWITCH
IIA	FILOME	FILOMENA
PAL	FINMAR	FINMARKEN
L	FINNEY	FINNEY
L6	FISHER	FISHER
H	FLEMIN	FLEMING
H3.4	FLOREN	FLORENCE
IIID	FOELLI	FOELLINGE
L	FOREBU	FORESTBURG
H5	FORECI	FOREST CITY
H4	FOREVA	FOREST VALE
L6	FORKSV	FORKSVILLE
H6	FORRES	FORREST
L	FORRLA	FORREST LAKES
H	FORSBA	FORSBACH
L6	FORSYT	FORSYTH
IIA	FORSYY	FORSYTH COUNTY
IA	FOSSIL	FOSSIL SPRINGS
IB	FOURCO	FOUR CORNERS
IIIA	FRANCE	FRANCEVILLE
IIIA	FRANFI	FRANKFORT (IRON)
HOW	FRANFS	FRANKFORT (STONE)
H	FRANKL	FRANKLIN
IIID	FREDA+	FREDA
L4	FREMON	FREMONT BUTTE
H3.4	FRENCH	FRENCHMAN BAY
L5	FUKUTO	FUKUTOMI
L6	FUTTEH	FUTTEHPUR
H	GAIL++	GAIL
H	GALAPI	GALAPIAN
LL	GALIM+	GALIM
IVB	GALLEG	GALLEGUILLOS
L6	GAMBAT	GAMBAT
H	GAOUPP	GAO (UPPER VOLTA)
IRANOM	GARDEN	GARDEN HEAD
DIO	GARLAN	GARLAND
H4	GARNET	GARNETT
L6	GARRAF	GARRAF
IRANOM	GAYGUL	GAY GULCH (KLONDIKE
H5	GEIDAM	GEIDAM
H	GEORGE	GEORGETOWN
H	GERONA	GERONA

CLASS	ABBREV	N A M E
L	GHUB54	GHUBARA (1954)
H4	GHUB62	GHUBARA (1962)
IVA	GIBEON	GIBEON
L6	GIFU++	GIFU
H5	GILGOI	GILGOIN
L6	GIRGEN	GIRGENTI
PAL	GIROUX	GIROUX
L6	GITGIT	GIT-GIT
IA	GLADIR	GLADSTONE (IRON)
H6	GLADST	GLADSTONE (STONE)
H	GLASAT	GLASATOVO
IIIA	GLASGO	GLASGOW
IRANOM	GLENOR	GLENORMISTON
PAL-AN	GLORIE	GLORIETA MOUNTAIN
H5	GNADEN	GNADENFREI
URE	GOALPA	GOALPARA
L4	GOODLA	GOODLAND
IA-AN	GOOSEL	GOOSE LAKE
H	GOPALP	GOPALPUR
L	GRAD33	GRADY (1933)
H3	GRAD37	GRADY (1937)
PAL	GRANCH	GRAN CHACO (PALLASI
IRANOM	GRANDR	GRAND RAPIDS
IIIB	GRANT+	GRANT
L	GRANTC	GRANT COUNTY
L	GRASSL	GRASSLAND
H	GREATB	GREAT BEAR LAKE
IIIA	GREENB	GREENBRIER COUNTY
IIA	GRESSK	GRESSK
L5	GRETNA	GRETNA
CV3	GROSNA	GROSNAJA
H5	GROSSD	GROSS-DIVINA
L6	GROSSL	GROSSLIEBENTHAL
H4	GRUENE	GRUENEBERG
H4	GRUVER	GRUVER
H	GUALEG	GUALEGUAYCHU
H6	GUAREN	GUARENA
H	GUETER	GUETERSLOH
IRANOM	GUFFEY	GUFFEY
LL	GUIDDE	GUIDDER
H	GUMOSC	GUMOSCHNIK
IRANOM	GUNCRE	GUN CREEK
IIIA	GUNDAR	GUNDARING
H4	GUNNAD	GUNNADORAH
L6	GURRAM	GURRAM KONDA
IIIA	HAIG++	HAIG
H	HAINAU	HAINAUT
MES	HAINHO	HAINHOLZ
L	HALEC1	HALE CENTER (NO.1)
H4	HALEC2	HALE CENTER (NO.2)
L3	HALLIN	HALLINGEBERG
LL3,4	HAMLET	HAMLET
IRANOM	HAMMON	HAMMOND
H4	HAMMOW	HAMMOND DOWNS
EUC	HARAIY	HARAIYA
L	HARDIN	HARDING COUNTY
L4	HARDWI	HARDWICK
CM2	HARIPU	HARIPURA
L6	HARLET	HARLETON
IVA	HARROF	HARRIMAN (OF)
IIIA	HARROM	HARRIMAN (OM)
L6	HARRTY	HARRISON COUNTY
L6	HARRVI	HARRISONVILLE
MEANOM	HARVAR	HARVARD UNIVERSITY
L	HASKEL	HASKELL
H	HASSAY	HASSAYAMPA
IIIC	HASSIJ	HASSI-JEKNA
H	HATCRE	HAT CREEK
IIIC	HAVANA	HAVANA
H	HAVEN+	HAVEN
URE	HAVERO	HAVEROE
H	HAVILA	HAVILAND (STONE)
H	HAWKSP	HAWK SPRINGS
L	HAYESC	HAYES CENTER
H	HEDESK	HEDESKOGA
L3	HEDJAZ	HEDJAZ
IA	HELTTO	HELT TOWNSHIP
IIIA	HENBUR	HENBURY
L5	HENDER	HENDERSONVILLE
H5	HEREDI	HEREDIA
L6	HERMIT	HERMITAGE PLAINS
H5	HESSLE	HESSLE
L6	HESSTO	HESSTON
IIA	HEXRIV	HEX RIVER MOUNTAINS
H5	HIGASH	HIGASHI-KOEN
L	HIGHPO	HIGH POSSIL
L	HILDRE	HILDRETH
IVA	HILLCI	HILL CITY
L	HINOJA	HINOJAL
H	HINOJO	HINOJO
IVB	HOBA++	HOBA
H4	HOBBS+	HOBBS
L4	HOEKMA	HOEKMARK
L6	HOLBRO	HOLBROOK
IIA	HOLLAN	HOLLAND≠S STORE
H	HOLLY+	HOLLY
LL	HOLMAN	HOLMAN ISLAND
H4	HOLYOK	HOLYOKE
L5	HOMEST	HOMESTEAD
L5	HONOLU	HONOLULU
IA	HOPE++	HOPE (BOAZ)
PAL	HOPEWE	HOPEWELL MOUNDS
IIIB	HOPPER	HOPPER
H	HORAC1	HORACE (NO.1)
L	HORAC2	HORACE (NO.2)
MEANOM	HORSEC	HORSE CREEK
IA	HOUCK+	HOUCK
H	HOWE++	HOWE

CLASS	ABBREV	N A M E
IID	HRASCH	HRASCHINA
PAL	HUCKIT	HUCKITTA
H	HUGOST	HUGO (STONE)
H5	HUGOTO	HUGOTON
IVA	HUIZOP	HUIZOPA
H6	HUNGEN	HUNGEN
E6	HVITTI	HVITTIS
DIO	IBBENB	IBBENBUEREN
EUC	IBITIR	IBITIRA
H	ICHKAL	ICHKALA
IA	IDAHO+	IDAHO
IIIA	IDER++	IDER
H	IDUTYW	I≠DUTYWA
IIIA	ILIMIR	ILIMAES (IRON)
PAL	ILIMPA	ILIMAES (PALLASITE)
IIIA	ILINSK	ILINSKAYA STANITZA
IRANOM	ILLIGU	ILLINOIS GULCH
PAL	IMILAC	IMILAC
H4	IMPERI	IMPERIAL
E4	INDARC	INDARCH
L5	INDIAN	INDIANOLA
IIA	INDIAV	INDIAN VALLEY
H6	INDIOR	INDIO RICO
H	INGALL	INGALLS
L3	IOKA++	IOKA
IVB	IQUIQU	IQUIQUE
IIB	IREDEL	IREDELL
IIIA	IRONCR	IRON CREEK
IVA	IRONRI	IRON RIVER
H	ISHING	ISHINGA
L	ISOULA	ISOULANE-N-AMAHAR
H	ISTHIL	ISTHILART
H	ITAPIC	ITAPICURU-MIRIM
PAL-AN	ITZAWI	ITZAWISIS
IIIA	IVANPA	IVANPAH
CI1	IVUNA+	IVUNA
L6	JACKAL	JACKALSFONTEIN
E6	JAJHDE	JAJH DEH KOT LALU
IVA	JAMEST	JAMESTOWN
H6	JAMKHE	JAMKHEIR
LL6	JELICA	JELICA
L6	JEMLAP	JEMLAPUR
IA	JENKIN	JENKINS
IA	JENNYS	JENNY≠S CREEK
L	JEROID	JEROME (IDAHO)
L4	JEROKA	JEROME (KANSAS)
L5	JHUNG+	JHUNG
HOW	JODZIE	JODZIE
IIIA	JOELSI	JOEL≠S IRON
IIIB	JOEWRI	JOE WRIGHT MOUNTAIN
L6	JOHNSO	JOHNSON CITY
DIO	JOHNST	JOHNSTOWN
EUC	JONZAC	JONZAC
L	JUAREZ	JUAREZ
H6	JUDESE	JUDESEGERI
IIIA	JUNCAL	JUNCAL
L	JUNCTI	JUNCTION
IIIA	JUROME	JUROMENHA
EUC	JUVINA	JUVINAS
CV2	KABA++	KABA
H6	KADONA	KADONAH
H	KAEE++	KAEE
L	KAGARL	KAGARLYK
CO3-AN	KAINSA	KAINSAZ
CHANOM	KAKANG	KAKANGARI
L6	KAKOWA	KAKOWA
H	KALABA	KALABA
H	KALDOO	KALDOONERA HILL
IIIA	KALKAS	KALKASKA
L6	KALUMB	KALUMBI
L	KAMALP	KAMALPUR
L	KAMSAG	KAMSAGAR
L	KANDAF	KANDAHAR (AFGANISTA
H5	KANGRA	KANGRA VALLEY
H5	KANS03	KANSAS CITY (1903)
HOW	KAPOET	KAPOETA
H6	KAPPAK	KAPPAKOOLA
L	KAPTAL	KAPTAL-ARYK
L	KARAGA	KARAGAI
LL	KARAKO	KARAKOL
LL6	KARATU	KARATU
IA-AN	KAREEK	KAREE KLOOF
L6	KAREWA	KAREWAR
H	KARGAP	KARGAPOLE
L	KARKH+	KARKH
L6	KARLOO	KARLOOWALA
CO4	KAROON	KAROONDA
H	KARVAL	KARVAL
H	KASAMA	KASAMATSU
L	KAUFMA	KAUFMAN
IIIA	KAYAKE	KAYAKENT
H	KEARNE	KEARNEY
IIA	KEENMO	KEEN MOUNTAIN
LL4	KELLY+	KELLY
MEANOM	KENDAL	KENDALL COUNTY
L	KENDLE	KENDLETON
H5	KENNAR	KENNARD
IIIA	KENTON	KENTON COUNTY
H	KERILI	KERILIS
L6	KERMIC	KERMICHEL
H6	KERNOU	KERNOUVE
H4	KESEN+	KESEN
L6	KEYES+	KEYES
E6	KHAIRP	KHAIRPUR
LL5	KHANPU	KHANPUR
LL	KHARKO	KHARKOV
L6	KHERAG	KHERAGUR
H	KHETRI	KHETRI

CLASS	ABBREV	N A M E
L	KHMELE	KHMELEVKA
L3	KHOHAR	KHOHAR
AUB	KHORTE	KHOR TEMIKI
L	KIEL++	KIEL
H	KIELPA	KIELPA
H6	KIKINO	KIKINO
H5	KILBOU	KILBOURNE
H	KILLET	KILLETER
H6	KIMBLE	KIMBLE COUNTY
L	KINGFI	KINGFISHER
L	KINGOO	KINGOONYA
IRANOM	KINGST	KINGSTON
EUC	KIRBYV	KIRBYVILLE
H	KISSIJ	KISSIJ
L	KISVAR	KISVARSANY
H6	KLEINW	KLEIN-WENDEN
IIIB	KNOWLE	KNOWLES
L5	KNYAHI	KNYAHINYA
IIE-AN	KODAIK	KODAIKANAL
IRANOM	KOFA++	KOFA
IVB	KOKOMO	KOKOMO
IIIE	KOKSTA	KOKSTAD
IIA	KOPJES	KOPJES VLEI
L	KORALE	KORALEIGH
E4	KOTAKO	KOTA-KOTA
IIIB	KOUGAM	KOUGA MOUNTAINS
LL5	KRAEHE	KRAEHENBERG
L	KRASNI	KRASNOI-UGOL
PAL	KRASNJ	KRASNOJARSK
L3	KRYMKA	KRYMKA
L	KUKSCH	KUKSCHIN
L5	KULAK+	KULAK
L	KULESC	KULESCHOVKA
L	KULNIN	KULNINE
H	KULP++	KULP
IIC	KUMERI	KUMERINA
L6	KUNASH	KUNASHAK
L	KUSIAL	KUSIALI
L	KUTTIP	KUTTIPPURAM
L	KUZNET	KUZNETZOVO
IIIA	KYANCU	KYANCUTTA
L	KYBUNG	KYBUNGA
L6	KYLE++	KYLE
L6	KYUSHU	KYUSHU
L6	LABECA	LA BECASSE
H5	LABORE	LABOREL
IRANOM	LACAIL	LA CAILLE
H	LACOLI	LA COLINA
L6	LADDER	LADDER CREEK
ACANOM	LAFAST	LA FAYETTE (STONE)
IVA	LAGRAN	LA GRANGE
L6	LAIGLE	L≠AIGLE
EUC	LAKANG	LAKANGAON
L	LAKEBO	LAKE BONNEY
L6	LAKEBR	LAKE BROWN
L	LAKEGR	LAKE GRACE
LL6	LAKELA	LAKE LABYRINTH
L	LAKEMO	LAKE MOORE (LAKE B
IIB	LAKEMU	LAKE MURRAY
L	LAKETO	LAKETON
L	LAKEWO	LAKEWOOD
L5	LALAND	LA LANDE
L	LALITP	LALITPUR
CO3	LANCE+	LANCE
H6	LANCON	LANCON
IA	LANDES	LANDES
L6	LANGHA	LANGHALSEN
IIIA	LANTON	LANTON
L	LANZEN	LANZENKIRCHEN
IIIA	LAPORT	LA PORTE
IRANOM	LAPRIM	LA PRIMITIVA
IA	LASVEG	LAS VEGAS
H4	LAUNDE	LAUNDRY EAST
H5	LAUNDR	LAUNDRY ROCKHOLE
L5	LAUNDW	LAUNDRY WEST
L6	LAUNTO	LAUNTON
IRANOM	LAUREN	LAURENS COUNTY
H	LAVILL	LA VILLA
L	LAVREN	LAVRENTIEVKA
L6	LAWREN	LAWRENCE
L6	LEEDEY	LEEDEY
IA	LEEDS+	LEEDS
L6	LEEUWF	LEEUWFONTEIN
H5	LEIGHT	LEIGHTON
IIIA	LENART	LENARTO
H	LEON++	LEON
L	LEONOV	LEONOVKA
CV3	LEOVIL	LEOVILLE
L6	LEPRES	LE PRESSOIR
L6	LESORM	LES ORMES
L	LESVES	LESVES
HOW	LETEIL	LE TEILLEUL
IA	LEXING	LEXINGTON COUNTY
IIIA	LEXIPS	LEXINGTON COUNTY,PS
H	LILLAV	LILLAVERKE
IRANOM	LIMECR	LIME CREEK
H	LIMERI	LIMERICK
L	LINCOL	LINCOLN COUNTY
L	LINUM+	LINUM
IA	LINWOO	LINWOOD
PAL	LIPOVS	LIPOVSKY
L6	LISSA+	LISSA
L5	LITTPI	LITTLE PINEY
H	LITTRI	LITTLE RIVER
IIIA	LIVIMO	LIVINGSTON (MONTANA
IRANOM	LIVITE	LIVINGSTON (TENNESS
H4	LIXNA+	LIXNA
L	LOCKNE	LOCKNEY

CLASS	ABBREV	N A M E
IIA	LOCUST	LOCUST GROVE
MEANOM	LODRAN	LODRAN
H	LOGAN+	LOGAN
IIA	LOMBAR	LOMBARD
IIE	LONACO	LONACONING
L6	LONGIS	LONG ISLAND
L	LOOMIS	LOOMIS
IRANOM	LOONGA	LOONGANA STATION
L	LOOP++	LOOP
IIIA	LORETO	LORETO
IIIB	LOSREY	LOS REYES
H5	LOSTCI	LOST CITY
L	LOSTLA	LOST LAKE
MES	LOWICZ	LOWICZ
L	LOYOLA	LOYOLA
L5	LUA+++	LUA
L	LUBBOC	LUBBOCK
L	LUCE++	LUCE
IIIB	LUISLO	LUIS LOPEZ
H6	LUMPKI	LUMPKIN
L6	LUNDSG	LUNDSGARD
HOW	LUOTOL	LUOTOLAX
H	LUPONN	LUPONNAS
L6	LUTSCH	LUTSCHAUNIG#S STONE
L	MABWEK	MABWE-KHOYWA
H5	MACAU+	MACAU
EUC	MACIBI	MACIBINI
IIIA	MADOC+	MADOC
L	MADRID	MADRID
HOW	MAESSI	MAESSING
L	MAFRA+	MAFRA
IIIC-AN	MAGNES	MAGNESIA
IA	MAGURA	MAGURA
L5	MAINZ+	MAINZ
L6	MAKARE	MAKAREWA
L	MALAKA	MALAKAL
H	MALAMP	MALAMPAKA
H	MALOTA	MALOTAS
HOW	MALVER	MALVERN
L	MAMRAS	MAMRA SPRINGS
LL6	MANBHO	MANBHOOM
DIO	MANEGA	MANEGAON
LL6	MANGWE	MANGWENDI
IVA	MANTOS	MANTOS BLANCOS
L3	MANYCH	MANYCH
IIIA	MAPLET	MAPLETON
PAL	MARBUR	MARBURG
H5	MARDAN	MARDAN
IVA	MARIAE	MARIA ELENA (1935)
H6	MARIDI	MARIDI
H4	MARILI	MARILIA
L6	MARIOI	MARION (IOWA)
L	MARIOK	MARION (KANSAS)
PAL	MARJAL	MARJALAHTI
L	MARLOW	MARLOW
L5	MARMAN	MARMANDE
IIIA	MARSHA	MARSHALL COUNTY
H	MARSLA	MARSLAND
IVA	MART++	MART
L	MASCOM	MASCOMBES
H	MASSEN	MASSENYA
IIIE	MATATI	MATATIELE
L6	MAUERK	MAUERKIRCHEN
L	MAURIT	MAURITIUS
H	MAYDAY	MAYDAY
IA	MAYERT	MAYERTHORPE
IIA	MAYODA	MAYODAN
IA	MAZAPI	MAZAPIL
L	MAZIBA	MAZIBA
IRANOM	MBOSI+	MBOSI
L	MCADDO	MCADDO
IA	MCCAME	MCCAME
IVA	MCDOWE	MCDOWELL COUNTY
L4	MCKINN	MCKINNEY
H	MCLEAN	MCLEAN
HOW	MEDANI	MEDANITOS
LL6	MELLEN	MELLENBYE
L5	MELROS	MELROSE
H4	MENOW+	MENOW
IIIA	MERCED	MERCEDITAS
L6	MERN++	MERN
IA-AN	MERTZO	MERTZON
LL	MERU++	MERU
H	MERUA+	MERUA
IB	MESAVE	MESA VERDE PARK
L	MESSIN	MESSINA
H4	METSAE	METSAEKYLAE
L	MEUSEL	MEUSELBACH
L	MEZEL+	MEZEL
L3	MEZOEM	MEZOE-MADARAS
L6	MHOW++	MHOW
H	MIAMI+	MIAMI
IA	MICHIG	MICHIGAN IRON
L6	MIDDLE	MIDDLESBROUGH
CM2	MIGHEI	MIGHEI
L	MIKE++	MIKE
L6	MILENA	MILENA
H	MILLKA	MILLER (KANSAS)
IIIA	MILLYM	MILLY MILLY
L	MINASG	MINAS GERAIS
MES	MINCY+	MINCY
L5	MIRZAP	MIRZAPUR
H4,5	MISSHO	MISSHOF
L	MISSIO	MISSION
IA	MISTEC	MISTECA
H	MJELLE	MJELLEIM
IA	MOAB++	MOAB
L6	MOCS++	MOCS

CLASS	ABBREV	N A M E
IA	MOCTEZ	MOCTEZUMA
L6	MODO05	MODOC (1905)
H	MOD048	MODOC (1948)
CV2	MOKOIA	MOKOIA
L5	MOLINA	MOLINA
PAL	MOLONG	MOLONG
HOW	MOLTEN	MOLTENO
IRANOM	MONAHA	MONAHANS
H4	MONROE	MONROE
L	MONTEC	MONTE COLINA
L	MONTED	MONTE DAS FORTES
L	MONTEM	MONTE MILONE
L	MONTLI	MONTLIVAULT
IA	MONUME	MONUMENT ROCK
L6	MONZE+	MONZE
IIIF	MOONBI	MOONBI
IA	MOORAN	MOORANOPPIN
EUC	MOOREC	MOORE COUNTY
H5	MOORES	MOORESFORT
L	MOORLE	MOORLEAH
IIIA	MOORUM	MOORUMBUNNA
L6	MORADA	MORADABAD
IA	MORDEN	MORDEN
IIIA	MORITO	MORITO
H6	MORLAN	MORLAND
H5	MORNAN	MORNANS
IRANOM	MORRAD	MORRADAL
IA-AN	MORRIL	MORRILL
MES	MORRIS	MORRISTOWN
H	MORVEN	MORVEN
L6	MOSCA+	MOSCA
H	MOSHES	MOSHESH LOCATION
H4	MOSQUE	MOSQUERO
H6	MOTIKA	MOTI-KA-NAGLA
L	MOTPEN	MOTPENA
H	MOTTAD	MOTTA DI CONTI
IA	MOUNAY	MOUNT AYLIFF
H6	MOUNBR	MOUNT BROWNE
IRANOM	MOUNDO	MOUNT DOOLING
PAL	MOUNDY	MOUNT DYRRING
IIIB	MOUNED	MOUNT EDITH
MEANOM	MOUNEG	MOUNT EGERTON
IIB	MOUNJO	MOUNT JOY
IRANOM	MOUNMA	MOUNT MAGNET
H	MOUNMN	MOUNT MORRIS (NEW Y
CHANOM	MOUNMW	MOUNT MORRIS (WISCO
IID	MOUNOU	MOUNT OURAY
MES	MOUNPA	MOUNT PADBURY
IA	MOUNSI	MOUNT SIR CHARLES
IA	MOUNST	MOUNT STIRLING
IVA	MOUNTA	MOUNT TABBY
PAL	MOUNVE	MOUNT VERNON
L	MUDDOO	MUDDOOR
L6	MUIZEN	MUIZENBERG

CLASS	ABBREV	N A M E
H	MULGNO	MULGA (NORTH)
H	MULGSO	MULGA (SOUTH)
L	MULLET	MULLETIWU
IRANOM	MUNDRA	MUNDRABILLA
IIIC	MUNGIN	MUNGINDI
IVA	MUONIO	MUONIONALUSTA
L	MURAID	MURAID
CM2	MURCHI	MURCHISON
IRANOM	MURFRE	MURFREESBORO
IRANOM	MURNPE	MURNPEOWIE
L	MUROC+	MUROC
L	MUROCD	MUROC DRY LAKE
IIA	MURPHY	MURPHY
CM2	MURRAY	MURRAY
H	NADIAB	NADIABONDI
LL6	NAES++	NAES
EUC	NAGARI	NAGARIA
L	NAGYBO	NAGY-BOROVE
IA	NAGYVA	NAGY-VAZSONY
ACANOM	NAKHLA	NAKHLA
L	NAKHON	NAKHON PATHOM
H	NALLAH	NALLAH
H5	NAMMIA	NAMMIANTHAL
H6	NANJEM	NANJEMOY
L	NANYAN	NAN YANG PAO
H	NAOKI+	NAOKI
H	NARD01	NARDOO NO.1
L	NARD02	NARDOO NO.2
L	NARELL	NARELLAN
L	NARETH	NARETHA
IIIB	NARRAB	NARRABURRA
H	NARUNA	NARUNA
L6	NASHST	NASHVILLE (STONE)
H	NASSIR	NASSIRAH
IIB	NAVAJO	NAVAJO
CM2	NAWAPA	NAWAPALI
IIIA	NAZARI	NAZARETH (IRON)
H	NAZARS	NAZARETH (STONE)
IRANOM	NEDAGO	NEDAGOLLA
IID	NEEDLE	NEEDLES
L6	NEENAC	NEENACH
IIA	NEGRIL	NEGRILLOS
IIIA	NEJED+	NEJED
L6	NEJO++	NEJO
IIIF	NELSON	NELSON COUNTY
IA	NEPTUN	NEPTUNE MOUNTAINS
L6	NERFT+	NERFT
H	NESS38	NESS COUNTY (1938)
L6	NESS94	NESS COUNTY (1894)
IIE-AN	NETSCH	NETSCHAEVO
L	NEWALM	NEW ALMELO
IRANOM	NEWBAL	NEW BALTIMORE
L6	NEWCON	NEW CONCORD
IA	NEWLEI	NEW LEIPZIG

CLASS	ABBREV	N A M E
PAL	NEWPOR	NEWPORT
L	NEWSOM	NEWSOM
IVA	NEWWES	NEW WESTVILLE
LL3	NGAWI+	NGAWI
IRANOM	NGOURE	N#GOUREYMA
IVA	NICO++	NICO
IA	NIEDFI	NIEDER FINOW
H	NIKOLA	NIKOLAEVKA
L	NIKOLS	NIKOLSKOE
H	NIO+++	NIO
IID	NKANDH	N#KANDHLA
EUC	NOBLEB	NOBLEBOROUGH
IRANOM	NOCOLE	NOCOLECHE
CM2	NOGOYA	NOGOYA
L	NORACR	NORA CREINA
L	NORCAT	NORCATEUR
IRANOM	NORDHE	NORDHEIM
IIIA	NORFOL	NORFOLK
IIIA	NORFOR	NORFORK
IIIB	NORRIS	NORRISTOWN
L4	NORTHE	NORTH EAST REID
L4	NORTHF	NORTH FORREST
URE	NORTHH	NORTH HAIG
IIB	NORTHP	NORTH PORTUGAL
LL5	NORTHR	NORTH REID
AUB	NORTON	NORTON COUNTY
IVA	NOVORY	NOVORYBINSKOE
URE	NOVOUR	NOVO-UREI
L	NOYANB	NOYAN-BOGDO
EUC	NUEVOL	NUEVO LAREDO
IIIA	NULERI	NULERI
H	NULLES	NULLES
IIIA	NUTWOO	NUTWOOD DOWNS
LL	NYIRAB	NYIRABRANY
L5	OAK+++	OAK
IIIF	OAKLIR	OAKLEY (IRON)
H6	OAKLST	OAKLEY (STONE)
LL5	OBERLI	OBERLIN
IVA	OBERNK	OBERNKIRCHEN
H4	OCHANS	OCHANSK
H	OCZERE	OCZERETNA
IA	ODESIR	ODESSA (IRON)
H	OESEDE	OESEDE
L6	OESEL+	OESEL
L	OFEHER	OFEHERTO
IA	OGALLA	OGALLALA
H6	OGI+++	OGI
H5	OHABA+	OHABA
L	OHUMA+	OHUMA
L	OJUELO	OJUELOS ALTOS
H	OKABE+	OKABE
IIA	OKAHAN	OKAHANDJA
IIA	OKANO+	OKANO
L4	OKECHO	OKECHOBEE

CLASS	ABBREV	N A M E
H	OKIRAI	OKIRAI
LL	OKNINY	OKNINY
L	OLDE30	OLDENBURG (1930)
LL5	OLIVEN	OLIVENZA
PAL	OLLAGU	OLLAGUE
H	OLMEDI	OLMEDILLA DE ALARCO
IIIB	ORANIR	ORANGE RIVER (IRON)
CI1	ORGUEI	ORGUEIL
H	ORLOVK	ORLOVKA
CO3	ORNANS	ORNANS
H5	OROGRA	ORO GRANDE
IIIB	OROVIL	OROVILLE
L6	ORVINI	ORVINIO
IA	OSCURO	OSCURO MOUNTAINS
H	OSHKOS	OSHKOSH
IA	OSSEO+	OSSEO
IVA	OTCHIN	OTCHINJAU
L6	OTEROY	OTEROY
L6	OTIS++	OTIS
H	OTOMI+	OTOMI
LL6	OTTAWA	OTTAWA
H	OUALLE	OUALLEN
LL	OUBARI	OUBARI
H	OVID++	OVID
L	OVIEDO	OVIEDO
IIIB	OWENSV	OWENS VALLEY
H	OZONA+	OZONA
L6	PACULA	PACULA
HOW	PADVAR	PADVARNINKAI
H	PALOLO	PALOLO VALLEY
L	PAMPAA	PAMPA DE AGUA BLANC
L	PAMPAI	PAMPA DEL INFIERNO
L5	PAMPAN	PAMPANGA
IA	PANDEA	PAN DE AZUCAR
L	PANNIK	PANNIKIN
H5	PANTAR	PANTAR
IVA	PARADE	PARA DE MINAS
LL	PARAGO	PARAGOULD
L6	PARANA	PARANAIBA
LL3	PARNAL	PARNALLEE
IIIA	PARRAL	PARRAL
EUC	PASAMO	PASAMONTE
L	PATRIM	PATRIMONIO
MES	PATWAR	PATWAR
PAL	PAVLDP	PAVLODAR (PALLASITE
H	PAVLDS	PAVLODAR (STONE)
L6	PAVLOG	PAVLOGRAD
HOW	PAVLOV	PAVLOVKA
L6	PEACER	PEACE RIVER
ACANOM	PECKEL	PECKELSHEIM
L5	PECKSS	PECK#S SPRING
L6	PEETZ+	PEETZ
AUB	PENABL	PENA BLANCA SPRINGS
H	PENOKE	PENOKEE

CLASS	ABBREV	N A M E	CLASS	ABBREV	N A M E
EUC	PERAMI	PERAMIHO	H	PULSOR	PULSORA
L6	PERPET	PERPETI	H5	PULTUS	PULTUSK
IIC	PERRYV	PERRYVILLE	IID	PUQUIO	PUQUIOS
IB	PERSIM	PERSIMMON CREEK	IIA	PURIPI	PURIPICA
LL5	PERTH+	PERTH	L6	PUTING	PUTINGA
L	PERVOM	PERVOMAISKY	IVA	PUTNAM	PUTNAM COUNTY
AUB	PESYAN	PESYANOE	IA	QUAIRA	QUAIRADING
EUC	PETERS	PETERSBURG	IA	QUEELA	QUEENSLAND
H	PETRKA	PETROPAVLOVKA	H6	QUEEME	QUEEN*S MERCY
LL	PEVENS	PEVENSEY	H4	QUENGG	QUENGGOUK
L	PHIL01	PHILLIPS COUNTY (19	IIA	QUILLA	QUILLAGUA
PAL	PHIL35	PHILLIPS COUNTY (19	L	QUINCA	QUINCAY
H	PHUHON	PHU HONG	IIIA	QUINNC	QUINN CANYON
L3	PIANCA	PIANCALDOLI	H	RACO++	RACO
IIIA	PICACH	PICACHO	IRANOM	RAFRUE	RAFRUETI
H	PICKEN	PICKENS COUNTY	IVA	RAILWA	RAILWAY
IRANOM	PIEDAD	PIEDADE DO BAGRE	L6	RAKOVK	RAKOVKA
L	PIERST	PIERCEVILLE (STONE)	L	RAMNAG	RAMNAGAR
E6	PILLIS	PILLISTFER	LL	RAMPUR	RAMPURHAT
IIA	PIMACO	PIMA COUNTY	L	RAMSDO	RAMSDORF
IA	PINERI	PINE RIVER	H	RANCHA	RANCHAPUR
MES	PINNAR	PINNAROO	IIIA	RANCPI	RANCHO DE LA PILA
IRANOM	PINON+	PINON	H	RANCPR	RANCHO DE LA PRESA
L	PINTOM	PINTO MOUNTAINS	L	RANGAL	RANGALA
H6	PIPECR	PIPE CREEK	H4	RANSOM	RANSOM
H	PIQUET	PIQUETBERG	H	RASTAN	RAS TANURA
L6	PIRGUN	PIRGUNJE	IIIA	RATELD	RATELDRAAI
IB	PITTS+	PITTS	PAL	RAWLPA	RAWLINNA (PALLASITE
H	PLAINS	PLAINS	H	RAWLST	RAWLINNA (STONE)
H5	PLAN17	PLAINVIEW (1917)	L	REAGER	REAGER
H	PLAN50	PLAINVIEW (1950)	IRANOM	REDFIE	REDFIELDS
H6	PLANTE	PLANTERSVILLE	IIIA	REDRIV	RED RIVER
H5	PLEASA	PLEASANTON	IRANOM	REEDCI	REED CITY
L	PLOSCH	PLOSCHKOVITZ	L3	REID++	REID
IIIA	PLYMOU	PLYMOUTH	L	RELIEG	RELIEGOS
L	PNOMPE	PNOMPEHN	IVA?	REMBAN	REMBANG
L5	POHLIT	POHLITZ	CV2-AN	RENAZZ	RENAZZO
IIIA	POINIR	POINT OF ROCKS (IRO	L	RENCA+	RENCA
L	POINST	POINT OF ROCKS (STO	IRANOM	REPEEV	REPEEV KHUTOR
PAL	POJOAQ	POJOAQUE	CI1	REVELS	REVELSTOKE
H5	POKHRA	POKHRA	IIIE	RHINEV	RHINE VILLA
CM2	POLLEN	POLLEN	H5	RICHAR	RICHARDTON
EUC	POMOZD	POMOZDINO	IIA	RICHLA	RICHLAND
IIB	PONCAC	PONCA CREEK	L5	RICHMO	RICHMOND
PAL	PORTOR	PORT ORFORD	L6	RICHMU	RICH MOUNTAIN
L6	POTTER	POTTER	IA	RIFLE+	RIFLE
H3	PRAIRI	PRAIRIE DOG CREEK	IIA	RIOLOA	RIO LOA
IRANOM	PREMIE	PREMIER DOWNS	L3,4	RIONEG	RIO NEGRO
H5	PRIBRA	PRIBRAM	L	RIVER+	RIVER
L6	PRICET	PRICETOWN	H	ROCHES	ROCHESTER
IIIA	PROVID	PROVIDENCE	DIO	RODA++	RODA
IIIA	PUENDE	PUENTE DEL ZACATE	IID	RODEO+	RODEO
L	PUENLA	PUENTE-LADRON	IIIA	ROEBOU	ROEBOURNE
IA	PULASK	PULASKI COUNTY	H	ROLLA+	ROLLA

CLASS	ABBREV	N A M E	CLASS	ABBREV	N A M E
H	ROMERO	ROMERO	H	SANLUI	SAN LUIS
IIIB	ROPERR	ROPER RIVER	IIA	SANMAR	SAN MARTIN
L	ROSAMO	ROSAMOND DRY LAKE	L	SANPED	SAN PEDRO SPRINGS
IA	ROSARI	ROSARIO	IIIA	SANTAP	SANTA APOLONIA
H	ROSEBU	ROSEBUD	L	SANTBA	SANTA BARBARA
H	ROSECI	ROSE CITY	IRANOM	SANTCA	SANTA CATHARINA
H	ROWENA	ROWENA	CM2	SANTCR	SANTA CRUZ
IIIA	ROWTON	ROWTON	PAL	SANTFE	SANTA FE
L5	ROY+33	ROY (1933)	L6	SANTIS	SANTA ISABEL
L6	ROY+34	ROY (1934)	IIB	SANTLU	SANTA LUZIA
IIIA	RUFFSM	RUFF#S MOUNTAIN	IRANOM	SANTRO	SANTA ROSA
L4	RUPOTA	RUPOTA	PAL	SANTRS	SANTA ROSALIA
H	RUSHCO	RUSH COUNTY	IRANOM	SANTSQ	SANTIAGO PAPASQUIER
L6	RUSHCR	RUSH CREEK	H	SAOJOS	SAO JOSE DO RIO PRE
L5	RUSHVI	RUSHVILLE	IIB	SAOJUL	SAO JULIAO DE MOREI
IIIA	RUSSEL	RUSSEL GULCH	L4	SARATO	SARATOV
L	RYECHK	RYECHKI	IA	SARDIS	SARDIS
IIIA	SACRAM	SACRAMENTO MOUNTAIN	IA	SAREPT	SAREPTA
H	SAIANN	ST. ANN	H	SASAGA	SASAGASE
L	SAICAP	ST. CAPRAIS-DE-QUIN	L6	SAUGUI	SAUGUIS
L	SAICHR	ST. CHRISTOPHE-LA-C	IIIA	SAVANN	SAVANNAH
L	SAIDEN	ST. DENIS WESTREM	LL	SAVTSC	SAVTSCHENSKOJE
IRANOM	SAIFRA	ST. FRANCOIS COUNTY	H	SCHAAP	SCHAAP-KOOI
IIIF	SAIGEN	ST. GENEVIEVE COUNT	L	SCHELL	SCHELLIN
H	SAIGER	ST. GERMAIN-DU-PINE	H4	SCHENE	SCHENECTADY
LL	SAILAW	ST. LAWRENCE	L6	SCHOEN	SCHOENENBERG
H3,4	SAILOU	ST. LOUIS	IIIA	SCHWET	SCHWETZ
H	SAIMAG	ST. MARGUERITE	H5	SCOTTC	SCOTT CITY
E5	SAIMAK	ST. MARK#S	IIA	SCOTTS	SCOTTSVILLE
LL3	SAIMAR	ST. MARY#S COUNTY	H5	SCURRY	SCURRY
LL6	SAIMES	ST. MESMIN	H4	SEAGRA	SEAGRAVES
L6	SAIMIC	ST. MICHEL	H5	SEARSM	SEARSMONT
L5	SAIPET	ST. PETER	IA	SEELAE	SEELAESGEN
E5	SAISAU	SAINT-SAUVEUR	L6	SEGOWL	SEGOWLIE
LL6	SAISEV	ST. SEVERIN	H	SEGUIN	SEGUIN
H5	SALINE	SALINE	H	SEIBER	SEIBERT
L	SALLA+	SALLA	H	SELDEB	SELDEBOURAK
H6	SALLES	SALLES	IA	SELIGM	SELIGMAN
PAL	SALTA+	SALTA	H4	SELMA+	SELMA
H5	SALTLA	SALT LAKE CITY	LL3	SEMARK	SEMARKONA
IIC	SALTRI	SALT RIVER	H4	SEMINO	SEMINOLE
IIIB	SAMSVA	SAMS VALLEY	H4	SENA++	SENA
IIIA	SANANG	SAN ANGELO	H	SENECA	SENECA
H	SANCAR	SAN CARLOS	IIIA	SENEFA	SENECA FALLS
IIA	SANCHE	SANCHEZ ESTATE	IVA	SENETO	SENECA TOWNSHIP
IB	SANCRI	SAN CRISTOBAL	H6	SEONI+	SEONI
IIIB	SANDER	SANDERSON	H3,4	SERES+	SERES
IIB	SANDIA	SANDIA MOUNTAINS	EUC	SERRAD	SERRA DE MAGE
IIIA	SANDTO	SANDTOWN	IVA	SERRAN	SERRANIA DE VARAS
H4	SANEMI	SAN EMIGDIO	H	SETELA	SETE LAGOAS
IIA	SANFDE	SAN FRANCISCO DEL M	LL	SEVILL	SEVILLA
IVA	SANFMO	SAN FRANCISCO MOUNT	L	SEVRUK	SEVRUKOVO
H	SANJOS	SAN JOSE	IA	SEYMOU	SEYMOUR
H6	SANJUA	SAN JUAN CAPISTRANO	H	SHAFTE	SHAFTER LAKE

CLASS	ABBREV	N A M E
DIO	SHALKA	SHALKA
AUB	SHALLO	SHALLOWATER
H3	SHARPS	SHARPS
L6-AN	SHAW++	SHAW
L5	SHELBU	SHELBURNE
ACANOM	SHERGO	SHERGOTTY
L	SHIKAR	SHIKARPUR
IRANOM	SHINGL	SHINGLE SPRINGS
IVA	SHIRAH	SHIRAHAGI
IA	SHREWS	SHREWSBURY
H	SHUPIY	SHUPIYAN
L6	SHYTAL	SHYTAL
LL	SIENA+	SIENA
H	SIERCO	SIERRA COUNTY
IIA	SIERGO	SIERRA GORDA
IIIA	SIERSA	SIERRA SANDON
IVA	SIGNAL	SIGNAL MOUNTAIN
IIB-AN	SIKHOT	SIKHOTE-ALIN
IIB	SILVBE	SILVER BELL
IA	SILVCR	SILVER CROWN
L6	SILVNE	SILVERTON (NEW SOUT
H	SILVTE	SILVERTON (TEXAS)
H	SIMMER	SIMMERN
MES	SIMOND	SIMONDIUM
L6	SINAI+	SINAI
H5	SINDHR	SINDHRI
H	SINNAI	SINNAI
EUC	SIOUXC	SIOUX COUNTY
H5	SITATH	SITATHALI
L6	SKI+++	SKI
IVB	SKOOKU	SKOOKUM (KLONDIKE)
H4	SLAVET	SLAVETIC
L	SLEEPE	SLEEPER CAMP
L4	SLOBOD	SLOBODKA
L	SMITHC	SMITH CENTER
IVA	SMITHL	SMITHLAND
IIA	SMITHO	SMITHONIA
IIIB	SMITHS	SMITHS MOUNTAIN
IIB	SMITHT	SMITHSONIAN IRON
IA	SMITHV	SMITHVILLE
IVA	SOCIAL	SOCIAL CIRCLE
LL4	SOKOBA	SOKO-BANJA
PAL	SOMERV	SOMERVELL COUNTY
H	SONE++	SONE
IRANOM	SOPER+	SOPER
MEANOM	SOROTI	SOROTI
PAL	SOUTBE	SOUTH BEND
IRANOM	SOUTBY	SOUTH BYRON
IA	SOUTER	SOUTHERN ARIZONA
E4	SOUTOM	SOUTH OMAN
IIIA	SPEARM	SPEARMAN
H	SPRINE	SPRINGER
L	SPRINF	SPRINGFIELD
PAL	SPRINW	SPRINGWATER

CLASS	ABBREV	N A M E
IIIA	SSYROM	SSYROMOLOTOVO
H5	STAELL	STAELLDALEN
EUC	STANNE	STANNERN
IIIE	STAUNT	STAUNTON
L6	STAVRO	STAVROPOL
IVA-AN	STEINB	STEINBACH
H	STONIN	STONINGTON
L	STRATH	STRATHMORE
L	SUBLET	SUBLETTE
L	SUCCES	SUCCESS
L6	SULTAN	SULTANPUR
IIB	SUMMIT	SUMMIT
H	SUNGAC	SUNGACH
H6	SUPUHE	SUPUHEE
IA	SURPRI	SURPRISE SPRINGS
IIIA	SUSUMA	SUSUMAN
H	SUTTON	SUTTON
L	SUWAAD	SUWAHIB (ADRAJ)
H	SUWAAI	SUWAHIB (AIN SALA)
H3	SUWABU	SUWAHIB (BUWAH)
H	SWEETW	SWEETWATER
H	SYLACA	SYLACAUGA
H5	TABOR+	TABOR
IA	TACUBA	TACUBAYA
L5	TADJER	TADJERA
H	TAFOYA	TAFOYA
L5	TAIBAN	TAIBAN
H	TAIGA+	TAIGA
H	TAKENO	TAKENOUCHI
IIIA	TAMARU	TAMARUGAL
IIIB	TAMBOQ	TAMBO QUEMADO
IIIA	TAMENT	TAMENTIT
IIIE	TANAKA	TANAKAMI MOUNTAIN
L5	TANE++	TANE
IRANOM	TARAPH	TARAPACA (H)
IIIA	TARAPO	TARAPACA (OM)
L	TARBAG	TARBAGATAI
L6	TARFA+	TARFA
DIO	TATAHO	TATAHOUINE
L	TATHLI	TATHLITH
H	TATUM+	TATUM
L6	TAUQ++	TAUQ
IVB	TAWALL	TAWALLAH VALLEY
IIID	TAZEWE	TAZEWELL
H	TELL++	TELL
IIIB	TEMORA	TEMORA
L	TEMPLE	TEMPLE
L6	TENHAM	TENHAM
L4	TENNAS	TENNASILM
IVB	TERNER	TERNERA
H	TEXLIN	TEXLINE
H6	THAL++	THAL
PAL	THIELM	THIEL MOUNTAINS
L6	THOMSO	THOMSON

CLASS	ABBREV	N A M E
IA	THOREA	THOREAU
IIIA	THULE+	THULE
IIIA	THUNDA	THUNDA
IIIB	THURLO	THURLOW
L	TIBERR	TIBERRHAMINE
IIIB	TIERAC	TIERACO CREEK
H3	TIESCH	TIESCHITZ
L6	TILDEN	TILDEN
L5	TIMMER	TIMMERSOI
H5	TIMOCH	TIMOCHIN
H	TIRUPA	TIRUPATI
H	TJABE+	TJABE
L5	TJEREB	TJEREBON
IVB	TLACOT	TLACOTEPEC
H4	TOBE++	TOBE
IIE	TOBYCH	TOBYCHAN
IIC	TOCAVI	TOCAVITA
IIA	TOCOPI	TOCOPILLA
IA	TOLUCA	TOLUCA
IIIE	TOLUPS	TOLUCA, PSEUDO-
LL	TOMAKO	TOMAKOVKA
H6	TOMATL	TOMATLAN
IRANOM	TOMBIG	TOMBIGBEE RIVER
H5	TOMHAN	TOMHANNOCK CREEK
L	TOMITA	TOMITA
CI1	TONK++	TONK
IIIA	TONOPA	TONOPAH
H	TORRIN	TORRINGTON
H	TOSTAD	TOSTADO
IIIA	TOUBIL	TOUBIL RIVER
H6	TOULOU	TOULOUSE
L6	TOURIN	TOURINNES-LA-GROSSE
H5	TRAVIS	TRAVIS COUNTY
IIIA	TRENTO	TRENTON
H6	TRENZA	TRENZANO
IIIB-AN	TREYSA	TREYSA
L	TRIFIR	TRIFIR
H	TROMOY	TROMOY
L6	TROUP+	TROUP
L	TRYON+	TRYON
L	TRYSIL	TRYSIL
L	TUANTU	TUAN TUC
MEANOM	TUCSON	TUCSON
H5	TULIA+	TULIA
IIIB	TURTLE	TURTLE RIVER
L6	TUZLA+	TUZLA
L	TWENTY	TWENTYNINE PALMS
IRANOM	TWINCI	TWIN CITY
H	TWOBUT	TWO BUTTES
H4	TYSNES	TYSNES ISLAND
H5	UBERAB	UBERABA
H5	UCERA+	UCERA
IA	UDEIST	UDEI STATION
LL6	UDEN++	UDEN
H5	UDIPI+	UDIPI
IIIA	UEGIT+	UEGIT
E6	UFANA+	UFANA
H	ULYSSE	ULYSSES
LL5	UMBALA	UMBALA
L6	UMMTIN	UMM TINA
IIA	UNION+	UNION
IRANOM	UNIONC	UNION COUNTY
IIC	UNTERM	UNTER-MAESSING
H	UTECRE	UTE CREEK
L6	UTRECH	UTRECHT
H	UTZENS	UTZENSTORF
H	UVALDE	UVALDE
IIA	UWET++	UWET
IIIA	UWHARR	UWHARRIE
IA	VAALBU	VAALBULT
MES	VACAMU	VACA MUERTA
H	VAGO++	VAGO
H	VALDAV	VALDAVUR
L6	VALDIN	VALDINIZZA
L6	VALKEA	VALKEALA
L	VALLEY	VALLEY WELLS
L6	VARPAI	VARPAISJAERVI
LL6	VAVILO	VAVILOVKA
IIIA	VELINI	VELIKO-NIKOLAEVSKY
H	VENGER	VENGEROVO
IRANOM	VENTUR	VENTURA
L4	VERA++	VERA
MES	VERAMI	VERAMIN
IIE	VERKDN	VERKHNE DNIEPROVSK
IIIB	VERKPS	V. DNIEPROVSK, PSEU
H	VERKTS	VERKHNE TSCHIRSKAIA
IIIA	VERKUD	VERKHNE UDINSK
H	VERNON	VERNON COUNTY
IRANOM	VICTOR	VICTORIA WEST
IIIA	VIEWHI	VIEW HILL
CV3	VIGARA	VIGARANO
H	VILLED	VILLEDIEU
L	VINCEN	VINCENT
L	VIRBA+	VIRBA
LL	VISHNU	VISHNUPUR
H	VISUNI	VISUNI
L	VOUILL	VOUILLE
H	VULCAN	VULCAN
IIIA	WABAR+	WABAR
L6	WACOND	WACONDA
H	WAIRAR	WAIRARAPA VALLEY
L	WALDO+	WALDO
IA	WALDRO	WALDRON RIDGE
IIA	WALKER	WALKER COUNTY
IID	WALLAP	WALLAPAI
L	WALLTO	WALLTOWN
L6	WALTER	WALTERS
IVB	WARBUR	WARBURTON RANGE

CLASS	ABBREV	N A M E
IIA	WARIAL	WARIALDA
CO3	WARREN	WARRENTON
IRANOM	WASHIN	WASHINGTON COUNTY
HOW	WASHOU	WASHOUGAL
IRANOM	WATERV	WATERVILLE
IIA	WATHEN	WATHENA
MEANOM	WEATHE	WEATHERFORD
IVB	WEAVER	WEAVER MOUNTAINS
L6	WEBB**	WEBB
IIID	WEDDER	WEDDERBURN
IIE	WEEKER	WEEKEROO STATION
H	WELDON	WELDONA
IIIA	WELLAN	WELLAND
H	WELLIN	WELLINGTON
H	WELLMA	WELLMAN
H	WESSEL	WESSELY
IVA	WESTER	WESTERN ARKANSAS
H4	WESTON	WESTON
H6	WESTRE	WEST REID
L	WETHER	WETHERSFIELD
H	WHITMA	WHITMAN
IA	WICHIT	WICHITA COUNTY
IA	WICKIR	WICKENBURG (IRON)
L5	WICKST	WICKENBURG (STONE)
L	WILBUR	WILBURTON
IIC-AN	WILEY*	WILEY
IIIA	WILLAM	WILLAMETTE
IIIA	WILLIA	WILLIAMSTOWN
IIIE	WILLOC	WILLOW CREEK
H	WILLOD	WILLOWDALE
H	WILMOT	WILMOT
H4	WILUNA	WILUNA
H	WINGEL	WINGELLINA
CHANOM	WINONA	WINONA
H5	WITKLI	WITKLIP FARM
LL	WITSAN	WITSAND FARM
L5	WITTEK	WITTEKRANTZ
L6	WOLDCO	WOLD COTTAGE
IIIB	WOLFCR	WOLF CREEK
IIIB	WONYUL	WONYULGUNNA
IB	WOODBI	WOODBINE
IVA	WOODSM	WOODS MOUNTAIN
H4	WOODWA	WOODWARD COUNTY
L	WOOLGO	WOOLGORONG
H	WRAY**	WRAY
LL	YALGOO	YALGOO
L	YANDAM	YANDAMA
IVA	YANHUI	YANHUITLAN
IRANOM	YARDEA	YARDEA
IA	YARDYM	YARDYMLY
IIIA	YARRI*	YARRI
IIA	YARROW	YARROWEYAH
H5	YATOOR	YATOOR
L	YAYJIN	YAYJINNA
IA	YENBER	YENBERRIE
E6	YILMIA	YILMIA
L	YOCEME	YOCEMENTO
H4,5	YONOZU	YONOZU
L	YORKST	YORK (STONE)
L	YORKTO	YORKTOWN
IIIA	YOUANM	YOUANMI
IA	YOUNDE	YOUNDEGIN
IRANOM	YSLETA	YSLETA
LL	YUKAN*	YUKAN
HOW	YURTUK	YURTUK
L	ZABORZ	ZABORZIKA
L6	ZABROD	ZABRODJE
IIIB	ZACA69	ZACATECAS (1969)
IA-AN	ZACA92	ZACATECAS (1792)
ACANOM	ZAGAMI	ZAGAMI
H	ZAISAN	ZAISAN LAKE
PAL	ZAISHO	ZAISHO
L	ZAVETN	ZAVETNOE
L6	ZAVID*	ZAVID
H	ZEBRAK	ZEBRAK
L6	ZEMAIT	ZEMAITKIEMIS
IA-AN	ZENDA*	ZENDA
IIIA-AN	ZERHAM	ZERHAMRA
H	ZHOVTN	ZHOVTNEVYI
HOW	ZMENJ*	ZMENJ
L6	ZOMBA*	ZOMBA
H	ZSADAN	ZSADANY
H	ZVONKO	ZVONKOV

B. Listing of Chondrites by Chemical Group and Petrologic Type

In the following list chondrites are ordered according to group and type, and alphabetically within each type. In those cases where the petrologic type was considered to be intermediate between two numbers (i.e., Florence, H 3, 4) the meteorite is listed in the lower of the two types.

The first column contains the abbreviated name; the full name can be obtained from Appendix II A. On the left side of the abbreviation is a "+" if the meteorite is an observed fall, and an O if the meteorite is an "anomalous member" of the group, as discussed in the previous section (e.g., Shaw is an anomalous L6 chondrite because of its low metal content). Literature sources for the classification are listed in the second column; if this column is blank, the classification was obtained from the silicate-composition reference.

The third column contains the composition of olivine expressed as mol % fayalite (FA) or of orthopyroxene expressed as mol % ferrosilite (FS). The only exceptions are the E4 and E5 chondrites, for which low-Ca clinopyroxene data are also included. The abbreviation UNEQU in this column indicates a compositional range of >2% for these minerals, as well as the fact that data defining the exact range were not available. References for the silicate-composition data are listed in the fourth column.

The final two columns list the sources of recent bulk-chemical analyses by reputable analysts (see Chapter IV). Note that these abbreviations refer to the name of the analyst, not to the first author of the paper (except for some joint papers by DYAKONOVA and KHARITONOVA). The "Remarks" column in this as well as in the following two lists gives

Table AII-2. Types of entries in the "Remarks" columns of Appendices IIB—IID, and abbreviations other than meteorite-group or element symbols found in the text of these columns

B	Minor component of polymict breccia; the abbreviated name of the dominant (host) component is given.
C	Classificational note—often a different classification proposed in the cited reference.
G	Genetically related to the listed meteorite. This remark is used only for anomalous meteorites which are so closely related to one another that they would be considered a group if there were at least five of them.
M	Mineralogical or petrological note—often in support of the given classification.
P	Paired with the listed meteorite (i.e., a fragment from the same fall). Data on these meteorites should be combined when preparing statistical compilations.
S	A synonym appearing in some publications is given.
U	Unpaired, i.e., distinct from the listed meteorite despite suggestions to the contrary in the literature.

Abbreviations in text:

KAM = kamacite	MET = metallic.
PC = per cent	POSS = possibly.

information of seven different sorts, as summarized in Table AII-2. Abbreviations which appear in the text of the "Remarks" columns are also encoded in Table AII-2.

At the end of this section are reference lists for 1. classification, silicate composition and remarks; and 2. bulk composition. These reference lists cover entries in Appendices IIB and IIC.

Appendix IIB

ABBREV	REF	SIL COMP	REF	REMARKS	REF	BULK ANALYSES
E4						
♦ABEE♦♦	VANS67A	FS 0.8	KEIL68A			VON69A
♦ADHIKO	BAED74A	FS 0.7	KEIL68A	C (E3)	VANS67A	
BETHUN	MASO66A					
♦INDARC	VANS67A	FS 1.1	KEIL68A			WII69A VON69A
KOTAKO	VANS67A	FS 1.2	KEIL68A			
SOUTOM	WASS74A			M 4 PC SI KAM	REED68A	
E5						
♦SAIMAK	VANS67A	FS 0.5	KEIL68A			WII69A VON69A
♦SAISAU	KEIL68A	FS 0.5	KEIL68A	C E4	VANS67A	
E6						
ATLANT	KEIL68A	FS 0.3	KEIL68A	C E5	VANS67A	WII69A VON69A
BLITHF	VANS67A	FS 0.04	KEIL68A			VON69A
♦DANIEL	VANS67A	FS 0.3	KEIL68A			
♦HVITTI	VANS67A	FS 0.2	KEIL68A			VON69A WII69A
♦JAJHDE	VANS67A	FS 0.2	KEIL68A			WII69A
♦KHAIRP	VANS67A	FS 0.3	KEIL68A			EAS67C
♦PILLIS	VANS67A	FS 0.4	KEIL68A			JAR69A DYA68A
♦UFANA♦	KEIL68A	FS 0.2	KEIL68A			
YILMIA	BUSE72A	FA 0.3	BUSE72A			
H3						
♦BREMER	VANS67A	FA15-19	DODD65A			WII69A
BROW37	VANS67A	FA19	MASO63A	C POSS L	VANS67A	
CLOVI1	VANS67A	UNEQU	DODD67A			JAR66A
DIMMIT	VANS67A	FA20	MASO63A	C TYPE 3,4	VANS67A	KHA69A
♦FLOREN	VANS67A	FA17.6	KEIL64A	C TYPE 3,4	VANS67A	
FRENCH	VANS73A	FA19	MASO67A	C TYPE 3,4	VANS73A	
GRAD37	VANS67A	UNEQU		C POSS L	VANS67A	JAR67A
PRAIRI	VANS67A	FA14-19	DODD65A			
♦SAILOU	VANS73A	FA19	MASO63A	C TYPE 3,4	VANS73A	
♦SERES♦	VANS73A	FA17	MASO63A	C TYPE 3,4	VANS73A	
♦SHARPS	VANS67A	UNEQU	DODD67A			JAR69B
SUWABU	VANS73A	FA14	MASO63A			
♦TIESCH	VANS67A	FA 0-43	DODD67A			WII69A

ABBREV	REF	SIL COMP	REF	REMARKS	REF	BULK ANALYSES
H4						
ADRIAN	VANS67A	FA19	MASO63A			
*AKBARP	VANS67A	FA19	MASO63A			
*ANKOBE	VANS73A	FA18	MASO67A	C TYPE UNSURE	VANS73A	
AURORA	VANS67A	FA19	MASO63A			
*BATH**	VANS67A	FA17.8	KEIL64A			WII69A
*BEAVER	VANS67A	FA18.3	KEIL64A			
*BIELOK	VANS67A	FA20	MASO63A			
BROKEN	VANS67A	FA20	MASO63A			
BUSHNE	VANS67A	FA19.4	KEIL64A			
CEDATE	VANS67A	FA18	MASO63A			
CULLIS	VANS67A	FA18	MASO63A			
CUSHIN	VANS67A	FA19	MASO63A			
ELMCRE	VANS67A	FA18	MASO63A			
ENIGMA	VANS73A	FA17	MASO67A			
EUSTIS	VANS67A	FA18	MASO63A			
*FARMVI	VANS67A	FA18	MASO63A			
*FOREVA	VANS67A	FA17.6	NOON72A			JAR67A DYA68A
GARNET	VANS67A	FA20	MASO63A			
GHUB62	VANS67A	FA18	MASO67A			
*GRUENE	VANS73A	FA17	MASO63A			
GRUVER	VANS67A	FA19	MASO63A			
GUNNAD	MCCA70A	FA18	MCCA70A			
HALEC2	VANS73A	FA17	MASO63A			
HAMMOW	MASO73A	FA18.8	MASO73A			JAR73A
HOBBS*	VANS67A	FA19	MASO63A			
HOLYOK	VANS67A	FA19	MASO63A			
IMPERI	VANS67A	FA18	MASO63A			
KESEN	VANS67A	FA16.1	KEIL64A			HAR62A VON69A
LAUNDE	MCCA70A	FA19	MCCA70A			
LIXNA	VANS73A	FA20	MASO63A			
*MARILI	AVAN73A	FA20	AVAN73A			
MENOW	VANS67A	FA17.4	KEIL64A			
METSAE	VANS67A	FA18	MASO63A			WII69A
*MISSHO	VANS67A	FA19	MASO63A	C TYPE 3,4	VANS67A	
*MONROE	VANS67A	FA17.6	KEIL64A			
MOSQUE	VANS73A	FA20	MASO67A			
*OCHANS	VANS67A	FA17.5	KEIL64A			WII69A DYA60A
*QUENGG	VANS73A	FA19	MASO63A			
RANSOM	READ72B	FA19	MASO63A			
SANEMI	VANS67A	FA20	MASO63A			
*SCHENE	FLEI70A	FA19	FLEI70A			
SEAGRA	VANS73A	FA18	MASO67A			
SELMA*	VANS67A	FA18.9	DODD65A			WII69A
SEMINO	HUSS72A	FA20	HUSS72A			
*SENA**	VANS67A	FA17	MASO63A			
*SLAVET	VANS73A	FA17	MASO63A			
TOBE**	VANS73A	FA20	MASO67A			
*TYSNES	VANS67A	FA20	MASO63A			
*WESTON	VANS67A	FA18.3	DODD67A			WII69A
*WILUNA	MCCA70B	FA19	MASO71A			
WOODWA	VANS73A	FA19	MASO63A			
*YONOZU	VANS67A	FA18	MASO63A	C TYPE 4,5	VANS67A	

ABBREV	REF	SIL COMP	REF	REMARKS	REF	BULK ANALYSES
H5						
ADAMSC	VANS67A	FA19	MASO63A			
*AGEN**	VANS67A	FA20	MASO63A	C L GROUP	BINN68A	
ALAMOG	VANS67A	FA19	MASO63A			
*ALLEGA	VANS67A	FA17.5	KEIL64A			JAR69A MAY67B
*AMBAPU	VANS67A	FA19	MASO63A			
ANTHON	VANS67A	FA19	MASO63A			
ASHMOR	CRAI71A	FA18.8	CRAI71A			
*ASSISI	VANS67A	FA17.9	KEIL64A			
*BARBOT	VANS67A	FA19	MASO67A			
*BEARDS	VANS67A	FA18.6	KEIL64A			VON69A
*BEDDGE	VANS67A	FA19	MASO63A			
BONITA	VANS67A	FA18	JARO66A			JAR66A
BURDET	FODO71B	FA18.0	FODO71B			JAR71G
*BURGHE	VANS67A	FA17.4	BINN68A			MAY67B
*CANGAS	VANS67A	FA18	MASO63A			
CARCOT	VANS73A	FA20	MASO63A			
*CASTAL	VANS67A	FA18-21	DODD65A			WII69A
*CERESE	VANS67A	FA19	MASO63A			
CHANNI	VANS67A	FA19	MASO63A			
COLBKA	VANS67A	FA18	MASO63A			
COLDST	VANS67A	FA17	MASO63A			
*COLLES	VANS67A	FA17.8	KEIL64A			WII69A
*COSINA	VANS67A	FA17.2	KEIL64A			
COVERT	VANS67A	FA18.5	JONE73A			
*CRONST	VANS67A	FA18	MASO63A			VON69A
*CROSSR	VANS67A	FA18	MASO63A			
CUERO*	VANS67A	FA19	MASO63A			
*DOKACH	VANS67A	FA18.0	KEIL64A			
*DUNDRU	VANS67A	FA19	MASO63A			
*EICHST	VANS67A	FA20	MASO63A			
FARLEY	VANS67A	FA18	MASO63A			
*FAVARS	VANS67A	FA18	MASO63A			
*FORECI	VANS67A	FA18.3	KEIL64A			WII69A VON69A
*GEIDAM	VANS67A	FA18.1	DODD67A			WII69A
GILGOI	VANS67A	FA17	MASO63A			
*GNADEN	VANS67A	FA18	MASO63A			
*GROSSD	VANS73A	FA19	MASO63A			
*HEREDI	VANS67A	FA18	MASO63A			
*HESSLE	VANS67A	FA17.9	KEIL64A			
*HIGASH	VANS67A	FA19	MASO63A			
HUGOTO	VANS67A	FA16	MASO63A			
*KANGRA	VANS67A	FA19	MASO63A			
KANS03	VANS67A	FA19	MASO63A			
KENNAR	VANS73A	FA19	MASO67A			
*KILBOU	VANS67A	FA19	MASO63A			
*LABORE	VANS67A	FA19	MASO63A	C TYPE UNSURE	VANS67A	
LAUNDR	MCCA70A	FA19	MCCA70A			
*LEIGHT	VANS67A	FA20	MASO63A			
*LOSTCI	CLAR71A	FA17.6	CLAR71A			JAR71B
MACAU	VANS67A	FA19	MASO63A			
*MARDAN	VANS67A	FA19	MASO63A			
MILLAR	VANS67A	FA19	MASO63A			WII69A
*MOORES	VANS67A	FA19	MASO63A	C L GROUP	KEIL64A	WII69A

ABBREV	REF	SIL COMP	REF	REMARKS	REF	BULK ANALYSES
H5						
+MORNAN	VANS67A	FA19	MASO63A			
+NAMMIA	VANS67A	FA19	MASO63A			
+OHABA+	VANS67A	FA20	MASO63A			
OROGRA	FODO72A	FA19.3	FODO72A			JAR72B
+PANTAR	VANS67A	FA18	MASO63A			KOE64A VON69A
PLAN17	VANS67A	FA18.2	BINN68A			VON69A
PLEASA	VANS67A	FA18	MASO63A			
+POKHRA	VANS67A	FA18	MASO63A			
+PRIBRA	CLAR71A	FA17.9	CLAR71A			JAR71C KHA65A
+PULTUS	VANS67A	FA18.2	KEIL64A			DYA61A VON69A
+RICHAR	VANS67A	FA16.2	KEIL64A			WII69A VON69A
SALINE	VANS67A	FA18	MASO63A			
SALTLA	VANS73A	FA19	MASO63A			
SCOTTC	VANS67A	FA20	MASO63A			
SCURRY	VANS67A	FA19	MASO63A			
+SEARSM	VANS67A	FA19	MASO63A			
+SINDHR	VANS67A	FA19.4	DODD67A			
+SITATH	VANS67A	FA19	MASO63A			
+STAELL	VANS67A	FA17.7	KEIL64A			
+TABOR+	VANS67A	FA16.5	KEIL64A			
+TIMOCH	VANS67A	FA20	MASO63A			KHA72A
TOMHAN	VANS67A	FA17.9	KEIL64A			WII69A
TRAVIS	VANS67A	FA18	MASO63A			
TULIA+	VANS67A	FA17.9	KEIL64A			
+UBERAB	VANS67A	FA19	MASO63A			
+UCERA+	CLAR71A	FA17.8	CLAR71A			JAR71C
+UDIPI+	VANS67A	FA18	MASO63A			
+WITKLI	VANS67A	FA18	MASO63A			
+YATOOR	VANS67A	FA19	MASO63A			
H6						
+BENLD+	VANS67A	FA20	MASO63A			
+BJELAJ	VANS67A	FA20	MASO63A			
+BLANSK	VANS67A	FA19	MASO63A			
BOERNE	VANS67A	FA20	MASO63A			
+BUTSUR	VANS67A	FA18.4	KEIL64A			
+CAPEGI	VANS67A	FA19	MASO63A			
CAVOUR	VANS67A	FA19	MASO63A			
CEDAKA	VANS67A	FA19.5	JONE70A			
+CHARSO	VANS67A	FA17.8	KEIL64A			
+CHARWA	VANS67A	FA19	MASO63A			
COBIJA	VANS67A	FA19	MASO63A			
+DJATIP	VANS67A	FA18.7	KEIL64A			
+DORONI	VANS67A	FA19	MASO63A			DYA72B
+ERXLEB	VANS67A	FA19	MASO63A			
ESTACA	VANS67A	FA18.4	KEIL64A			WII69A
FORRES	MCCA70A	FA19	MCCA70A			
GLADST	VANS67A	FA17	MASO63A			
+GUAREN	VANS67A	FA19	MASO63A			JAR69A
+HUNGEN	VANS73A	FA19	MASO63A			
INDIOR	VANS73A	FA18	MASO63A			
+JAMKHE	VANS67A	FA19	MASO63A			

ABBREV	REF	SIL COMP	REF	REMARKS	REF	BULK ANALYSES
H6						
+JUDESE	VANS73A	FA18	MAS063A			
+KADONA	VANS67A	FA17	MAS063A			
KAPPAK	VANS67A	FA17	MAS063A			
+KERNOU	VANS67A	FS16.5	KEIL64A	M NO OLIVINE	KEIL64A	
KIKINO	VANS67A	FA19	MAS063A			
KIMBLE	VANS67A	FA19	MAS063A			
+KLEINW	VANS67A	FA19	MAS063A			
+LANCON	VANS67A	FA19	MAS063A			DYA68A
+LUMPKI	VANS67A	FA19	MAS063A			
MARIDI	VANS67A	FA19	MAS063A			
MORLAN	VANS67A	FA19	MAS063A			
+MOTIKA	VANS67A	FA19	MAS063A			
+MOUNBR	VANS67A	FA17	MAS067A			
+NANJEM	VANS67A	FA18	MAS063A			
OAKLST	VANS67A	FA18.3	KEIL64A			WII69A EAS67C
+OGI+++	VANS67A	FA19	MAS063A			
PIPECR	VANS67A	FA19	MAS063A			
+PLANTE	VANS67A	FA20	MAS063A			
+QUEEME	VANS67A	FA18	MAS063A			
+SALLES	VANS67A	FA18	MAS063A			
+SEONI+	BUNC72A	FA19.7	BUNC72A			
+SUPUHE	VANS67A	FA19	MAS063A			
+THAL++	VANS67A	FA19	MAS063A			
+TOMATL	VANS67A	FA18	MAS063A			
+TOULOU	VANS67A	FA19	MAS063A			
+TRENZA	VANS67A	FA19	MAS063A			
WESTRE	MCCA70A	FA21	MCCA70A			
H (PETROLOGIC TYPE NOT KNOWN)						
+AARHUS		FA18	MAS063A			
ABBOTT		FA19	MAS067A			
ABO+++		FA19	MAS063A			
ACHILL		FA20	MAS063A			
ACME++		FA19	MAS063A			
ADDAHB		FA19	MAS071A			
AKR045		FA19	MAS063A			
AKR054		FA16	MAS063A			
+AKWANG		FA19	MAS063A			
+ALESSA		FA18	MAS063A			
+ALEXKY		FA18	MAS063A			DYA61A
ALGHST		FA19	MAS067A			
+ANDURA		FA20	MAS067A			
+ARCHIE		FA18.0	KEIL64A			
+ARROYO		FA20	MAS067A			
ASHSHA		FA19	MAS067A			
+AVANHA		FA18	MAS067A			
+AVILEZ		FA19	MAS063A			
BARWIS		FA18	FROS71B			
BELLYR		FA20	MAS063A			WII69A
BELMON		FA18	MAS067A			
+BENONI		FA18	HUTC71B			
+BETHLE		FA19	MAS063A			
BIRHAD		FA19	MAS067A			

ABBREV	REF	SIL COMP	REF	REMARKS	REF	BULK ANALYSES
H (PETROLOGIC TYPE NOT KNOWN)						
*BIRNIN		FA18	MASO63A			
BOGOSL		FA19	MASO63A			DYA61A
*BOLSHA		FA19	MASO63A			KHA65A
*BORODI		FA20	MASO63A			KHA68A
BOWDEN		FA18	MASO63A			
*BREITS		FA17.4	KEIL64A			
BROW64		FA19	MASO67A			
BURNAB		FA18	MASO67A			
CACILA		FA20	WLOT71A			
*CANELL		FA17	MASO63A			
CANY57		FA18	MASO67A			
CAPILL		FA19	MASO63A			
CAROLI		FA19	MASO63A			
CASHIO		FA18	MASO63A			
*CENTER		FA19	MASO67A			
CHAIL		FA18	MASO63A			
CHAMBE		FA18	MASO63A			
CHICOH		FA18	MASO67A			
COCKLE		FA18	MASO67A			
COMAST		FA19	MASO67A			
COOMAN		FA19	MASO63A			
COONAN		FA19	MASO67A			
COPE**		FA18	MASO63A			
CORTEZ		FA19	MASO63A			
COTTON		FA18	MASO63A			
CRANFI		FA18	MASO67A			
CROSBY		FA17	MASO63A			
CULBER		FA18	MASO63A			
DARMST		FA19	MASO63A			
DEATHV		FA20	MASO63A			
*DECEWS		FA18	MASO63A			
DEKALB		FA19	READ71A			
*DESURI		FA20	MASO67A			
DIMBOO		FA19	MASO63A			
DISPAT		FA18.5	JONE73A			
*DISTRI		FA18	MASO67A			
*DJERMA		FA19	MASO63A			
*DONGAK		FA19	MASO63A			
DOYLEV		FA20	MASO63A			
DRESKA		FA18	FROS71B			
*DRESON		FA20	MASO63A			
DUMAS*		FA16	MASO67A			
DUNCAN		FA19	MASO63A			
*DWALEN		FA20	MASO71A			
EHOLE		FA19.0	KEIL64A			JAR66A
EKEBY		FA18.9	KEIL64A			
*EKHKHE		FA18	MASO67A			
ELKHAR		FA18	MASO63A			
ELPERD		FA19	MASO63A			
ELSINO		FA17	MASO63A			
*EPINAL		FA19	MASO63A			
EROFEE		FA17	MASO63A			KHA65A
*ESNAND		FA18	MASO63A			

ABBREV	REF	SIL COMP	REF	REMARKS	REF	BULK ANALYSES
H (PETROLOGIC TYPE NOT KNOWN)						
+ESU+++		FA19	MASO63A			
FAUCET		FA18	MASO67A			JAR71C
+FAYETT		FA19	MASO63A			MUE66A
+FEIDCH		FA17	MASO63A			
FENBAR		FA18	MASO71A			
+FENGHS		FA18	MASO63A			
FERGSW		FA17	MASO63A			
FLEMIN		FA18	MASO63A			
+FORSBA		FA19	MASO63A			
FRANKL		FA19	MASO63A			
GAIL++		FA18	MASO63A			
+GALAPI		FA20	MASO63A			
+GAOUPP		FA18	MASO67A			
GEORGE		FA19	MASO67A			
GERONA		FA19	MASO63A			
+GLASAT		FA18	MASO63A			DYA61A
+GOPALP		FA20	MASO63A			
GREATB		FA19	MASO63A			
GUALEG		FA19	MASO63A			
+GUETER		FA20	MASO63A			
+GUMOSC		FA19	MASO63A			
+HAINAU		FA19	MASO63A			WII69A
HASSAY		FA18	MASO63A			
HATCRE		FA18	MASO63A			
HAVEN+		FA20	MASO63A			
HAVILA		FA19	MASO63A			
HAWKSP		FA19	MASO63A			
+HEDESK		FA17.6	KEIL64A			
HINOJO		FA20	MASO63A			
HOLLY+		FA18	MASO63A			
HORACI		FA19	MASO63A			
HOWE++		FA19	MASO63A			
HUGOST		FA17	MASO63A			
+ICHKAL		FA18	MASO63A			KHA65A
+IDUTYW		FA19	MASO63A			VON69A
INGALL		FA20	MASO63A			
+ISHING		FA18	MASO63A			
+ISTHIL		FA18	MASO63A			
+ITAPIC		FA18	MASO63A			
+KAEE++		FA19	MASO63A			
+KALABA		FA19	MASO71A			
KALDOO		FA19	MASO63A			
KARGAP	KHAR69A					KHA69A
KARVAL		FA19	MASO63A			
+KASAMA		FA18	MASO67A			HAR63D
KEARNE		FA20	MASO63A			
+KERILI		FA19	MASO63A			
+KHETRI		FA18	MASO63A			
KIELPA		FA17	FROS71B			
+KILLET		FA20	MASO63A			
KISSIJ		FA17	MASO63A			DYA64A
+KULP++		FA17	MASO67A			DYA64A
+LACOLI		FA17	MASO63A			

ABBREV	REF	SIL COMP	REF	REMARKS	REF	BULK ANALYSES
H (PETROLOGIC TYPE NOT KNOWN)						
LAVILL		FA19	MASO63A			
LEON++		FA18	MASO67A			
+LILLAV		FA18	MASO63A			
+LIMERI		FA19	MASO63A			
LITTRI		FA17	FROS71B			
LOGAN+		FA19	MASO63A			
+LUPONN		FA19	MASO63A			
+MALAMP		FA19	MASO63A			
+MALOTA		FA19	MASO63A			
MARSLA		FA16	MASO63A			
MASSEN		FA19.0	CAYE69A			
MAYDAY		FA19	MASO63A			
MCLEAN		FA19	MASO63A			
+MERUA+		FA16.5	KEIL64A			
MIAMI+		FA17	MASO63A			
MILLKA		FA19	MASO63A			
+MJELLE		FA19	MASO63A			
MOD048		FA19	MASO67A			
MORVEN		FA18	MASO63A			
MOSHES	VONM69A					VON69A
+MOTTAD		FA17	MASO63A			
MOUNMN		FA17	MASO63A			
MULGNO		FA18	MASO67A			
MULGSO		FA18	MCCA68A			
+NADIAB		FA18	MASO63A			
NALLAH		FA18	MASO71A			
+NAOKI+		FA20	MASO63A			
NARDO1		FA17	MASO63A	P ELSINORA	MASO63A	
NARUNA		FA19	MASO63A			
+NASSIR		FA19	MASO63A			
NAZARS		FA19	MASO63A			
NESS38		FA20	MASO63A			
+NIKOLA		FA19	MASO63A			DYA61A
+NIO+++		FA19	MASO67A			HAR67A
+NULLES		FA19	MASO63A			
OCZERE		FA19	MASO63A			
+OESEDE		FA19	MASO67A			
+OKABE+		FA19	MASO67A			HAR67A
OKIRAI		FA19	MASO63A			
+OLMEDI		FA18.3	BINN68A			
ORLOVK		FA19	MASO63A			DYA60A
OSHKOS		FA18	MASO63A			
+OTOMI+		FA18	MASO67A			
OUALLE		FA20	MASO63A			
OVID++		FA20	MASO63A			
OZONA+		FA19	MASO63A			
+PALOLO		FA19	MASO67A			
+PAVLDS		FA18	MASO67A			
PENOKE		FA19	MASO63A			
PETRKA		FA16	MASO63A			DYA60A
+PHUHON		FA18	MASO63A			
PICKEN		FA20	MASO63A			

ABBREV	REF	SIL COMP	REF	REMARKS	REF	BULK ANALYSES
H (PETROLOGIC TYPE NOT KNOWN)						
•PIQUET		FA17	MASO67A			
PLAINS		FA16	HUTC71B			
PLANSO		FA20	MASO63A			
•PULSOR		FA19	MASO63A			
•RACO••		FA18	MASO63A			
•RANCHA		FA17.9	DODD65A			
•RANCPR		FA19	MASO63A			
•RASTAN		FA20	MASO67A			
RAWLST		FA20	MASO63A			
•ROCHES		FA20	MASO63A			
ROLLA•		FA19	MASO63A	C ROLLAS PAIR	MASO63A	
ROMERO		FA19	MASO63A			
ROSEBU		FA19	MASO63A			
•ROSECI		FA19	MASO63A			WII69A
ROWENA		FA19	MASO67A			
RUSHCO		FA17	MASO63A			
SAIANN		FA18	MASO63A			
•SAIGER		FA18	MASO63A			
•SAIMAG		FA20	MASO63A			
SANCAR		FA18	MASO63A			
SANJOS		FA18	MASO63A			
SANLUI		FA19	MASO67A			
•SAOJOS		FA18	MASO67A			
•SASAGA		FA19	MASO71A			HAR62B
SCHAAP		FA19	MASO63A			VON69A
SEGUIN		FA20	MASO63A			
SEIBER		FA18	MASO63A			
•SELDEB		FA19	MASO63A			
SENECA		FA19	MASO63A			
•SETELA		FA19	MASO63A			
SHAFTE		FA19	MASO63A			
•SHUPIY		FA19	MASO63A			
SIERCO		FA19	MASO67A			
SILVTE		FA17	MASO63A			
•SIMMER		FA19	MASO63A			
•SINNAI		FA18	MASO67A			
•SONE••		FA18	MASO67A			HAR63B
SPRINE		FA19	MASO71A			
STONIN		FA19	MASO63A			
•SUNGAC		FA19	MASO67A			DYA64A
SUTTON		FA18	MASO67A			
SUWAAI		FA18	MASO63A			
SWEETW		FA19	MASO67A			
•SYLACA		FA20	MASO63A			
TAFOYA		FA20	MASO67A			
TAIGA•		FA19	MASO63A			
•TAKENO		FA19	MASO63A			
TATUM•		FA18	MASO63A			
TELL••		FA18	MASO67A			
TEXLIN		FA19	MASO63A			
•TIRUPA		FA18	MASO63A			
•TJABE•		FA19	MASO63A			

ABBREV	REF	SIL COMP	REF	REMARKS	REF	BULK ANALYSES
H (PETROLOGIC TYPE NOT KNOWN)						
•TORRIN		FA18	MASO63A			
TOSTAD		FA19	MASO63A			
•TROMOY		FA18	MASO63A			WII69A
TWOBUT		FA19	HUTC71B			
ULYSSE		FA17	MASO63A			
UTECRE		FA19	MASO63A			
•UTZENS		FA18	MASO63A			
UVALDE		FA19	MASO63A			
•VAGO••		FA19	MASO63A			
•VALDAV		FA20	MASO63A			
•VENGER		FA19	MASO63A			DYA72B
•VERKTS		FA18	MASO63A			
•VERNON		FA19	MASO63A			
VILLED		FA19	MASO63A			
•VISUNI		FA20	MASO67A			
VULCAN		FA20	MASO67A			
WAIRAR		FA18	MASO63A			
WELDON		FA19	MASO63A			
WELLIN		FA18	HUTC71B			
WELLMA		FA18	MASO63A			
•WESSEL		FA18	MASO63A			
WHITMA		FA19	MASO63A			
WILLOD		FA18	JONE73A			
WILMOT		FA18	MASO63A			
WINGEL		FA19	MASO63A			
WRAY••		FA15	MASO63A			
•ZAISAN		FA18	MASO67A			
•ZEBRAK		FA18	MASO63A			
•ZHOVTN		FA19.0	KEIL64A			DYA60A
•ZSADAN		FA18	MASO63A			
•ZVONKO		FA19	MASO63A			
L3						
•BISHUN	VANS67A	FA 0-40	DODD67A	C POSS LL	DODD67A	JAR66A
CARRAW	VANS67A	UNEQU	DODD67A			WII69A
•HALLIN	VANS67A	FA 0-37	DODD67A			JAR66A
•HEDJAZ	VANS67A	FA24	KRAU71A	C TYPE 3-6	KRAU71A	
IOKA••	VANS67A	UNEQU	DODD67A			JAR66A
•KHOHAR	VANS67A	FA 7-26	DODD67A			JAR66A
•KRYMKA	VANS67A	FA 0-38	DODD67A	C POSS LL	DODD67A	DYA60A
•MANYCH	VANS67A	UNEQU	DODD67A	C POSS LL	DODD67A	DYA64A
•MEZOEM	VANS67A	FA16-29	DODD65A	C POSS LL	DODD67A	JAR67A EAS68A
•PIANCA	CARA71A	FA 2-34	CARA71A			
REID••	MCCA70A	UNEQU	MCCA70A			
•RIONEG	VANS67A	FA25	MASO63A	C TYPE 3,4	VANS67A	
L4						
•ALBARE	VANS67A	FA24.4	KEIL64A			
•ATARRA	VANS67A	FA23	MASO63A			
•AWERE•	HUTC71A	FA25	HUTC71A			
•BALDMO	VANS67A	FA22.0	KEIL64A			
BARRAT	VANS67A	FA14-28	DODD67A			WII69A

ABBREV	REF	SIL COMP	REF	REMARKS	REF	BULK ANALYSES
L4						
•BJURBO	VANS67A	FA24.5	KEIL64A			VON69A
•CYNTHI	VANS67A	FA25	DODD67A			MAY67B
DALGET	VANS73A	FA24	MASO63A			
FREMON	VANS67A	FA24	MASO67A			
GOODLA	VANS67A	UNEQU	DODD67A			JAR67A
HARDWI	VANS67A	FA24	MASO63A			
•HOEKMA	VANS67A	FA22.7	KEIL64A			WII69A
JEROKA	VANS67A	FA19	MASO63A	C H GROUP	MASO63A	
MCKINN	VANS67A	FA24	MASO63A			WII69A VON69A
NORTHE	MCCA70A	FA24	MCCA70A			
NORTHF	MCCA70A	FA24	MCCA70A			
OKECHO	VANS67A	FA24	MASO63A			
•RUPOTA	JARO69A	FA24.0	FRED68A			JAR69A
•SARATO	VANS67A	FA24	MASO63A			
•SLOBOD	VANS67A	FA23.3	FRED68A			DYA61A
•TENNAS	VANS67A	FA21-27	DODD65A			WII69A
VERA••	VANS67A	FA24	MASO63A			
L5						
ADELIE	VANS67A	FA24	MASO63A			
ARAPAH	VANS67A	FA23.6	DODD65A			
ARRIBA	VANS67A	FA24.2	KEIL64A			
ASSAM•	VANS67A	FA24	MASO63A			
•AUSSON	VANS67A	FA23.8	KEIL64A			
BEENHA	VANS67A	FA23	MASO63A			
•BEUSTE	VANS73A	FA25	MASO63A			
•BLACWE	VANS73A	FA26	MASO67A			
BLUFF•	VANS67A	FA25	MASO63A			WII69A
•BORKUT	VANS67A	FA26	MASO63A			
•CHANDA	VANS67A	FA24	MASO63A			
•CHERVE	VANS67A	FA26	MASO63A			
COORAR	MCCA70A	FA25	MCCA70A			
•CRUMLI	VANS67A	FA24	MASO63A			
•ERGHEO	VANS67A	FA23.7	KEIL64A			
•FARMIN	VANS67A	FA23.8	KEIL64A			WII69A VON69A
FARNUM	VANS67A	FA25	MASO63A			
•FUKUTO	VANS67A	FA21.6	KEIL64A			
GRETNA	VANS67A	FA25	MASO63A			
HENDER	VANS67A	FA24	MASO63A			
•HOMEST	VANS67A	FA22.0	KEIL64A			
•HONOLU	VANS67A	FA24	MASO63A			DYA68A
INDIAN	VANS67A	FA24.1	BINN68A			
•JHUNG•	VANS67A	FA25	MASO63A			
•KNYAHI	VANS67A	FA24.4	FRED68A			WII69A
•KULAK•	GRAH73A	FA24.0	GRAH73A			
LALAND	VANS67A	FA24	MASO63A			
LAUNDW	MCCA70A	FA25	MCCA70A			
•LITTPI	VANS67A	FA23.4	BINN68A			
•LUA•••	VANS67A	FA25.1	DODD67A			
MAINZ•	VANS67A	FA25	MASO63A			
•MARMAN	VANS73A	FA24	MASO63A			
MELROS	VANS67A	FA23	MASO63A			

ABBREV	REF	SIL COMP	REF	REMARKS	REF	BULK ANALYSES
L5						
*MIRZAP	VANS67A	FA24	MASO63A			
*MOLINA	VANS67A	FA24	MASO63A			
OAK***	MCCA70A	FA25	MCCA70A			
*PAMPAN	VANS73A	FA24	MASO63A			
PECKSS	VANS67A	FA25	MASO63A			
*POHLIT	VANS67A	FA25	MASO63A			
*RICHMO	VANS67A	FA26	MASO63A			
ROY*33	VANS67A	FA25	MASO63A			
RUSHVI	VANS67A	FA24	MASO63A			
SAIPET	VANS67A	FA25	MASO63A			
*SHELBU	VANS67A	FA24	MASO63A			
*TADJER	VANS67A	FA25	MASO63A			WII69A
TAIBAN	VANS67A	FA23.8	KEIL64A			
*TANE**	VANS67A	FA25.0	FRED68A			HAR67A
TIMMER	SMIT72A	FA24	SMIT72A			
*TJEREB	VANS67A	FA24	MASO63A			
WICKST	VANS67A	FA23	MASO63A			
*WITTEK	VANS67A	FA23	MASO63A			VON69A
L6						
ABERNA	VANS67A	FA23	MASO63A			
*ACHIRA	VANS67A	FA25	MASO63A			
*AGUADA	VANS67A	FA25	MASO63A			
AKABA	VANS67A	FA24	MASO63A			
*ALEPPO	VANS67A	FA24	MASO63A			
*ALFIAN	VANS67A	FA23.7	KEIL64A			KHA68A VON69A
*ANDOVE	VANS67A	FA25	MASO63A			
*ANGERS	VANS67A	FA25	MASO63A			
*APT***	VANS67A	FA24	MASO63A			
ARTRAC	VANS67A	FA24	MASO63A			
*ASHDON	VANS67A	FA25	MASO63A			
ATWOOD	VANS67A	FA25	MASO67A			
*AUMALE	VANS67A	FA25	MASO63A			
*AUMIER	VANS67A	FA24	MASO63A			
AZTEC	VANS73A	FA22.9	KEIL64A			
*BACHMU	VANS67A	FA24	MASO63A			
*BALDWY	VANS67A	FA25	MASO63A			
*BAROTI	VANS67A	FA24.8	FRED68A			
*BATHFU	VANS67A	FA24.0	KEIL64A			
*BERLAN	VANS67A	FA25	MASO63A			
*BHAGUR	VANS67A	FA25	MASO63A			
*BHERAI	VANS67A	FA24	MASO63A			
*BLANKE	VANS67A	FA23.5	KEIL64A			
BOCAS	VANS67A	FA26	MASO63A			
*BORI**	VANS67A	FA25	MASO63A			
BREWST	VANS67A	FA24	MASO63A			
*BRUDER	VANS67A	FA23.8	KEIL64A			KOE64B MAY61A
*BUSCHH	VANS67A	FA24	MASO63A			
*CABEZO	VANS67A	FA24	MASO63A			
*CASTIN	VANS67A	FA25	MASO63A			
*CHANDP	VANS67A	FA24	MASO63A			
*CHANTO	VANS67A	FA23.3	KEIL64A			

ABBREV	REF	SIL COMP	REF	REMARKS	REF	BULK ANALYSES
L6						
♦CHATEA	VANS67A	FA23.7	KEIL64A			WII69A
♦COLBWI	VANS67A	FA23.4	KEIL64A			VON69A
CONCHO	VANS67A	FA23	MASO63A			
COONBU	VANS67A	FA24.3	KEIL64A			
♦CRANGA	VANS67A	FA24	MASO63A			
♦DANDAP	VANS67A	FA25	MASO63A			
♦DANVIL	VANS67A	FA23	MASO63A			
DENS79	VANS67A	FA23	MASO63A			
♦DENVER	MASO68A	FA24	MASO68A			JAR68A
♦DRAKEC	VANS67A	FA25	MASO63A			VON69A
♦DURALA	VANS67A	FA25	MASO63A			
♦DURUMA	VANS67A	FA25	MASO63A			
DWIGHT	VANS67A	FA25	MASO63A			
ELIELW	VANS67A	FA23.5	KEIL64A			
♦FISHER	VANS67A	FA23	MASO63A			
♦FORKSV	VANS67A	FA23.9	KEIL64A			
♦FORSYT	VANS67A	FA23	MASO63A			
♦FUTTEH	VANS67A	FA24.3	KEIL64A			
♦GAMBAT	VANS67A	FA23	MASO63A			
GARRAF	VANS67A	FA24	MASO63A			
♦GIFU♦♦	VANS67A	FA25	MASO63A			HAR62C
♦GIRGEN	VANS67A	FA24.3	KEIL64A			JAR72C
♦GITGIT	VANS67A	FA24	MASO63A			
♦GROSSL	VANS67A	FA25	MASO63A			DYA72B
♦GURRAM	VANS67A	FA24	MASO63A			
♦HARLET	VANS67A	FA24.1	KEIL64A			
♦HARRTY	VANS67A	FA26	MASO63A			
HARRVI	VANS67A	FA23.6	KEIL64A			
HERMIT	VANS67A	FA24	MASO63A			
HESSTO	VANS67A	FA24	MASO63A			
♦HOLBRO	VANS67A	FA24.6	KEIL64A			WII69A VON69A
♦JACKAL	VANS67A	FA24	MASO63A			VON69A
♦JEMLAP	VANS67A	FA25	MASO63A			
JOHNSO	VANS67A	FA22.7	KEIL64A			
♦KAKOWA	VANS67A	FA23	MASO63A			
♦KALUMB	VANS67A	FA24	MASO63A			
♦KAREWA	VANS67A	FA24	MASO63A			
♦KARLOO	VANS67A	FA23	MASO63A			
KERMIC	VANS67A	FA24	MASO63A			
KEYES♦	VANS67A	FA23	MASO63A			
♦KHERAG	VANS67A	FA24	MASO63A			
♦KUNASH	VANS67A	FA24	MASO63A			DYA61A
KYLE♦♦	FODO71A	FA26.2	FODO71A			JAR71A
♦KYUSHU	VANS67A	FA23.3	KEIL64A			WII69A VON69A
♦LABECA	VANS67A	FA23	MASO63A			
LADDER	VANS67A	FA22.6	KEIL64A			
♦LAIGLE	VANS67A	FA22.9	KEIL64A			
LAKEBR	VANS67A	FA25	MASO63A			
♦LANGHA	VANS67A	FA24.2	FRED68A			WII69A
♦LAUNTO	VANS67A	FA23	MASO63A			
LAWREN	VANS67A	FA23	MASO63A			
♦LEEDEY	VANS67A	FA24.4	FRED68A			WII69A JAR67A
♦LEEUWF	VANS67A	FA24	MASO63A			

ABBREV	REF	SIL COMP	REF	REMARKS	REF	BULK ANALYSES
L6						
+LEPRES	VANS67A	FA24	MASO63A			
+LESORM	VANS67A	FA24	MASO63A			
+LISSA+	VANS67A	FA23	MASO63A			
LONGIS	VANS67A	FA25	MASO63A			
+LUNDSG	VANS67A	FA24.2	FRED68A			
LUTSCH	VANS67A	FA24	MASO63A			
MAKARE	VANS67A	FA24	MASO63A			
+MARIOI	VANS67A	FA24	MASO63A			
+MAUERK	VANS67A	FA24	MASO63A			
+MERN++	VANS67A	FA24	MASO63A			
+MHOW++	VANS67A	FA24	MASO63A			
+MIDDLE	VANS67A	FA23	MASO63A			
+MILENA	VANS67A	FA24	MASO63A			
+MOCS++	VANS67A	FA24.4	KEIL64A			WII69A KHA65A
+MODO05	VANS67A	FA25.6	DODD67A			WII69A VON69A
+MONZE+	VANS67A	FA25	MASO63A			VON69A
+MORADA	VANS67A	FA24	MASO63A			
MOSCA+	VANS73A	FA24	MASO67A			
MUIZEN	VANS67A	FA24	MASO63A			VON69A
NASHST	VANS67A	FA22	MASO63A			
NEENAC	VANS67A	FA24	MASO63A			
+NEJO++	CLAR72A	FA25	CLAR72A			JAR72A
+NERFT+	VANS67A	FA23	MASO63A			DYA72B
NESS94	VANS67A	FA25	MASO63A			
+NEWCON	VANS67A	FA23.6	KEIL64A			WII69A VON69A
+OESEL+	VANS67A	FA25	MASO63A			KHA69A
+ORVINI	VANS67A	FA23	MASO63A			
+OTEROY	VANS67A	FA25	MASO63A			
OTIS++	VANS67A	FA24	MASO63A			
+PACULA	VANS67A	FA24	MASO63A			
PARANA	VANS73A	FA23	MASO67A			
+PAVLOG	VANS67A	FA25	MASO63A			DYA60A
+PEACER	VANS67A	FA23	MASO67A			
PEETZ+	VANS67A	FA26	MASO63A			
+PERPET	VANS67A	FA25	MASO63A			
+PIRGUN	VANS67A	FA25	MASO63A			
POTTER	VANS67A	FA23	MASO63A			
+PRICET	VANS67A	FA24	MASO63A			
+PUTING	SYME70A	FA27	SYME70A			
+RAKOVK	VANS67A	FA24	MASO63A			
+RICHMU	VANS67A	FA24	MASO63A			
ROY+34	VANS67A	FA24	MASO63A			
RUSHCR	VANS67A	FA26	MASO63A			
+SAIMIC	VANS67A	FA23.9	KEIL64A			VON69A
+SANTIS	VANS67A	FA24	MASO63A			
+SAUGUI	VANS67A	FA25	MASO63A			
+SCHOEN	VANS67A	FA25	MASO63A			
+SEGOWL	VANS67A	FA25	MASO63A			
OSHAW++	VANS67A	FA23.7	FRED67A	M MET FE 1.76	FRED67A	MAY67C
+SHYTAL	VANS67A	FA25	MASO63A			
SILVNE	VANS67A	FA25	MASO63A			
+SINAI+	VANS67A	FA23	MASO63A			
+SKI+++	VANS67A	FA24	MASO63A			

ABBREV	REF	SIL COMP	REF	REMARKS	REF	BULK ANALYSES
L6						
♦STAVRO	VANS67A	FA24	MASO63A			KHA65A
♦SULTAN	VANS67A	FA26	MASO63A			
TARFA♦	VANS67A	FA25	MASO63A			
♦TAUQ♦♦	VANS67A	FA25	MASO63A			
♦TENHAM	VANS67A	FA24	MASO63A			
THOMSO	VANS67A	FA24	MASO63A			
♦TILDEN	VANS67A	FA26	MASO63A			
♦TOURIN	VANS67A	FA25	MASO63A			
♦TROUP♦	VANS67A	FA25	MASO63A			
TUZLA♦	VANS67A	FA25	MASO63A			
♦UMMTIN	VANS67A	FA23	MASO63A			
♦UTRECH	VANS67A	FA24	MASO63A			
♦VALDIN	VANS67A	FA24.7	FRED68A			JAR71D
VALKEA	VANS67A	FA24	MASO63A			WII69A
VARPAI	VANS67A	FA25	MASO63A			WII69A
WACOND	VANS67A	FA25	MASO63A			
WALTER	VANS67A	FA24.4	KEIL64A			
WEBB♦♦	MCCA70A	FA23	MCCA70A			
♦WOLDCO	VANS67A	FA24.4	MOSS67A			EAS67C
♦ZABROD	VANS73A	FA25	MASO63A			
♦ZAVID♦	VANS67A	FA24	MASO63A			
♦ZEMAIT	VANS67A	FA24	MASO63A			KHA65A
♦ZOMBA♦	VANS67A	FA23	MASO63A			
L (PETROLOGIC TYPE NOT KNOWN)						
ACCALA		FA24	MASO63A			
♦AIR♦♦♦		FA23	MASO63A			
AKRO61		FA25	MASO67A			
ALAMOS		FA25	MASO63A			
ALBIST		FA22	MASO63A			
AMBER♦		FA24	MASO63A			
AMHERS		FA25	MASO63A			
ANDRYU		FA23	MASO67A			
ARMEL♦		FA25	MASO71A			
♦ASCO♦♦		FA26	MASO63A			
♦ATEMAJ		FA23	MASO63A			
♦ATOKA♦		FA24	MASO63A			
♦BANSWA		FA24	MASO63A			
♦BARWEL	MASO67A	FA23.7	JOBB66A			EAS66A EAS67C
♦BAXTER		FA23.6	KEIL64A			
BELLEP		FA24	MASO63A			
♦BEYROU		FA26	MASO63A			
BILLYG		FA25	MASO67A			
♦BLACMP		FA24	MASO63A			
♦BOTSCH		FA26	MASO63A			KHA65A
♦BOVEDY		UNEQU	HUTC71B	C POSS LL	HUTC71B	
BOWESM		FA25	MASO67A			
BRADY♦		FA24.1	KEIL64A			
BRISCO		FA26	MASO63A			
BURRIK		FA24	MASO67A			
♦BURSA♦		FA24	MASO67A			
CADELL		FA24	MASO63A			
CALLIH		FA23.8	KEIL64A			

ABBREV	REF	SIL COMP	REF	REMARKS	REF	BULK ANALYSES
L (PETROLOGIC TYPE NOT KNOWN)						
•CANAKK		FA25	MASO67A			
CARDAN		FA24	MASO67A			
CARTOO		FA24	MASO63A			
CLARET		FA25	MASO63A			
CLAYTO		FA24	HUTC71B			
•CLOHAR		FA25	MASO63A			
CLOVI2		FA24	MASO67A			
COCKBU		FA25	JOHN67A			
COCUND		FA24	MASO63A			
COOLAM		FA25	MASO67A			
COTESF		FA23	MASO63A			
CREDO•		FA24	MASO71A			
DALEDR		FA24	MASO63A			
DAOURA		FA23	MASO63A			
DAVY••		FA23	MASO63A			
•DEAL••		FA25	MASO63A			
•DELHI•		FA24	MASO67A			
•DEMINA		FA23	MASO63A			DYA61A
DENOVA		FA23	MASO63A			
•DIEPRI		FA25	MASO63A			VON69A
DIX•••		FA23	MASO63A			
•DOLGOV		FA25	MASO63A			
•DOMANI		FA24	MASO63A			
•DOSSO•		FA25	MASO63A			
•ELENOV		FA25	MASO63A			
ELLERS		FA25	MASO63A			
ERIE••		FA24	MASO67A			
FINNEY		FA24	MASO67A			
FOREBU		FA24	MASO63A			
FORRLA		FA26	MASO67A			
GHUB54		FA24	MASO67A	P OTHER L GHU	MASO67A	DYA68A EAS68A
GRAD33		FA25	MASO63A			
GRANTC		FA26	MASO63A			
GRASSL		FA23	MASO67A			
HALECI		FA25	MASO63A			
HARDIN		FA26	MASO63A			
HASKEL		FA23	MASO63A			
HAYESC		FA23	MASO63A			
•HIGHPO		FA25	MASO63A			
HILDRE		FA25	MASO63A			
HINOJA		FA26	MASO67A			
HORAC2		FA25	MASO63A			
ISOULA		FA25	MASO63A			
JEROID		FA24	MASO63A			
JUAREZ		FA24	MASO63A			
JUNCTI		FA24	MASO63A			
•KAGARL		FA23	MASO63A			
•KAMALP		FA25	MASO67A			
•KAMSAG		FA24	MASO63A			
•KANDAF		FA23.8	KEIL64A			
•KAPTAL		FA23	MASO63A			DYA61A
KARAGA		FA23	MASO63A			
•KARKH•		FA25	MASO63A	C H6	VANS67A	

ABBREV	REF	SIL COMP	REF	REMARKS	REF	BULK ANALYSES
L (PETROLOGIC TYPE NOT KNOWN)						
KAUFMA		FA23	MASO63A			
♦KENDLE		FA25	MASO63A			
♦KHMELE		FA23	MASO63A			DYA64A
♦KIEL♦♦		FA25	MASO67A			KOE64B
KINGFI		FA23	MASO63A			
KINGOO		FA24	MASO63A			
♦KISVAR		FA23	MASO67A			WII69A
KORALE		FA24	MASO63A			
♦KRASNI		FA24	MASO63A			DYA61A
♦KUKSCH		FA24	MASO63A			
♦KULESC		FA25	MASO63A			
KULNIN		FA24	MASO63A			
♦KUSIAL		FA24	MASO63A			
♦KUTTIP		FA25	MASO63A			
♦KUZNET		FA23	MASO63A			DYA61A
KYBUNG		FA25	MASO71A			
LAKEBO		FA25	MASO67A			
LAKEGR		FA24	MASO67A			
LAKEMO		FA25	MASO63A			
LAKETO		FA25	MASO63A			
LAKEWO		FA24	MASO71A			
♦LALITP		FA24	MASO63A			
♦LANZEN		FA24.0	KURA65A			
♦LAVREN		FA24	MASO63A			KHA68A
♦LEONOV		FA24	MASO63A			
♦LESVES		FA25	MASO63A			
LINCOL		FA24	MASO63A			
♦LINUM♦		FA23	MASO63A			WII69A
LOCKNE		FA23	MASO63A			
LOOMIS		FA23	MASO63A			
LOOP♦♦		FA24	MASO67A			
LOSTLA		FA25	MASO63A			
LOYOLA		FA23	MASO63A	P HOMESTEAD	MASO63A	
LUBBOC		FA24	MASO63A			
♦LUCE♦♦		FA24	MASO63A			
♦MABWEK		FA24	MASO67A			
♦MADRID		FA24	MASO63A			
♦MAFRA♦		FA25	MASO67A			
♦MALAKA		FA24.5	HUTC71B			
♦MAMRAS		FA24	MASO63A			
MARIOK		FA24	MASO63A			
MARLOW		FA24	MASO63A			
♦MASCOM		FA24	MASO63A			
♦MAURIT		FA26	MASO63A			
♦MAZIBA		FA25	MASO63A			
MCADDO		FA25	MASO63A			
♦MESSIN		FA24	MASO67A			
♦MEUSEL		FA24	MASO63A			
♦MEZEL♦		FA24	MASO63A			
♦MIKE♦♦		FA24	MASO63A			
MINASG		FA25	MASO63A			
MISSIO		FA23	MASO63A			
MONTEC		FA25	MASO67A			

ABBREV	REF	SIL COMP	REF	REMARKS	REF	BULK ANALYSES
L (PETROLOGIC TYPE NOT KNOWN)						
•MONTED		FA24	MASO63A			WII69A
•MONTEM		FA25	MASO63A			
•MONTLI		FA24	MASO63A			
•MOORLE		FA23	MASO63A			
MOTPEN		FA24	MASO71A			
•MUDDOO		FA23	MASO67A			
•MULLET		FA25	MASO63A			
•MURAID		FA26	MASO67A			
MUROC•		FA25	MASO63A			
MUROCD		FA25	MASO63A			
•NAGYBO		FA24	MASO63A			
•NAKHON		FA24	NELE72A			NEL72A
•NANYAN		FA24	MASO63A			
NARDO2		FA25	MASO63A			KHA68A
•NARELL		FA25	MASO63A			
NARETH		FA25	MASO67A			
NEWALM		FA24	MASO63A			KHA69A
NEWSOM		FA24	MASO63A			
•NIKOLS		FA24	MASO63A			DYA60A
NORACR		FA24	MASO67A			
NORCAT		FA25	MASO63A			
•NOYANB		FA23	MASO67A			
•OFEHER		FA23	MASO63A			
•OHUMA•		FA23.5	MOSS67A			EAS67C
•OJUELO		FA25	MASO63A			
•OLDE30		FA24	MASO67A			
•OVIEDO		FA25	MASO63A			
PAMPAA		FA24	MASO63A			
PAMPAI		FA24	MASO63A			
PANNIK		FA24	MASO67A			
•PATRIM		FA25	MASO63A			
•PERVOM		FA24	MASO63A			DYA61A
•PHIL01		FA23	HUTC71B			
PIERST		FA24	MASO63A			
PINTOM		FA24	MASO63A			DYA68A
•PLOSCH		FA24	MASO63A			
•PNOMPE		FA24	MASO63A			
POINST		FA24	MASO67A			
•PUENLA		FA24	MASO63A			
•QUINCA		FA24	MASO63A			
•RAMNAG		FA25	MASO67A			
•RAMSDO		FA21.6	KEIL64A	M CLINOPYROXE	KEIL64A	
•RANGAL		FA25	MASO63A			
REAGER		FA25	MASO63A			
•RELIEG		FA23	MASO63A			
•RENCA•		FA24	MASO63A			
RIVER•		FA25	MCCA68A			
ROSAMO		FA24	MASO63A			
•RYECHK		FA23	MASO63A			
•SAICAP		FA25	MASO63A			
•SAICHR		FA25	MASO63A			
•SAIDEN		FA25	MASO63A			
SALLA•		FA24	MASO67A			

ABBREV	REF	SIL COMP	REF	REMARKS	REF	BULK ANALYSES
L (PETROLOGIC TYPE NOT KNOWN)						
SANPED		FA23	MAS063A			
+SANTBA		FA25	MAS063A			
+SCHELL		FA25	MAS063A			
+SEVRUK		FA25	MAS063A			DYA60A
+SHIKAR		FA25	MAS067A			
SLEEPE		FA25	MAS067A			
SMITHC		FA24	MAS063A			
SPRINF		FA23	MAS063A			
+STRATH		FA25	MAS063A			
SUBLET		FA24	MAS063A			
+SUCCES		FA24	MAS063A			
+SUWAAD		FA23	MAS063A			
TARBAG		FA23	MAS063A			DYA61A
+TATHLI		FA26	MAS071A			
TEMPLE		FA24.5	KEIL64A			
TIBERR		FA24.5	CHRI70A			WII70A
+TOMITA		FA23	MAS067A			HAR63C
+TRIFIR		FA24	MAS067A			
TRYON+		FA24	MAS063A			
+TRYSIL		FA25	MAS063A			
+TUANTU		FA24	MAS063A			
TWENTY		FA26	MAS063A			
+VALLEY		FA24	MAS063A			
VINCEN		FA24	MAS063A			
+VIRBA+		FA25	MAS063A			
+VOUILL		FA24	MAS063A			
WALDO+		FA25	MAS063A			
WALLTO		FA23	MAS067A			
+WETHER		FA25	MAS071A			
WILBUR		FA25	MAS063A			
+WOOLGO		FA25	MAS067A			
YANDAM		FA25	MAS063A			
YAYJIN		FA25	MAS067A			
YOCEME		FA24	HUTC71B			
YORKST		FA24	MAS063A			
+YORKTO		FA24	MAS063A	M OTHERS FA19	MAS063A	
+ZABORZ		FA25	MAS063A			DYA72B
+ZAVETN		FA24	MAS063A			DYA60A
LL3						
+CHAINP	VANS67A	FA 0-37	DODD65A	C POSS L	DODD67A	WII69A
+HAMLET	VANS67A	UNEQU	DODD67A	C TYPE 3,4	VANS67A	WII69A
+NGAWI+	VANS67A	UNEQU	DODD67A	C POSS L	DODD67A	
+PARNAL	VANS67A	FA 0-47	DODD67A	C POSS L	VANS67A	WII69A EAS67A
+SAIMAR	NOON73A	FA 0-25	NOON73A			JAR73B
+SEMARK	VANS67A	UNEQU	DODD67A	C POSS L	DODD67A	JAR66A
LL4						
KELLY+	VANS67A	FA29.0	FRED68A			
+SOKOBA	VANS67A	FA28.8	FRED68A			

ABBREV	REF	SIL COMP	REF	REMARKS	REF	BULK ANALYSES
LL5						
•ALDSWO	VANS67A	FA28	MASO63A			
•CHEROK	JARO69A	FA29.0	FRED68A	C LL6	VANS67A	JAR69A
•KHANPU	VANS67A	FA29	MASO63A			
•KRAEHE	VANS67A	FA27	MASO63A			MUE69A
NORTHR	MCCA70A	FA28	MCCA70A			
OBERLI	VANS67A	FA27.2	FRED68A			
•OLIVEN	VANS67A	FA29.8	FRED68A			
•PERTH•	VANS67A	FA28	MASO63A			
•UMBALA	VANS67A	FA28	MASO63A			
LL6						
•APPLEY	VANS67A	FA31.1	FRED68A			WII69A
ARCADI	VANS67A	FA29	MASO63A			
•ATHENS	VANS67A	FA31	MASO63A			
•BANDON	VANS67A	FA29.4	BINN68A			
•BENARE	VANS67A	FA28	MASO63A			
•BENTON	VANS67A	FA30.5	FRED68A			
•BORGOS	VANS67A	FA29	MASO63A			
•CARATA	VANS67A	FA29	MASO63A			
•CHICOR	VANS67A	FA29	MASO63A			
•DHURMS	VANS67A	FA27.2	FRED68A			VON69A
•DOUARM	VANS67A	FA29.8	FRED68A			
•ENSISH	VANS67A	FA29.2	FRED68A			
•JELICA	VANS67A	FA32.3	FRED68A			
•KARATU	VANS73A	FA30.8	FRED68A			JAR66A
LAKELA	VANS67A	FA28.8	BINN68A			DYA68A
•MANBHO	VANS67A	FA31.1	FRED68A			WII69A
•MANGWE	VANS67A	FA29.5	FRED68A			
MELLEN	VANS67A	FA27	MASO63A			
NAES••	VANS67A	FA30.6	FRED68A			WII69A
•OTTAWA	VANS67A	FA29.3	FRED68A			WII69A
•SAIMES	VANS67A	FA29.0	FRED68A	M UNEQU SIL	FRED68A	
•SAISEV	JARO69A	FA29.6	FRED68A			JAR69A MUE68A
•UDEN••	VANS73A	FA30.3	FRED68A			
•VAVILO	VANS73A	FA30.6	FRED68A			DYA61A
LL (PETROLOGIC TYPE NOT KNOWN)						
BEELER		FA30	BILD73A	S KANSADA	READ72A	
BERDYA		FA31	MASO63A			KHA65A
•BHOLA•		FA27	MASO63A			
BOELUS		FA30	MASO63A			
CHICO•		FA27	MASO63A			KHA69A
•GALIM•		FA29	MASO63A			
•GUIDDE		FA27.8	FRED68A			
HOLMAN		FA29	MASO63A			MUE66A
•KARAKO		FA28	MASO63A			
•KHARKO	KHAR72A			M FA 24	MASO63A	KHA72A
•MERU••		FA28	MASO63A			
•NYIRAB		FA28.7	FRED68A			WII69A
•OKNINY		FA27	MASO63A			
OUBARI		FA28.0	FRED68A	C L4	VANS67A	
•PARAGO		FA27.6	FRED68A			

ABBREV	REF	SIL COMP	REF	REMARKS	REF	BULK ANALYSES	
L (PETROLOGIC TYPE NOT KNOWN)							
PEVENS		FA28	MASO63A				
•RAMPUR		FA27	MASO67A				
SAILAW		FA30	MASO71A				
•SAVTSC		FA28.0	FRED68A			DYA61A	
•SEVILL		FA28	MASO63A				
•SIENA•		FA28.7	KURA69A			WII69A	
•TOMAKO		FA30	MASO63A				
•VISHNU		FA30	MASO63A				
•WITSAN		FA27	MASO63A				
YALGOO		FA27	MASO67A				
•YUKAN•		FA29	MASO63A				
CV2							
●ALRAIS	VANS73B	FA 0-51	WOOD67A			WII69A	
•KABA••	VANS69A	FA 0-99	VANS69A	C CV3	MASO71A		
•MOKOIA	VANS69A	FA 0-42	WOOD67A	C CV3	MASO71B	WII69A	VON69A
●RENAZZ	VANS73B	FA 0-42	WOOD67A	M 12 PC METAL	MASO71B	WII69A	
CV3							
•ALLEND	MASO71B	FA 0-45	CLAR70A			JAR70B	MCC72A
•BALI••	MASO71B	UNEQU					
EFREMO	VANS69A	UNEQU	VANS69A			WIL73A	
•GROSNA	VANS69A	FA 0-72	VANS69A			WII69A	
LEOVIL	MASO71B	UNEQU				WII69A	MCC72A
•VIGARA	VANS69A	UNEQU	VANS69A			WII69A	VON69A
CV4							
COOLID	VANS69A	FA14.2	VANS69A			JAR66A	MCC72A
CO3							
•FELIX•	VANS69A	UNEQU	VANS69A			WII69A	VON69A
●KAINSA	VANS69A	FA 0-60	VANS69A	M 16 PC METAL	DYAK64A	DYA64A	WIL73A
•LANCE•	VANS69A	UNEQU				WII69A	VON69A
•ORNANS	VANS69A	FA 0-62	VANS69A			WII69A	
•WARREN	VANS69A	FA 0-58	VANS69A			WII69A	MCC72A
CO4							
•KAROON	VANS67A	FA33.4	VANS69A	C CO3	MASO71B	WII69A	
CM2							
•BELLS•	VANS67A	UNEQU					
•BORISK	VANS67A	UNEQU				WII69A	
•COLDBO	VANS67A	FA 0-54	WOOD67A			WII69A	VON69A
•CRESCE	VANS67A	UNEQU					
•ERAKOT	VANS67A	UNEQU				WII69A	
•ESSEBI	MASO71B	UNEQU				WII69A	
•HARIPU	VANS67A	FA 0-52	WOOD67A			WII69A	
•MIGHEI	VANS67A	FA 0-69	WOOD67A			WII69A	VON69A
•MURCHI	MASO71B	UNEQU				JAR71E	
•MURRAY	VANS67A	FA 0-51	WOOD67A			WII69A	VON69A
•NAWAPA	VANS67A	UNEQU				WII69A	
•NOGOYA	VANS67A	FA 0-44	WOOD67A				
•POLLEN	VANS67A	FA 0-52	WOOD67A				
•SANTCR	VANS67A	UNEQU				WII69A	

ABBREV	REF	SIL COMP	REF	REMARKS	REF	BULK ANALYSES
CI1						
ALAIS	VANS67A	UNEQU	VANS73B			WII69A
IVUNA	VANS67A	UNEQU	VANS73B			WII69A
*ORGUEI	VANS67A	FA 0-13	REID70A			WII69A VON69A
*REVELS	MASO71B	UNEQU				
*TONK**	VANS67A	UNEQU				
ANOMALOUS CHONDRITES						
*CUMBEC		FS 2-17	BINN69A	B CUMBEA	BINN69A	JAR67A EAS69A
*KAKANG	MASO66B	UNEQU	MASO66B			WII69A
MOUNMW	MASO67C	FA 5	MASO67C	C TYPE 6**	VANS73A	
WINONA		FA 5	MASO67A	C TYPE 6**	VANS73A	JAR67B

References for Classification, Silicate Composition, and Remarks

Appendices II B and II C

AVAN73A AVANZO, P. E., LEVI-DONATI, G. R., SIGHINOLFI, G. P.: Meteoritics **8,** 141 (1973).

BAED74A BAEDECKER, P. A., WASSON, J. T.: Geochim. Cosmochim. Acta **38** (1974).

BILD73A BILD, R. W.: Unpublished data, 1973.

BINN67A BINNS, R. A.: Nature **213,** 1111 (1967).

BINN68A BINNS, R. A.: Geochim. Cosmochim. Acta **32,** 299 (1968).

BINN69A BINNS, R. A.: In: Millman, P. M., (Ed.): Meteorite Research, p. 696. Dordrecht: Reidel 1969.

BOST66A BOSTROM, K., FREDRIKSSON, K.: Smithsonian Misc. Coll. **151,** 39 pp. (1966).

BUNC69A BUNCH, T. E., FUCHS, L. H.: Am. Mineralogist **54,** 1509 (1969).

BUNC70A BUNCH, T. E., KEIL, K., OLSEN, E.: Contrib. Mineral. Petrol. **25,** 297 (1970).

BUNC72A BUNCH, T. E., MALL, A. P., LEWIS, C. F.: Meteoritics **7,** 87 (1972).

BUSE69A BUSECK, P. R., GOLDSTEIN, J. I.: Bull. Geol. Soc. Amer. **80,** 2141 (1969).

BUSE72A BUSECK, P. R., HOLDSWORTH, E. F.: Meteoritics **7,** 429 (1972).

CARA71A CARAPEZZA, M., NUCCIO, M.: Meteoritics **6,** 255 (1971).

CAYE69A CAYÉ, R., GIRAUD, R., SANDREA, A.: In: Millman, P. M., (Ed.): Meteorite Research, p. 657. Dordrecht: Reidel 1969.

CHRI70A CHRISTOPHE MICHEL-LÉVY, M., LÉVY, C., LEFRANC, J.-P., WIIK, H. B.: Bull. Soc. Fr. Mineral. Cristallogr. **93,** 114 (1970).

CLAR70A CLARKE, R. S., JAROSEWICH, E., MASON, B., NELEN, J., GOMEZ, M., HYDE, J. R.: Smithson. Contrib. Earth Sci. **5,** 1 (1970).

CLAR71A CLARKE, R. S., JAROSEWICH, E., NELEN, J.: J. Geophys. Res. **76,** 4135 (1971).

CLAR72A CLARKE, R. S., JAROSEWICH, E., NELEN, J.: Smithson. Contrib. Earth Sci. **9,** 67 (1972).

CRAI71A CRAIG, J. R., HOWE, D. A., HARRIS, R. L., KULLERUD, G., BRYAN, W. B.: Meteoritics **6,** 33 (1971).

DODD65A DODD, R. T., VAN SCHMUS, R.: J. Geophys. Res. **70,** 3801 (1965).

DODD67A DODD, R. T., VAN SCHMUS, W. R., KOFFMAN, D. M.: Geochim. Cosmochim. Acta **31,** 921 (1967).

DODD68A DODD, R. T.: Geochim. Cosmochim. Acta **32,** 1111 (1968).

DOER65A DÖRFLER, G., HECHT, F., PLÖCKINGER, E.: Tschmermaks Mineral. Petrogr. Mitt. **10,** 413 (1965).

DUKE67A DUKE, M. B., SILVER, L. T.: Geochim. Cosmochim. Acta **31,** 1637 (1967).

DYAK 60A DYAKONOVA, M. I., KHARITONOVA, V. YA.: Meteoritika **18,** 48 (1960).
DYAK 61A DYAKONOVA, M. I., KHARITONOVA, V. YA.: Meteoritika **21,** 52 (1961).
DYAK 64A DYAKONOVA, M. I.: Meteoritika **25,** 129 (1964).
FLEI 70A FLEISCHER, R. L., LIFSHIN, E., PRICE, P. B., WOODS, R. T., CARTER, R. W., FIREMAN, E. L.: Icarus **12,** 402 (1970).
FODO71A FODOR, R. V., KEIL, K., JAROSEWICH, E., HUSS, G. I.: Meteoritics **6,** 71 (1971).
FODO71B FODOR, R. V., KEIL, K., JAROSEWICH, E., HUSS, G. I.: Chem. Erde **30,** 103 (1971).
FODO72A FODOR, R. V., KEIL, K., JAROSEWICH, E.: Meteoritics **7,** 495 (1972).
FRED63A FREDRIKSSON, K., KEIL, K.: Geochim. Cosmochim. Acta **27,** 717 (1963).
FRED67A FREDRIKSSON, K., MASON, B.: Geochim. Cosmochim. Acta **31,** 1705 (1967).
FRED67B FREDRIKSSON, K., KRAUT, F.: Geochim. Cosmochim. Acta **31,** 1701 (1967).
FRED68A FREDRIKSSON, K., NELEN, J., FREDRIKSSON, B. J.: In: AHRENS, L. H., (Ed.): Origin and Distribution of the Elements, p. 457. New York-London: Pergamon 1968.
FROS71A FROST, M. J.: Mineral. Mag. **38,** 89 (1971).
FROS71B FROST, M. J.: In: Letter from HUTCHISON, R., dated 23 Nov. 1971.
GRAH73A GRAHAM, A. L.: Meteoritics **8,** 181 (1973).
HESS49A HESS, H. H., HENDERSON, E. P.: Am. Mineralogist **34,** 494 (1949).
HUSS72A HUSS, G. I., BUSECK, P. R., MOORE, C. B.: Meteoritics **7,** 463 (1972).
HUTC71A HUTCHISON, R.: Meteoritics **6,** 53 (1971).
HUTC71B HUTCHISON, R.: Letter dated 23 Nov. 1971.
HUTC72A HUTCHISON, R.: Nature Phys. Sci. **240,** 58 (1972).
JARO66A JAROSEWICH, E.: Geochim. Cosmochim. Acta **30,** 1261 (1966).
JARO67A JAROSEWICH, E.: Geochim. Cosmochim. Acta **31,** 1103 (1967).
JARO69A JAROSEWICH, E., MASON, B.: Geochim. Cosmochim. Acta **33,** 411 (1969).
JARO73A JAROSEWICH, E.: Letter dated 9 August 1973.
JERO72A JÉROME, D. Y., CHRISTOPHE MICHEL-LÉVY, M.: Meteoritics **7,** 449 (1972).
JOBB66A JOBBINS, E. A., DIMES, F. G., BINNS, R. A., HEY, M. H., REED, S. J. B.: Mineral. Mag. **35,** 881 (1966).
JOHN67A JOHNSON, J. E., MCCALL, D. H.: Trans. Roy. Soc. S. Aust. **91,** 37 (1967).
JONE73A JONES, R. E.: Unpublished data 1973.
KEIL64A KEIL, K., FREDRIKSSON, K.: J. Geophys. Res. **69,** 3487 (1964).
KEIL68A KEIL, K.: J. Geophys. Res. **73,** 6945 (1968).
KHAR69A KHARITONOVA, V. YA.: Meteoritika **29,** 91 (1969)
KHAR72A KHARITONOVA, V. YA.: Meteoritika **31,** 116 (1972).
KRAU71A KRAUT, F., FREDRIKSSON, K.: Meteoritics **6,** 284 (1971).
KURA69A KURAT, G., FREDRIKSSON, K., NELEN, J.: Geochim. Cosmochim. Acta **33,** 765 (1969).
KVAS72A KVASHA, L. G., DYAKONOVA, M. I.: Meteoritika **31,** 109 (1972).
LOVE62A LOVERING, J. F.: In: MOORE, C. B., (Ed.): Researches on Meteorites, p. 179. New York: Wiley 1962.
MARV72A MARVIN, U. B., WOOD, J. A.: Meteoritics **7,** 601 (1972).
MASO62A MASON, B.: Meteorites, p. 274. New York: Wiley 1962.
MASO63A MASON, B.: Geochim. Cosmochim. Acta **27,** 1011 (1963).

MASO63B Mason, B.: Am. Mus. Novitates **2155,** 13 (1963).
MASO66A Mason, B.: Geochim. Cosmochim. Acta **30,** 23 (1966).
MASO66B Mason, B., Wiik, H. B.: Am. Mus. Novitates **2272,** 1 (1966).
MASO66C Mason, B.: Geochim.Mason, B., Wiik, H. B.: Am. Mus. Novitates **2273,** 1 (1966).
MASO67A Mason, B.: Geochim. Cosmochim. Acta **31,** 1100 (1967).
MASO67B Mason, B.: Geochim. Cosmochim. Acta **31,** 107 (1967).
MASO67C Mason, B., Jarosewich, E.: Geochim. Cosmochim. Acta **31,** 1097 (1967).
MASO68A Mason, B., Jarosewich, E.: Science **160,** 878 (1968).
MASO68B Mason, B., Nelen, J.: Geochim. Cosmochim. Acta **32,** 661 (1968).
MASO71A Mason, B.: Letter, Sept. 21, 1971.
MASO71B Mason, B.: Meteoritics **6,** 59 (1971).
MASO73A Mason, B.: Meteoritics **8,** 1 (1973).
MASO73B Mason, B., Jarosewich, E.: Mineral. Mag. **39,** 204 (1973).
MCCA68A McCall, G. J. H., Cleverly, W. H.: Mineral. Mag. **36,** 691 (1968).
MCCA70A McCall, G. J. H., Cleverly, W. H.: J. Roy. Soc. Western Australia **53,** 69 (1970).
MCCA70B McCall, G. J. H., Jeffery, P. M.: Mineral. Mag. **37,** 880 (1970).
VONM69A von Michaelis, H., Ahrens, L. H., Willis, J. P.: Earth Planet. Sci. Letters **5,** 387 (1969).
MOSS67A Moss, A. A., Hey, M. H., Elliott, C. J., Easton, A. J.: Mineral. Mag. **36,** 101 (1967).
MUEL66A Müller, O., Zähringer, J.: Earth Planet. Sci. Letters **1,** 25 (1966).
NELE72A Nelen, J. A., Fredriksson, K.:Smithson. Contrib. Earth Sci. **9,** 69 (1972).
NOON72A Noonan, A. F., Fredriksson, K. A., Nelen, J.: Smithson. Contrib. Earth Sci. **9,** 57 (1972).
NOON73A Noonan, A. F., Jarosewich, E., Clarke, R. S.: Meteoritics **8,** 61 (1973).
OLSE70A Olsen, E., Jarosewich, E.: Earth Planet. Sci. Letters **8,** 261 (1970).
POWE71A Powell, B. N.: Geochim. Cosmochim. Acta **35,** 5 (1971).
RAMD67A Ramdohr, P.: Chem. Erde **26,** 1 (1967).
RAMD69A Ramdohr, P., El Goresy, A.: Naturwissenschaften **56,** 512 (1969).
READ71A Read, W. F.: Meteoritics **6,** 105 (1971).
READ72A Read, W. F.: Meteoritics **7,** 417 (1972).
READ72B Read, W. F.: Meteoritics **7,** 509 (1972).
REED68A Reed, S. J. B.: Mineral. Mag. **36,** 850 (1968).
REID67A Reid, A. M., Cohen, A. J.: Geochim. Cosmochim. Acta **31,** 661 (1967).
REID70A Reid, A. M., Bass, M. N., Fujita, H., Kerridge, J. F., Fredriksson, K.:Geochim. Cosmochim. Acta **34,** 1253 (1970).
SCHA72A Schaudy, R., Wasson, J. T., Buchwald, V. F.: Icarus **17,** 174 (1972).
SMIT72A Smith, P. G. W., Bell, J. D., Frisch, T.: Meteoritics **7,** 1 (1972).
SYME70A Symes, R. F., Hutchison, R.: Mineral. Mag. **37,** 221 (1970).
TERA70A Tera, F., Eugster, O., Burnett, D. S., Wasserburg, G. J.: Geochim. Cosmochim. Acta, Suppl. I, 1637 (1970).
VANS67A van Schmus, W. R., Wood, J. A.: Geochim. Cosmochim. Acta **31,** 747 (1967).
VANS69A van Schmus, W. R.: In: Millman, P., (Ed.): Meteorite Research, p. 480. Dordrecht: Reidel 1969.
VANS73A van Schmus, W. R.: Private communication 1973.

VANS73B van Schmus, W. R., Hayes, J. M.: Geochim. Cosmochim. Acta **38**, 47—64 (1973).

WASS67A Wasson, J. T., Kimberlin, J.: Geochim. Cosmochim. Acta **31**, 2065 (1967).

WASS69B Wasson, J. T., Sedwick, S. P.: Nature **222**, 22 (1969).

WASS70B Wasson; J. T.: Geochim. Cosmochim. Acta **34**, 957 (1970).

WASS70C Wasson, J. T., Wai, C. M.: Geochim. Cosmochim. Acta **34**, 169 (1970).

WASS74A Wasson, J. T.: This work (1974).

WIIK69A Wiik, H. B.: Commun. Phys.-Math. (Helsinki) **34**, 135 (1969).

WIIK72A Wiik, H. B.: Meteoritics **7**, 553 (1972).

WLOT71A Wlotzka, F.: Unpublished data 1971.

WOOD67A Wood, J. A.: Geochim. Cosmochim. Acta **31**, 2095 (1967).

YODE57A Yoder, H. S., Sahama, T. G.: Am. Mineralogist **42**, 475 (1957).

References for Bulk Analyses Entries

Appendices II B and II C

DYA 60A Dyakonova, M.I., Kharitonova, V.Y.: Meteoritika **18,** 48 (1960).

DYA 60A Dyakonova, M.I., Kharitonova, V.Y.: Meteoritika **21,** 52 (1961) (in Russian).

DYA 64A Dyakonova, M.I.: Meteoritika **25,** 129 (1964).

DYA 68A Dyakonova, M.I.: Meteoritika **28,** 131 (1968).

DYA 72A Kvasha, L.G., Dyakonova, M.I.: Meteoritika **31,** 109 (1972).

DYA 72B Dyakonova, M.I.: Meteoritika **31,** 119 (1972).

EAS 63A Easton, A.J., Lovering, J.F.: Geochim. Cosmochim. Acta **27,** 753 (1963).

EAS 66A Jobbins, E.A., Dimes, F.G., Binns, R.A., Hey, M.H., Reed, S.J.B.: Mineral. Mag. **35,** 881 (1966).

EAS 67A Hey, M.H., Easton, A.J.: Geochim. Cosmochim. Acta **31**, 1789 (1967).

EAS 67B Binns, R.A.: Mineral. Mag. **36,** 319 (1967).

EAS 67C Moss, A.A., Hey, M.H., Elliott, C.J., Easton, A.J.: Mineral. Mag. **36,** 101 (1967).

EAS 68A Binns, R.A.: Geochim. Cosmochim. Acta **32,** 299 (1968).

EAS 69A Binns, R.A.: In: Millman, P.M., (Ed.): Meteorite Research, p.696. Dordrecht: Reidel 1969.

EAS 70A Symes, R.F., Hutchison, R.: Mineral. Mag. **37,** 221 (1970).

HAR 62A Miyashiro, A.: Jap. J. Geol. Geogr. **33,** 73 (1962).

HAR 62B Miyashiro, A., Murayama, S., Haramura, H.: Jap. J. Geol. Geogr. **33,** 239 (1962).

HAR 62C Miyashiro, A.: Jap. J. Geol. Geogr. **33,** 125 (1962).

HAR 63A Mason, B.: Am. Mus. Novitates **2155,** 13 (1963).

HAR 63B Miyashiro, A., Murayama, S., Haramura, H.: Bull. Nat. Sci. Mus. Tokyo **6,** 352 (1963).

HAR 63C Miyashiro, A., Murayama, S., Haramura, H.: Jap. J. Geol. Geogr. **34,** 63 (1963).

HAR 63D Miyashiro, A., Murayama, S., Haramura, H.: Jap. J. Geol. Geogr. **34,** 193 (1963).

HAR 67A Miyashiro, A., Murayama, S.: Chem. Erde **26,** 219 (1967).

JAR 66A Jarosewich, E.: Geochim. Cosmochim. Acta **30,** 1261 (1966).

JAR 67A Jarosewich, E.: Geochim. Cosmochim. Acta **31,** 1103 (1967).

JAR 67B Mason, B., Jarosewich, E.: Geochim. Cosmochim. Acta **31,** 1097 (1967).

JAR 68A Mason, B., Jarosewich, E.: Science **160,** 878 (1968).

JAR 68B McCall, G.J.H., Cleverly, W.H.: Mineral. Mag. **36,** 691 (1968).

JAR 68C McCall, G.J.H.: Mineral. Mag. **36,** 726 (1968).

JAR69A JAROSEWICH, E., MASON, B.: Geochim. Cosmochim. Acta **33**, 411 (1969).

JAR69B FREDRIKSSON, K., JAROSEWICH, E., NELEN, J.: In: MILLMAN, P.M., (Ed.): Meteorite Research, p. 155. Dordrecht: Reidel 1969.

JAR70A BUNCH, T.E., KEIL, K., OLSEN, E.: Contrib. Mineral. Petrol. **25**, 297 (1970).

JAR70B CLARKE, R.S., JAROSEWICH, E., MASON, B., NELEN, J., GOMEZ, M., HYDE, J.R.: Smithson. Contrib. Earth Sci. **5**, 1 (1970).

JAR70C OLSEN, E., JAROSEWICH, E.: Earth Planet. Sci. Letters **8**, 261 (1970).

JAR71A FODOR, R.V., KEIL, K., JAROSEWICH, E., HUSS, G.I.: Meteoritics **6**, 71 (1971).

JAR71B CLARKE, R.S., JAROSEWICH, E., NELEN, J.: J. Geophys. Res. **76**, 4135 (1971).

JAR71C KING, E.A., JAROSEWICH, E., BROOKINS, D.G.: Meteoritics **6**, 280 (1971).

JAR71D LEVI-DONATI, G.R., JAROSEWICH, E.: Meteoritics **6**, 1 (1971).

JAR71E JAROSEWICH, E.: Meteoritics **6**, 49 (1971).

JAR71F OLSEN, E., JAROSEWICH, E.: Science **174**, 583 (1971).

JAR71G FODOR, R.V., KEIL, K., JAROSEWICH, E., HUSS, G.I.: Chem. Erde **30**, 103 (1971).

JAR71H MASON, B., JAROSEWICH, E.: Meteoritics **6**, 241 (1971).

JAR72A CLARKE, R.S., JAROSEWICH, E., NELEN, J.: Smithson. Contrib. Earth Sci. **9**, 67 (1972).

JAR72B FODOR, R.V., KEIL, K., JAROSEWICH, E.: Meteoritics **7**, 495 (1972).

JAR72C LEVI-DONATI, G.R., JAROSEWICH, E.: Meteoritics **7**, 109 (1972).

JAR73A MASON, B.: Meteoritics **8**, 1 (1973).

JAR73B NOONAN, A.F., JAROSEWICH, E., CLARKE, R.S.: Meteoritics **8**, 61 (1973).

JAR73C MASON, B., JAROSEWICH, E.: Mineral. Mag. **39**, 204 (1973).

JAR73D JAROSEWICH, E.: Letter dated 9 August 1973.

KHA65A KHARITONOVA, V.Y.: Meteoritika **26**, 146 (1965).

KHA68A KHARITONOVA, V.Y.: Meteoritika **28**, 138 (1968).

KHA69A KHARITONOVA, V.Y.: Meteoritika **29**, 91 (1969).

KHA72A KHARITONOVA, V.Y.: Meteoritika **31**, 116 (1972).

KOE64A KÖNIG, H.: Geochim. Cosmochim. Acta **28**, 1397 (1964).

KOE64B KÖNIG, H.: Geochim. Cosmochim. Acta **28**, 1697 (1964).

MAY61A DUKE, M., MAYNES, D., BROWN, H.: J. Geophys. Res. **66**, 3557 (1961).

MAY67A DUKE, M.B., SILVER, L.T.: Geochim. Cosmochim. Acta **31**, 1637 (1967).

MAY67B MASON, B., MAYNES, A.D.: Proc. U.S. Nat. Mus. **124**, No. 3624, 12 pp. (1967).

MAY67C FREDRIKSSON, K., MASON, B.: Geochim. Cosmochim. Acta **31**, 1705 (1967).

MCC72A MCCARTHY, T., AHRENS, L.: Earth Planet. Sci. Letters **14**, 97 (1972).

MUE66A MÜLLER, O., ZÄHRINGER, J.: Earth Planet. Sci. Letters **1**, 25 (1966).

MUE68A MÜLLER, O.: Z. Naturforsch. **23a**, 48 (1968).

MUE69A KEMPE, W., MÜLLER, O.: In: MILLMAN, P.M., (Ed.): Meteorite Research, p. 418. Dordrecht: Reidel 1969.

NEL68A MASON, B., NELEN, J.: Geochim. Cosmochim. Acta **32**, 661 (1968).

NEL72A NELEN, J.A., FREDRIKSSON, K.: Smithsonian Contrib. Earth Sci. **9**, 69 (1972).

NEL72B NELEN, J., MASON, B.: Smithsonian Contrib. Earth. Sci. **9**, 55 (1972).

POW71A	POWELL, B. N.: Geochim. Cosmochim. Acta **35**, 5 (1971).
VON69A	VON MICHAELIS, H., AHRENS, L. H., WILLIS, J. P.: Earth Planet. Sci. Letters **5**, 387 (1969).
WII63A	MASON, B.: Am. Mus. Novitates **2163**, 19 (1963).
WII69A	WIIK, H. B.: Commun. Phys. Math. (Helsinki) **34**, 135 (1969).
WII70A	CHRISTOPHE MICHEL-LEVY, M., LEVY, C., LEFRANC, J.-P., WIIK, H. B.: Bull. Soc. Fr. Mineral. Cristallogr. **93**, 114 (1970).
WII72A	WIIK, H. B.: Meteoritics **7**, 553 (1972).
WIL73A	AHRENS, L. H., WILLIS, J. P., ERLANK, A. J.: Meteoritics **8**, 133 (1973).

C. Listing of Differentiated Silicate-Rich Meteorites by Chemical Group

In the following list achondrites and other differentiated meteorites with appreciable silicate contents are ordered according to group, and alphabetically within each group. The first two columns are as described in Appendix II B. In the third column a number preceded by FA is the olivine composition in mol % fayalite; a number preceded by FS is generally an orthopyroxene or low-Ca clinopyroxene composition in mol % ferrosilite. However, in all eucrites and in a few anomalous meteorites the analyzed pyroxene is pigeonite, and the FS value is not mol % ferrosilite, but rather the Fe/(Fe + Mg) ratio in units of mol %. The fourth column contains the source of the data on silicate-mineral composition.

The Ca/Mg ratios in the fifth column are in units of atoms/100 atoms. The sixth column (labelled FEMET) gives whole-rock (= bulk) metallic Fe contents in wt %. The CA/MG and FEMET data are taken from the reference cited in the first column under Bulk Analyses. If two analyses are cited, the name of the second analyst is truncated to two letters, but inspection of the reference list will show that the designations are still unique.

The meanings of the abbreviations used in the Remarks column are given in Table AII-2. Because of a shortage of space, no references are given for the remarks. Generally these points are discussed in the references cited here for classification or, for the pallasites or iron meteorites, in Appendix II D.

This section contains the complete listing of mesosiderites and pallasites. Appendix II D contains the complete listing of the silicate-bearing iron meteorite groups I A, I B, II E, and IV A.

Separate reference lists for 1. classification and silicate composition, and 2. bulk analytical data are given at the end of Appendix II B.

Appendix II C

ABBREV	REF	SIL COMP	REF	CA/MG	FEMET	BULK ANALYSES	REMARKS
AUB							
*AUBRES	REID67A	FS 0.02	REID67A				
*BISHOV	REID67A	FS 0.03	WASS70C				
*BUSTEE	REID67A	FS<0.03	REID67A				
*CUMBEA	REID67A	FS 0.05	WASS70C				
*KHORTE	REID67A	FS<0.03	REID67A	1.6	0.22	EAS67B	
*NORTON	REID67A	FS 0.02	WASS70C	2.0	1.00	WII69A VO69A	
*PENABL	REID67A	FS 0.08	REID67A				
*PESYAN	REID67A	FS 0.04	REID67A	1.9	0.64	DYA60A MU66A	
SHALLO	REID67A	FS<0.02	WASS70C				
DIO							
*ELLEME	MASO63B	FS26	MASO63B				
*GARLAN	MASO63B	FS25	MASO63B				
*IBBENB	MASO63B	FS26	MASO63B				
*JOHNST	MASO63B	FS27	MASO63B	3.5	---	JAR71H VO69A	
*MANEGA	MASO63B	FS25	MASO63B				
*RODA**	MASO63B	FS27	MASO63B				
*SHALKA	MASO63B	FS26	MASO63B	2.1	---	VON69A	
*TATAHO	MASO63B	FS25	MASO63B				
URE							
DINGOP	MCCA68A	FA10	MCCA68A	3.3	0.36	JAR68B	
*DYALPU	MASO62A			2.5	3.07	WII69A	
GOALPA	MASO62A			0.9	5.86	WII69A	
*HAVERO	MARV72A	FA 5-17	MARV72A	0.2	3.6	WII72A	
NORTHH	MCCA68A	FA 0-30	MCCA68A	2.6	0.34	JAR68B	
*NOVOUR	MASO62A	FA21	MASO62A	1.4	2.90	WII69A DY64A	
EUC							
ADALIA	MASO67B						
*BEREBA	MASO67B						
*BIALYS	MASO67B	FS60-70	DUKE67A				
CACHAR	MASO67B	FS47	FRED67B				
*CHERVO	MASO67B						
*EMMAVI	MASO67B						
*HARAIY	MASO67B						
*IBITIR	MASO67B						
*JONZAC	MASO67B						
*JUVINA	MASO67B	FS61	DUKE67A	109.	0.04	MAY67A	
*KIRBYV	MASO67B						
*LAKANG	MASO67B						
*MACIBI	MASO67B						
*MOOREC	MASO67B	FS50	DUKE67A	81.	0.67		
*NAGARI	MASO67B						
*NOBLEB	MASO67B						
NUEVOL	MASO67B	FS67	DUKE67A	117.			
*PASAMO	MASO67B	FS48-70	DUKE67A	109.	---	MAY67A VO69A	
*PERAMI	MASO67B						
*PETERS	MASO67B	FS58	DUKE67A				
POMOZD	KVAS72A			78.	0.90	DYA72A	

ABBREV	REF	SIL COMP	REF	CA/MG	FEMET	BULK ANALYSES	REMARKS
EUC							
•SERRAD	MASO67B	FS44	DUKE67A				
•SIOUXC	MASO67B	FS59	DUKE67A	103.	0.03	MAY67A VO69A	
•STANNE	MASO67B	FS62	DUKE67A	116.	0.02	MAY67A	
HOW							
•BHOLGH	MASO67B						
BINDA•	MASO67B	FS32	DUKE67A				
•BRIENT	MASO67B			55.	0.13	DYA61A	
•BUNUNU	MASO67B	FS22-35	MASO67B	34.	1.01	JAR67A	
•CHAVES	MASO67B						
•FRANFS	MASO67B	FS22-35	MASO66B	14.	0.00	WII69A	
•JODZIE	MASO67B						
•KAPOET	MASO67B	FS15-61	FRED63A	48.	0.00	WII69A MU66A	
•LETEIL	MASO67B						
•LUOTOL	DUKE67A			33.	0.07	WII69A	
•MAESSI	MASO67B						
•MALVER	VONM69A			50.	---	VON69A	
•MEDANI	SYME70A	FS41	SYME70A	59.	0.22	EAS70A	
•MOLTEN	FROS71A						
•PADVAR	BINN67A	FS36-38	BINN67A				
•PAVLOV	MASO67B						
•WASHOU	MASO67B	FS20-47	JERO72A				
•YURTUK	MASO67B						
•ZMENJ•	MASO67B			22.	0.25	WII69A	
MES							
•BAREA•	MASO73B	FS23-37	MASO73B	28.	50.3	JAR73C	
BONDOC	POWE71A	FS24-31	POWE71A	11.	40.0	POW71A	
BUDULA	POWE71A						
CHINGU	POWE71A						
CLOVER	POWE71A			39.	35.59	WII69A	
CRABOR	POWE71A	FS26-36	POWE71A	34.	49.1	POW71A	
DALGAR	POWE71A						
•DYARRL	MASO73B	FS34-51	MASO73B	36.	15.8	JAR73C	
EMERY•	MASO73B	FS30-37	MASO73B	38.	45.1	JAR73C	
•ESTHER	POWE71A	FS16-28	DUKE67A	20.	51.0	POW71A NE72A	
HAINHO	POWE71A	FS19-28	POWE71A	31.	48.3	POW71A	
•LOWICZ	POWE71A	FS20-37	POWE71A	23.	54.5	POW71A	
MINCY•	POWE71A	FS22-31	POWE71A				
MORRIS	POWE71A	FS33-39	POWE71A	39.	45.0	POW71A	
MOUNPA	POWE71A			19.		WII69A	
•PATWAR	POWE71A	FS21-41	POWE71A	42.	32.0	JAR69A PO71A	
PINNAR	POWE71A						
SIMOND	POWE71A						
VACAMU	POWE71A	FS23-44	POWE71A	44.	42.8	POW71A	
•VERAMI	POWE71A	FS19-27	POWE71A	12.	44.4	POW71A	
PAL							
ADMIRE	BUSE69A	FA12.1	BUSE69A				
AHUMAD	BUSE69A	FA11.6	BUSE69A				
ALBIPA	BUSE69A	FA12.5	BUSE69A				

ABBREV	REF	SIL COMP	REF	CA/MG	FEMET	BULK ANALYSES	REMARKS
PAL							
ANDERS	BUSE69A	FA12.3	BUSE69A				P BRENHAM
ANTOFA	BUSE69A	FA12.6	BUSE69A				P IMILAC
ARGONI	BUSE69A	FA13.5	BUSE69A				
BRAHIN	BUSE69A	FA11.7	BUSE69A				
BRENHA	BUSE69A	FA12.4	BUSE69A				
COLDBA	BUSE69A	FA19.3	BUSE69A				
OEAGLES	BUSE69A	FA20.0	BUSE69A				
ESQUEL	BUSE69A	FA11.7	BUSE69A				
FINMAR	BUSE69A	FA12.7	BUSE69A				
GIROUX	BUSE69A	FA10.8	BUSE69A				
OGLORIE	BUSE69A	FA13.2	BUSE69A				
GRANCH	MASO63C	FA11.	MASO63C				P IMILAC
HOPEWE	WASS69B						P BRENHAM
HUCKIT	BUSE69A	FA12.7	BUSE69A				
ILIMPA	BUSE69A	FA12.6	BUSE69A				P IMILAC
IMILAC	BUSE69A	FA12.3	BUSE69A				
OITZAWI	BUSE69A	FA19.2	BUSE69A				
KRASNJ	BUSE69A	FA12.2	BUSE69A				
LIPOVS	BUSE69A	FA11.6	BUSE69A				
MARBUR	BUSE69A	FA12.3	BUSE69A				
*MARJAL	BUSE69A	FA11.9	BUSE69A				
MOLONG	BUSE69A	FA11.3	BUSE69A				
MOUNDY	BUSE69A	FA13.2	BUSE69A				
MOUNVE	BUSE69A	FA11.9	BUSE69A				
NEWPOR	BUSE69A	FA11.8	BUSE69A				
OLLAGU	BUSE69A	FA12.4	BUSE69A				
PAVLDP	BUSE69A	FA12.7	BUSE69A				
PHIL35	BUSE69A	FA17.9	BUSE69A				
POJOAQ	BUSE69A	FA12.9	BUSE69A				P GLORIETA
PORTOR	BUSE69A	FA12.3	BUSE69A				
RAWLPA	BUSE69A	FA15.9	BUSE69A				
SALTA*	BUSE69A	FA12.5	BUSE69A				
SANTRS	BUSE69A	FA12.1	BUSE69A				
SOMERV	BUSE69A	FA12.6	BUSE69A				
SOUTBE	BUSE69A	FA12.0	BUSE69A				
SPRINW	BUSE69A	FA18.0	BUSE69A	0.0	---	WII63A	
THIELM	BUSE69A	FA12.7	BUSE69A				
*ZAISHO	MASO63C	FA19.	MASO63C				
IA							
CAMPOD	WASS70B	FS 6.2	BUNC70A	3.8	---	JAR70A	
COPIAP	WASS70B	FS 6.7	BUNC70A				
LINWOO	WASS70B	FS 7.2	BUNC70A				
ODESIR	WASS70B	FS 6.6	BUNC70A				
PINERI	WASS70B	FS 3.6	BUNC70A				
TACUBA	WASS70B	FS 6.9	BUNC70A				
TOLUCA	WASS70B	FS 6.6	BUNC70A				
*UDEIST	WASS70B	FS 8.3	BUNC70A				
IB							
FOURCO	WASS70B	FS 6.5	BUNC70A				
PERSIM	WASS70B	FS 7.2	BUNC70A				
PITTS	WASS70B	FS 6.9	BUNC70A				
WOODBI	WASS70B	FS 7.7	BUNC70A	4.7	73.06	JAR67A	

ABBREV	REF	SIL COMP	REF	CA/MG	FEMET	BULK ANALYSES	REMARKS
IIE							
COLOME	WASS70B	FS23.5	BUNC70A				
ELGA++	WASS70B						
OKODAIK	WASS70B	FS16.6	BUNC70A				
ONETSCH	WASS70B	FS13.6	BUNC70A	6.0		JAR71F	
WEEKER	WASS70B	FS21.2	BUNC70A	23.		JAR70C	
IVA							
OSTEINB	SCHA72A	FS10.3	DOER65A				
ANOMALOUS SILICATE-RICH METEORITES							
+ANGRST	MASO62A	FA46.5	HUTC72A				
BENCUB	WASS70C	FS 3.6	WASS70C			JAR68C EA63A	G WEATHERF
+CHASSI	MASO62A	FA33	MASO62A	1.2	≤0.1	DYA60A	
ENON++	BUNC70A	FS10.4	BUNC70A				
HARVAR	RAMD67A	FA18	RAMD67A				G LODRAN
KENDAL	WASS70B	FS 0.7	BUNC70A				
LAFAST	MASO62A						G NAKHLA
+LODRAN	MASO62A						G HARVARD
MOUNEG	WASS70C	FS<0.02	WASS70C				
+NAKHLA	MASO62A	FA66	MASO62A				G LAFAYETT
+PECKEL	RAMD69A	FS14.4	RAMD69A				
+SHERGO	WASS74A						G ZAGAMI
TUCSON	WASS70B	FS 0.4	BUNC69A				
WEATHE	MASO68B	FS 3.1	MASO68B	6.0	48.	NEL68A	G BENCUBBI
+ZAGAMI	WASS74A						G SHERGOTT

D. Listing of Differentiated Metal-Rich Meteorites by Chemical Group

In the following list iron meteorites and other differentiated meteorites with appreciable metal contents are ordered according to group. As discussed in Chapter II, mean Ge content generally decreases with increasing group numeral, and for the same numeral, with increasing alphabetical position of the suffix letter. Within each group the meteorites are listed in order of increasing Ni content. As a result, this listing lends itself to the selection of a suite of meteorites spanning the entire fractionation range found within a single chemical group. It should be noted that all groups are independent except for the following four pairs—IA—IB, IIA—IIB, IIIA—IIIB, and IIIC—IIID. The anomalous metal-rich meteorites are listed in order of decreasing Ge content.

The name abbreviation and the use of + for an observed fall and O for an "anomalous member" of a group are as described in Appendix IIB. Columns two through five give Ni (%), Ga (ppm), Ge (ppm) and Ir (ppm) data taken from the reference cited in column six. The seventh column gives the kamacite bandwidth of octahedrites and ataxites in units of mm, and the eighth column gives the structural class according to the BUCHWALD (1974) system (see Table II-6 for the meaning of the structural class symbols). If no reference is given for the bandwidth and structural class, these are taken from the reference for the compositional data.

The meanings of the abbreviations found in the "Remarks" column are given in Table A II-2. Because of a shortage of space, no references are given for the remarks, but with rare exceptions, all points are discussed either in the reference for the analytical data or in BUCHWALD (1974).

These listings of iron-meteorite groups are complete. Complete listings of mesosiderites and pallasites are included in Appendix II C.

Appendix II D

ABBREV	NI	GA	GE	IR	REF	BW	STRC	REF	REMARKS
IA									
OZACA92	5.88	83.8	307.	2.2	W70A	---	ANO	B74	
CORRIZ	6.1	81.2	337.	1.0	S71A	1.9	OG	B74	
OBALLER	6.19	84.5	326.	2.1	W70A	2.6	OG	B74	
LANDES	6.31	88.7	414.	2.9	W74A		ANO		
QUAIRA					B74A		OG		P YOUNDEGIN
YOUNDE	6.38	90.8	383.	2.0	W70A	2.3	OG	B74	
COOKEV	6.4	94.1	384.	2.1	W70A	2.4	OG		
LINWOO	6.4	90.4	374.	2.7	W70A	2.8	OG	B74	
OSSEO*	6.44	91.7	450.	5.4	W70A	2.8	OG	B74	
SEELAE	6.47	96.8	493.	1.1	W70A	3.1	OG	B74	
GLADIR	6.53	93.7	418.	3.0	W70A	2.8	OG	B74	
SEYMOU	6.54	89.0	382.	1.7	W70A	2.2	OG	B74	
OSCURO	6.55	91.0	359.	2.6	W74A	1.75	OG	B74	
SAREPT	6.55	99.9	457.	3.4	W70A	2.2	OG	B74	
COSBYS	6.57	91.5	431.	2.9	W70A	2.5	OG	B74	
SARDIS	6.58	93.7	400.	1.3	W74A	2.5	OG	B74	
MORDEN	6.6	81.	329.		R72A				
BOLIVI	6.6	88.0	377.	1.4	W70A	2.7	OG	B74	
CAMPOD	6.62	90.0	392.	3.2	W70A	3.0	OG	B74	
MAGURA	6.67	94.6	483.	3.2	W70A	2.4	OG	B74	
HOPE**	6.68	88.3	388.	0.59	W70A	2.1	OG	B74	S BOAZ
LEXING	6.69	85.4	316.	2.3	W74A	2.1	OG	B74	
SELIGM	6.69	91.3	423.	2.8	W70A	2.3	OG	B74	
BURGAV	6.71	95.8	519.	1.1	W70A	2.6	OG	B74	
*YARDYM	6.71	88.2	387.	1.6	W70A	2.2	OG	B74	
YENBER	6.72	86.7	312.	2.9	W74A	2.1	OG	B74	
BLACMT	6.73	98.	460.	2.2	W74A	2.6	OG	B74	
QUEELA	6.78	86.9	380.	2.6	W70A		OG		P GLADSTONE
SMITHV	6.78	86.9	363.	2.0	W70A	2.2	OG	B74	
WICHIT	6.78	86.9	341.	2.1	W70A	2.4	OG	B74	
BLOODY	6.79	81.6	320.	2.0	W70A	2.0	OG		P CANYON DI
LASVEG	6.8	80.3	320.	1.6	W74A				P CANYON DI
MOUNSI	6.8	90.	394.		R72A				

ABBREV	NI	GA	GE	IR	REF	BW	STRC	REF	REMARKS
IA									
CRANBO	6.80	85.4	358.	1.8	W70A	2.2	OG	B74	
MONUME	6.84	80.9	320.	1.9	W70A	1.9	OG		P CANYON DI
PANDEA	6.84	82.1	308.	2.0	W74A	2.2	OG	B74	
VAALBU	6.84	84.3	323.	1.7	W76A				
JENKIN	6.85	86.2	353.	1.8	W70A	2.3	OG	B74	
BURKET	6.87	87.2	368.	2.1	W70A	2.0	OG	B74	
NEWLEI	6.88	93.1	445.	2.5	W74A	2.6	OG	B74	
DUNGAN	6.9	79.7	330.	2.1	W70A	2.0	OG	B74	
CASEYC	6.96	82.5	317.	1.1	W70A	2.2	OG	B74	
ASHFOR	6.97	83.2	330.	2.3	W70A	2.4	OG		P CANYON DI
EHRENB	6.98	79.9	314.	1.8	W70A	2.0	OG		P CANYON DI
MOAB++	6.98	83.5	330.	1.6	W74A				P CANYON DI
PULASK					B74A				P CANYON DI
FOSSIL	6.98	77.7	308.	1.7	W74A		OG		P CANYON DI
CANYD1	6.98	81.8	324.	1.9	W70A	2.0	OG	B74	
SILVCR	6.98	82.0	321.	1.6	W74A	2.1	OG	B74	
HELTTO					B74A				P CANYON DI
JENNYS	7.0	86.0	320.	2.3	W70A	2.2	OG	B74	
COPIAP	7.01	69.8	252.	2.5	W70A	1.5	ANO	B74	
CAMPVE	7.06	78.2	322.	2.0	W70A	2.0	OG		P CANYON DI
WICKIR	7.07	82.6	321.	1.8	W70A	2.0	OG		P CANYON DI
MCCAME	7.07	74.1	281.	2.0	W70A	1.5	OG		P ODESSA
FAIROA	7.08	81.8	318.	1.9	W70A	1.6	OG		P CANYON DI
IDAHO+	7.09	81.1	321.	2.07	W74A				
NEPTUN	7.1	73.9	269.	2.0	W74A	1.9	OG	B74	
HOUCK+	7.1	80.5	330.	1.8	W70A		OG		P CANYON DI
DEELFO	7.11	83.1	306.	1.4	W70A	1.75	OG	B74	
+BOGOU+	7.15	77.4	301.	1.4	W70A	1.90	OG	B74	
ROSARI	7.16	89.5	401.	1.5	W74A	1.70	OG	B74	
MAYERT	7.19	75.5	283.	2.4	W70A	2.0	OG		
ODESIR	7.20	74.7	285.	2.2	W70A	1.70	OG	B74	
RIFLE+	7.20	77.2	281.	1.7	W70A	1.7	OG		
BOHUMI	7.37	75.3	264.	1.8	W70A	1.90	OG	B74	
MOUNST	7.4	84.6	346.	1.5	W74A		OG		P YOUNDEGIN
COOLAC	7.4	91.5	423.	2.4	W74A	2.1	OG	B74	
PINERI	7.40	76.9	234.	2.6	W70A	1.2	ANO	B74	
THOREA	7.4	73.8	271.	1.9	W70A	1.8	OG	B74	P ODESSA
WALDRO	7.55	74.6	282.	2.0	W70A	1.5	OG	B74	
MOORAN	7.7	79.1	328.	1.3	W74A		OG		P YOUNDEGIN
CALIFO	7.7	66.7	253.	1.9	W74A				
OANNAHE	7.74	79.8	302.	3.5	W70A	1.8	OG	B74	
MOUNAY	7.76	70.0	250.	1.8	W70A	1.6	OG	B74	
OGALLA	7.85	66.7	266.	2.6	W70A	1.60	OG	B74	
BISCHT	7.88	68.4	238.	1.9	W70A	1.80	OG	B74	
OCANYD6	7.9	79.3	317.	1.9	W70A	1.1	OM		P CANYON DI
●BAHJOI	7.95	64.5	265.	1.9	W70A	1.50	OG	B74	
NAGYVA	7.98	68.9	237.	2.1	W74A	1.40	OG	B74	
MOCTEZ	7.98	67.2	244.	2.4	W74A	1.35	OG	B74	
LEEDS+	7.99	67.1	241.	2.1	W70A	1.30	OG	B74	
OGOOSEL	8.00	67.2	305.	2.3	W70A	1.25	OM	B74	
SOUTER	8.06	66.2	242.	1.9	W70A	1.5	OG	B74	
TOLUCA	8.07	70.6	246.	1.8	W70A	1.40	OG	B74	
AMATES	8.09	66.2	237.	1.8	W74A		OG		P TOLUCA

ABBREV	NI	GA	GE	IR	REF	BW	STRC	REF	REMARKS
IA									
COMAIR	8.1	76.1	262.	2.2	W70A	1.50	OG	B74	
OKAREEK	8.1	79.5	355.	1.5	W70A	1.6	OG	B74	
DEPORT	8.11	69.9	255.	2.2	W70A	1.3	OG	B74	
SURPRI	8.12	69.6	265.	2.0	W70A	1.4	OG	B74	
TACUBA	8.13	66.9	249.	1.9	W74A		OG		P TOLUCA
MICHIG	8.16	69.7	251.	2.2	W74A		OG		P TOLUCA
OCANYD9	8.2	80.1	332.	2.0	W70A	1.1	OM		P CANYON DI
NIEDFI	8.27	72.0	257.	2.6	W74A				
MISTEC	8.27	67.8	233.	1.6	W70A	1.35	OG	B74	
OMORRIL	8.38	58.0	296.	1.7	W70A	0.90	OM	B74	
BALFOU	8.39	56.4	194.	2.0	W70A	1.30	OG	B74	
SHREWS	8.42	62.0	204.	2.6	W70A	1.15	OM	B74	
OZENDA◆	8.5	54.7	214.	2.1	W74A	0.85	OM	B74	
◆MAZAPI	8.64	60.2	221.	5.5	W70A	1.20	OM	B74	
◆UDEIST	8.83	61.6	204.	0.51	W70A		ANO	B74	
OMERTZO	8.98	68.0	293.	2.4	W70A	0.80	OM	B74	
IB									
FOURCO	8.90	48.7	179.	2.0	W70A	0.80	OM	B74	
MESAVE	10.56	53.0	142.	1.8	W74A	0.60	OM	B74	
WOODBI	10.6	37.3	114.	1.4	W70A	0.3	ANO	B74	
COLFAX	10.84	52.8	153.	1.5	W70A	0.60	OM	B74	
BITBUR	12.4	34.8	140.	0.46	W74A				
◆PITTS◆	12.80	33.7	94.2	0.86	W70A	0.20	ANO	B74	
PERSIM	14.45	34.7	78.3	0.65	S73A	0.06	ANO	S73	
SANCRI	25.0	11.8	25.0	0.32	S73A	0.01	ANO	S73	
IIA									
SIERGO	5.27	57.4	170.	43.	W69A		H		
BENNET	5.28	59.1	179.	41.	W69A		H		
SCOTTS	5.31	60.4	172.	49.	W69A		H		
HOLLAN	5.35	60.9	184.	20.	W74A		H		
ARAGON					B74A		H		P CEDARTOWN
CEDATO	5.36	63.2	181.	8.2	W74A		H		
EDMOTC	5.37	60.4	172.	33.	W74A		H		
KEENMO	5.38	62.0	183.	12.	W74A		H		
RICHLA	5.40	60.6	182.	8.2	W74A		H		
NEGRIL	5.41	59.0	179.	59.	W69A		H		
BRUNO◆	5.41	61.3	185.	37.	W69A		H		
MURPHY	5.42	60.5	186.	34.	W74A		H		
◆BOGUSL	5.45	60.5	180.	24.	W69A		H		
CALICO	5.45	57.3	185.	8.6	W69A		H		
WALKER	5.46	58.2	189.	3.0	W74A		H		
INDIAV	5.48	62.5	174.	12.	W69A		H		
ANGRIR	5.48	57.5	188.	31.	W74A		H		
YARROW	5.48	59.3	171.	18.	W69A		H		
◆BRAUNA	5.49	61.3	183.	12.	W69A		H		
COAHUI	5.49	57.6	178.	16.	W69A		H		
◆AVCE◆◆	5.49	58.1	182.	57.	W74A		H		
CHICOM	5.50	59.3	176.	6.2	W74A		H		
FORSYY	5.50	60.8	176.	31.	W74A		H		
SANFDE	5.50	61.4	183.	21.	W74A		H		
SANMAR	5.51	59.4	177.	3.7	W69A		H		P TOCOPILLA

ABBREV	NI	GA	GE	IR	REF	BW	STRC	REF	REMARKS
IIA									
WATHEN	5.51	59.6	184.	7.0	W74A		H		
RIOLOA	5.52	59.5	181.	3.8	W69A		H		P TOCOPILLA
MAYODA	5.54	59.3	180.	14.	W69A		H		
TOCOPI	5.54	58.6	176.	3.5	W69A		H		
FILOME	5.54	58.7	176.	3.6	W69A		H		P TOCOPILLA
•OKANO•	5.55	59.9	180.	11.	W69A		H		
COYANO	5.55	58.9	174.	3.6	W69A		H		P TOCOPILLA
LOCUST	5.55	60.6	180.	7.5	W74A		H		
PIMACO	5.56	60.3	181.	8.9	W74A		H		
BINGER	5.58	59.7	185.	3.2	W69A		H		
SANCHE	5.58	60.7	189.	15.	W74A		H		P COAHUILA
BARRAB					B74A		H		P BINGERA
WARIAL					B74A		H		P BINGERA
PURIPI	5.58	59.5	174.	3.8	W69A		H		P TOCOPILLA
HEXRIV	5.59	60.7	181.	4.4	W69A		H		
LOMBAR	5.59	58.0	174.	2.3	W69A		H		
CHESTE	5.61	58.9	178.	1.8	W74A		H		
UWET••	5.61	62.3	182.	2.7	W69A		H		
GRESSK	5.61	62.1	177.	7.7	W69A		H		
QUILLA	5.62	57.5	184.	3.4	W69A		H		P TOCOPILLA
KOPJES	5.65	59.9	182.	3.1	W69A		H		
OKAHAN	5.74	56.3	186.	10.	W74A		H		
UNION•	5.79	60.5	187.	3.2	W74A		H		P TOCOPILLA
SMITHO	5.86	65.3	187.	34.	W69A		H		
ELBERT					B74A		H		P SMITHONIA
IIB									
NAVAJO	5.5[illegible]	55.0	180.	0.46	W74A	10	OGG	B74	
SMITHT	5.64	55.3	165.	0.057	W74A		OGG		U COAHUILA
MOUNJO	5.68	59.1	183.	0.46	W69A	10	OGG	B74	
SANDIA	5.85	59.0	174.	0.14	W69A	10	OGG	B74	
•SIKHOT	5.87	51.8	161.	0.029	W69A	9.	OGG	B74	
AINSWO	5.9	55.7	144.	0.023	W69A	6.0	OGG	B74	P CENTRAL M
ELBURR	5.95	58.1	167.	0.059	W69A	10	OGG	B74	
IREDEL	6.0	58.1	163.	0.07	W74A	10	OGG	B74	
NORTHP					B74A		OGG		P SAO JULIA
SAOJUL	6.1	46.2	107.	0.012	W69A	6.	OGG	B74	
CENTRA	6.22	53.4	144.	0.025	W69A	6.	OGG	B74	
PONCAC	6.3	55.0	145.	0.041	W69A	6.	OGG	B74	P CENTRAL M
LAKEMU	6.3	53.9	141.	0.018	W69A	10	OGG	B74	
SANTLU	6.3	47.9	110.	0.010	W69A	5.	OGG	B74	
SILVBE	6.43	45.6	111.	0.012	W69A	5.	OGG	B74	
SUMMIT	6.56	50.5	115.4	0.025	W74A	6.	OGG	B74	
IIC									
CRATSO	8.97	36.3	91.4	9.5	W74A	0.06	OPL	B74	
PERRYV	9.27	37.0	88.0	10.	W69A	0.06	OPL	B74	
TOCAVI	9.63	38.7	96.1	6.8	W69A		OPL		P? SALT RIV
KUMERI	9.69	36.8	93.4	8.1	W69A	0.07	OPL	B74	
BALLOO	9.72	39.0	94.4	9.0	W69A	0.07	OPL	B74	
UNTERM	9.8[illegible]	37.1	101.	4.4	W69A		OPL		
SALTRI	10.02	37.8	100.	6.3	W69A	0.07	OPL	B74	
OWILEY•	11.5[illegible]	38.8	114.	6.2	W69A	0.035	OPL	B74	

ABBREV	NI	GA	GE	IR	REF	BW	STRC	REF	REMARKS
IID									
OARLTUN	9.64	77.1	83.1	17.	W74A	0.005	D	B74	
BRIDGE	9.8	81.0	82.0	10.	W74A	0.65	OM	B74	
*NKANDH	9.96	71.8	83.3	18.	W69A	0.85	OM	B74	
CARBO*	10.02	70.0	87.2	13.	W69A	0.85	OM	B74	
PUQUIO	10.08	77.0	87.9	13.	W69A	0.75	OM	B74	
MOUNOU	10.13	71.4	84.3	15.	W74A	0.80	OM	B74	
ELBOGE	10.2	74.5	87.0	14.	W74A	0.75	OM	B74	
RODEO*	10.2	82.1	93.0	8.0	W69A	0.65	OM	B74	
NEEDLE	10.3	77.2	92.7	4.8	W69A	0.47	OF	B74	
BROW66	10.32	77.8	85.2	10.	W74A				
*HRASCH	10.6	74.5	89.4	13.	W74A	0.75	OM	B74	
WALLAP	11.3	82.9	98.3	3.5	W69A	0.43	OF	B74	
IIE									
WEEKER	7.51	28.2	67.0	2.8	S73A		ANO	S73	
COLOME	7.86	28.4	74.6	7.9	S73A		ANO	S73	
TOBYCH	7.91	27.8	75.5	5.7	W74A				
ELGA**	7.98	24.1	72.4	4.	S73A		ANO	S73	
BARRAN	8.07	22.1	63.9	4.9	S73A		ANO	S73	
OKODAIK	8.22	20.7	65.6	5.2	S73A	0.15	ANO	S73	
ARLING	8.42	21.8	64.9	5.8	S73A	0.80	OM	S73	
ONETSCH	8.6	24.8	66.	1.8	S73A	1.25	ANO	S73	
VERKDN	8.78	22.8	70.4	6.1	W74A		ANO		
LONACO	9.7	23.5	62.1	0.9	W74A	2.0	OG	B74	
IIIA									
PICACH	7.08	19.1	33.9	19.	S73A	1.0	OM	S73	
LIVIMO	7.22	19.6	35.4	9.3	S73A	0.95	OM	S73	
HAIG**	7.24	18.8	33.2	10.	S73A	0.90	OM	S73	
RATELD	7.28	18.5	32.5	12.	S73A	0.9	OM	S73	
DAVISM	7.29	17.4	33.7	14.	S73A	0.95	OM	S73	
DALTON	7.35	18.4	33.1	9.6	S73A	1.10	OM	S73	
GREENB	7.38	18.1	33.3	10.	S73A	1.00	OM	S73	
KENTON	7.38	18.2	35.0	14.	S73A	0.90	OM	S73	
MORITO	7.38	18.7	35.8	9.2	S73A	1.05	OM	S73	
PARRAL					B74A		OM		P MORITO
KALKAS	7.39	18.1	33.5	11.	S73A	1.00	OM	S73	
NEJED*	7.4[illegible]	19.9	38.1	8.7	S73A		OM		P WABAR
UEGIT*	7.4	17.	39.		R72A				
HARROM	7.41	19.2	36.8	10.	S73A	0.95	OM	S73	
ANGELI	7.42	18.3	34.9	9.3	S73A	1.20	OM	S73	
NULERI	7.44	18.0	36.7	9.3	W74A				
SCHWET	7.44	18.3	33.5	11.	S73A	1.00	OM	S73	
NORFOL	7.45	20.2	38.1	10.	S73A	1.00	OM	S73	
BASEDO	7.45	19.1	34.5	11.	S73A		OM		P HENBURY
VERKUD	7.46	19.0	39.8	3.3	S73A	1.15	OM	S73	
CHULAF	7.47	17.8	33.7	5.5	W74A	1.10	OM	B74	
HENBUR	7.47	17.7	33.7	13.	S73A	0.95	OM	S73	
COSTIL	7.49	18.7	33.6	14.	S73A	1.00	OM	S73	
WILLIA	7.50	18.2	32.9	15.	S73A		OM		P KENTON CO
IVANPA	7.51	21.1	38.1	3.8	S73A	1.05	OM	S73	
DUKETO	7.52	19.8	38.1	4.	S73A	1.0	OM	S73	
MADOC*	7.52	19.4	36.4	6.8	S73A	0.95	OM	S73	
RUSSEL	7.52	19.1	35.6	7.2	S73A	0.90	OM	S73	

ABBREV	NI	GA	GE	IR	REF	BW	STRC	REF	REMARKS
IIIA									
SANANG	7.52	19.4	37.6	8.0	S73A	0.95	OM	S73	
SANTAP	7.52	19.5	35.8	8.3	S73A	0.95	OM	S73	
CHAMBO	7.53	18.4	35.0	10.	S73A	0.9	OM	S73	
CACARI	7.56	19.1	35.6	8.7	S73A	1.20	OM	S73	
AKPOHO					B74A		OM		P CAPE YORK
CANTON	7.58	18.6	35.9	9.3	S73A	1.05	OM	S73	
CAPEYO	7.58	19.2	36.0	5.0	S73A	1.20	OM	S73	
CANYCI	7.58	19.8	36.8	11.	S73A	1.00	OM	S73	
TOUBIL	7.59	19.8	38.1	5.0	S73A	1.0	OM	S73	
GLASGO	7.60	20.6	38.8	5.0	S73A	1.05	OM	S73	
MILLYM	7.62	19.1	38.6	2.8	S73A	1.00	OM	S73	
WABAR*	7.62	21.3	38.4	6.0	S73A	0.95	OM	S73	
WILLAM	7.62	18.6	37.3	4.7	S73A	1.05	OM	S73	
DIMITR	7.64	20.3	40.2	3.0	S73A	1.05	OM	S73	
BOXHOL	7.64	18.1	37.2	8.2	S73A	1.00	OM	S73	
NUTWOO	7.66	18.5	35.2	11.	S73A		OM		P HENBURY
LORETO	7.67	19.3	38.3	3.8	S73A	1.15	OM	S73	
DEXTER	7.67	20.5	40.9	1.2	S73A	1.1	OM	S73	
REDRIV	7.70	19.7	38.5	4.4	S73A	1.05	OM	S73	
IRONCR	7.72	20.2	39.6	3.3	S73A	1.05	OM	S73	
MAPLET	7.73	20.3	40.6	1.4	S73A	1.00	OM	S73	
*NORFOR	7.75	20.3	40.1	3.0	S73A	1.05	OM	S73	
BILLIN	7.77	19.5	37.4	3.7	S73A	1.15	OM	S73	
CHILKO	7.77	20.0	39.3	1.8	S73A	1.00	OM	S73	
CASASG	7.77	19.9	37.4	5.1	S73A	1.15	OM	S73	
YARRI*	7.77	19.8	38.5	4.0	S73A	1.1	OM	S73	
*ROWTON	7.79	20.5	38.1	2.8	S73A	1.15	OM	S73	
DESCUB	7.79	20.5	39.7	2.3	S73A		OM		P CHARCAS
MERCED	7.82	19.5	38.9	3.6	S73A	1.00	OM	S73	
SACRAM	7.82	19.2	36.6	6.7	S73A	1.00	OM	S73	
UWHARR	7.83	20.6	39.0	3.6	S73A	1.15	OM	S73	
LEXIPS	7.85	21.7	42.6	1.1	S73A				
FRANFI	7.85	20.2	40.4	1.8	S73A	1.15	OM	S73	
SSYROM	7.85	19.6	40.9	3.3	W74A	0.95	OM	B74	
YOUANM	7.85	21.0	37.7	2.6	S73A	1.0	OM	S73	
CHARCA	7.86	19.4	41.4	1.9	S73A	1.05	OM	S73	
SUSUMA	7.86	20.5	41.0	2.2	S73A	1.00	OM	S73	
LAPORT	7.88	21.5	43.1	1.4	S73A	1.05	OM	S73	
MARSHA	7.92	21.1	44.3	2.6	S73A	1.20	OM	S73	
RANCPI	7.93	20.8	42.4	0.70	S73A	1.05	OM	S73	
CUMPAS	7.94	21.0	42.8	2.7	S73A	1.20	OM	S73	
ABAKAN	7.97	20.3	42.3	4.6	W74A		OM		P TOUBIL RI
SAVANN	7.99	21.1	44.2	0.6	S73A	1.20	OM	S73	
CHANAR	8.0	22.0	44.1	0.17	S73A	0.9	OM	S73	P? ILIMA IR
DURANG	8.00	20.4	40.2	1.0	S73A	1.15	OM	S73	
OZERHAM	8.00	18.2	33.5	10.	W74A	1.1	OM	B74	
BAGDAD	8.01	19.8	39.7	6.8	S73A	1.10	OM	S73	
ROEBOU	8.01	21.2	42.4	0.65	S73A	1.10	OM	S73	
JUNCAL	8.05	20.5	41.2	1.8	S73A	1.10	OM	S73	
KYANCU	8.06	19.9	39.5	1.7	S73A	1.05	OM	S73	
THUNDA	8.08	20.2	38.9	2.2	S73A	1.20	OM	S73	
SANDTO	8.09	21.0	41.4	1.4	S73A	1.20	OM	S73	
APOAPS	8.09	21.7	41.9	1.1	S73A	0.9	OM	S73	P DURANGO

ABBREV	NI	GA	GE	IR	REF	BW	STRC	REF	REMARKS
IIIA									
ILIMIR	8.1	21.2	43.5	0.17	S73A	0.9	OM	S73	
BRIGGS	8.17	20.1	40.7	0.72	S73A	1.25	OM	S73	
OAUGUCO	8.19	18.7	35.3	9.2	S73A	1.15	OM	S73	
PUENDE	8.20	20.6	40.5	1.4	S73A	1.05	OM	S73	
FRANCE	8.2	20.4	42.4	0.38	S73A	1.10	OM	S73	
COWELL	8.2	21.	38.		R72A				
DENTON	8.21	19.7	42.7	0.28	S73A	1.15	OM	S73	
ASWAN+	8.21	20.0	41.8	0.22	S73A	1.3	OM	S73	
DRUMMO	8.23	20.4	41.8	0.64	S73A	1.15	OM	S73	
CARTHA	8.24	21.5	43.7	0.57	S73A	1.25	OM	S73	
PROVID	8.25	20.2	41.5	0.39	S73A	1.15	OM	S73	
SENEFA	8.25	21.1	42.8	0.25	S73A	1.2	OM	S73	
GUNDAR	8.27	20.0	43.9	0.31	S73A	1.40	OG	S73	
LANTON	8.28	20.6	39.3	3.5	S73A	1.05	OM	S73	
IDER++	8.3	20.1	40.	2.8	S73A	1.20	OM	S73	
+KAYAKE	8.32	19.9	44.0	1.1	S73A	1.20	OM	S73	
ASARCO	8.32	21.2	44.4	0.23	S73A	1.1	OM	S73	
TRENTO	8.34	20.8	44.5	2.6	S73A	1.15	OM	S73	
TAMENT	8.37	20.3	42.7	2.5	S73A	1.2	OM	S73	
POINIR	8.4	21.1	41.4	0.46	S73A	1.0	OM	S73	
PLYMOU	8.4	23.1	42.4	0.66	S73A	1.30	OM	S73	
TONOPA					B74A		OM		P QUINN CAN
JOELSI	8.40	22.6	43.6	0.26	S73A	1.10	OM	S73	P? SIER SAN
CASIMI	8.40	20.9	41.0	0.25	S73A	1.3	OM	S73	
QUINNC	8.40	20.9	41.5	0.58	S73A	1.10	OM	S73	
TARAPO					B74A		OM		P TAMARUGAL
TAMARU	8.44	21.6	43.8	0.58	S73A	1.10	OM	S73	
AGGIEC	8.48	20.5	39.9	0.46	S73A	1.20	OM	S73	
THULE+	8.52	19.3	39.8	2.6	S73A		OM	S73	
SIERSA	8.55	20.8	43.8	0.28	S73A	1.00	OM	S73	
RUFFSM	8.56	21.5	46.9	0.47	S73A	1.25	OM	S73	
CAPERR	8.58	21.0	45.3	0.24	S73A	1.00	OM	S73	
SPEARM	8.63	20.2	46.0	0.71	S73A	1.15	OM	S73	
BARTLE	8.68	20.6	46.0	0.64	S73A	1.10	OM	S73	
VELINI	8.75	21.5	47.4	0.62	W74A	1.15	OM	B74	
BAQUED	8.76	20.4	43.1	0.10	S73A	1.20	OM	S73	
WELLAN	8.77	21.0	46.7	0.29	S73A	1.20	OM	S73	
+JUROME	8.81	21.2	40.3	0.24	S73A	---	ANO	S73	
LENART	8.85	21.7	43.5	0.33	S73A	1.15	OM	S73	
VIEWHI	8.87	21.0	42.6	0.27	S73A	0.80	OM	S73	
MOORUM	8.98	21.7	44.	0.26	S73A	0.9	OM	S73	
NAZARI	9.04	20.3	40.3	0.44	W74A	1.00	OM	B74	
AVOCWE	9.21	20.6	44.5	0.30	W74A		OM	B74	
ILINSK	9.28	19.5	39.2	0.29	S73A	0.70	OM	S73	
IIIB									
OROVIL	8.36	20.3	40.7	0.053	S73A	0.85	OM	S73	
ORANIR	8.46	21.2	43.7	0.12	S73A	1.20	OM	S73	
OWENSV	8.53	21.5	45.9	0.15	S73A	1.15	OM	S73	
ELCAPI	8.56	21.5	45.1	0.11	S73A	1.10	OM	S73	
LUISLO	8.64	20.1	41.9	0.15	S73A	1.15	OM	S73	
CAMPBE	8.65	20.4	43.8	0.09	S73A	1.25	OM	S73	
LOSREY	8.71	20.9	40.7	0.12	S73A	0.90	OM	S73	

ABBREV	NI	GA	GE	IR	REF	BW	STRC	REF	REMARKS
IIIB									
WONYUL	8.72	19.5	39.6	0.028	S73A	0.95	OM	S73	
TURTLE	8.80	20.5	41.4	0.057	S73A	1.05	OM	S73	
CLEVEL	8.85	21.0	41.9	0.094	S73A	1.0	OM	S73	
ZACA69	9.0	20.3	38.8	0.029	S73A	0.70	OM	S73	
JOEWRI	9.1[illegible]	20.1	35.5	0.015	S73A	0.85	OM	S73	
●TREYSA	9.1	20.4	43.1	1.2	S73A	0.85	OM	S73	
WOLFCR	9.22	18.4	37.3	0.036	S73A	0.85	OM	S73	
GRANT+	9.24	19.8	37.0	0.040	S73A	0.80	OM	S73	
BALDEA	9.25	18.1	37.1	0.018	S73A	0.65	OM	S73	
KOUGAM	9.29	18.0	35.3	0.022	S73A	0.9	OM	S73	
KNOWLE	9.35	18.5	31.6	0.023	S73A	0.75	OM	S73	
APOALA	9.39	18.4	35.7	0.016	S73A	0.65	OM	S73	
MOUNED	9.4	20.1	37.5	0.016	S73A	0.80	OM	S73	
BREECE	9.48	19.7	37.9	0.044	S73A		OM		P GRANT
ODELEGA	9.5	20.3	41.7	1.6	S73A	0.85	OM	S73	
AUGUST	9.56	17.8	37.6	0.03	S73A	0.8	OM	S73	
SMITHS	9.56	17.4	30.6	0.023	S73A	0.63	OM	S73	
HOPPER					B74A		OM		P SMITHS MO
VERKPS	9.6	18.4	33.5	0.027	S73A		OM		P AUGUSTINO
NORRIS	9.64	18.2	32.4	0.016	S73A	0.65	OM	S73	
SANDER	9.69	18.2	35.8	0.021	S73A	0.75	OM	S73	
CHUPAD	9.7	17.2	29.6	0.020	S73A	0.65	OM	S73	
SAMSVA	9.77	18.4	35.1	0.017	S73A	0.75	OM	S73	
BEARCR	9.8[illegible]	18.4	32.8	0.019	S73A	0.60	OM	S73	
ROPERR	9.8	18.1	33.9	0.04	S73A	0.6	OM	S73	
THURLO	9.9	15.9	27.3	0.017	S73A	0.65	OM	S73	
BELLAR	10.06	16.7	31.1	0.014	S73A	0.70	OM	S73	
ADARGA	10.1	17.6	28.7	0.015	S73A		OM		P CHUPADERO
TEMORA					B74A		OM		P NARRABURR
NARRAB	10.13	16.6	28.7	0.016	S73A	0.60	OM	S73	
TAMBOQ	10.2	17.9	31.5	0.039	S73A	0.75	OM	S73	
CUERNA	10.4	16.3	31.3	0.021	S73A		OM		P CHUPADERO
TIERAC	10.5	16.2	28.0	0.041	S73A	0.5	OM	S73	
IIIC									
+HASSIJ	10.5[illegible]	27.4	69.6	0.18	S73A	0.35	OF	S73	
OMAGNES	11.0	14.5	22.4	0.18	W74A		OM		
HAVANA	11.37	20.5	21.6	0.3	W71A	0.16	OFF	W71	
MUNGIN	11.5	19.4	22.1	0.47	W71A	0.40	OF	B74	
ANOKA+	11.95	17.2	15.7	0.16	W71A	0.34	OF	B74	
EDMOTK	12.7	25.4	34.6	0.55	S73A	0.32	OF	S73	
CARLTO	13.0	11.4	8.5	0.076	W71A	0.21	OF	B74	
IIID									
TAZEWE	16.64	4.69	3.79	0.063	W71A	0.045	OFF	B74	
DAYTON	17.02	5.16	3.52	0.028	W71A	0.045	OFF	B74	
FOELLI	18.13	4.02	3.15	0.072	W71A	0.04	OFF	B74	
WEDDER	22.36	1.51	1.47	0.052	W71A	0.01	D	B74	
FREDA+	22.57	2.09	2.24	0.021	W71A	0.015	D	B74	
IIIE									
BURLIN	8.15	16.9	34.9	0.45	S73A	1.3	OM	S73	
STAUNT	8.21	18.9	36.6	0.11	S73A	1.60	OG	S73	

ABBREV	NI	GA	GE	IR	REF	BW	STRC	REF	REMARKS
IIIE									
MATATI	8.32	17.7	34.4	0.54	S73A		OG		P KOKSTAD
KOKSTA	8.33	17.4	35.9	0.6	S73A	1.35	OG	S73	
COOPER	8.47	17.0	34.9	0.51	S73A	1.5	OG	S73	
RHINEV	8.63	18.8	36.3	0.12	S73A	1.4	OG	S73	
WILLOC	8.76	16.9	36.4	0.054	S73A	1.40	OG	S73	
TANAKA	8.94	18.2	34.6	0.22	S73A	1.50	OG	S73	
TOLUPS	8.98	17.0	34.1	0.37	W74A	1.5	OG	B74	
IIIF									
CLARKC	6.79	6.92	1.11	6.2	S72A	1.0	OM	S72	
NELSON	7.02	5.33	0.92	7.6	S72A	1-10	OGG	S72	
OAKLIR	7.32	7.20	1.13	5.5	W74A	1.40	OG	B74	
SAIGEN	7.68	6.86	0.781	2.0	S72A	0.49	OF	S72	
MOONBI	7.70	6.84	0.831	1.3	S72A	0.55	OM	S72	
ABANCA	7.92	6.59	0.779	1.8	S72A	0.49	OF	S72	P ST.GENEVI
IVA									
OBERNK	7.33	1.80	0.092	3.2	S72A	0.26	OF	S72	
LAGRAN	7.40	2.07	0.116	2.3	S72A	0.27	OF	S72	
JAMEST	7.45	1.80	0.093	3.5	S72A	0.26	OF	S72	
HUIZOP	7.48	2.22	0.120	2.2	S72A	0.28	OF	S72	
YANHUI	7.49	1.75	0.105	2.7	S72A	0.33	OF	S72	
BISHOC	7.59	2.20	0.111	2.6	S72A	0.30	OF	S72	
SANFMO	7.62	2.09	0.102	3.0	S72A	0.23	OF	S72	
MARIAE	7.64	1.72	0.096	3.1	S72A	0.30	OF	S72	
SOCIAL	7.65	1.63	0.092	3.6	S72A	0.30	OF	S72	
GIBEON	7.68	1.97	0.111	2.4	S72A	0.30	OF	S72	
NICO**					B74A		OF		P GIBEON
CRAT31	7.72	2.19	0.111	2.3	S72A	0.30	OF	S72	
WESTER	7.72	1.97	0.100	2.8	S72A	0.30	OF	S72	
OTCHIN	7.82	2.13	0.119	2.6	S72A	0.29	OF	S72	
RAILWA	7.82	1.99	0.119	2.1	S72A	0.30	OF	S72	P GIBEON
SIGNAL	7.84	2.11	0.121	2.5	S72A	0.28	OF	S72	
SHIRAH	7.86	2.19	0.120	2.4	W74A	0.25	OF	B74	
IRONRI	7.87	2.12	0.118	2.1	S72A		OF	S72	
BRISTO	7.90	2.13	0.125	1.6	S72A	0.30	OF	S72	
BODAIB	7.91	1.96	0.111	1.7	S72A	0.30	OF	S72	
HARROF	7.96	2.21	0.130	2.3	S72A	0.30	OF	S72	
PUTNAM	7.98	2.17	0.129	2.1	S72A	0.28	OF	S72	
*PARADE	7.99	2.21	0.125	2.3	S72A	0.33	OF	S72	
SERRAN	8.00	2.16	0.130	1.8	S72A	0.30	OF	S72	
*CHARLO	8.04	2.24	0.118	1.5	S72A	0.30	OF	S72	
MCDOWE					B74A		OF		P WOODS MTN
WOODSM	8.13	2.39	0.147	2.3	S72A	0.30	OF	S72	
ALTONA	8.17	2.33	0.123	1.5	S72A	0.28	OF	S72	
SENETO	8.42	2.17	0.124	1.8	S72A	0.28	OF	S72	
MUONIO	8.42	2.24	0.133	1.6	S72A	0.29	OF	S72	
*REMBAN	8.7	<3.	<3.		R72A				
BOOGAL	8.79	2.26	0.133	0.60	S72A	0.42	OF	S72	
BUSHMA	8.83	2.08	0.134	0.98	S72A	0.33	OF	S72	U GIBEON
MANTOS	8.89	2.42	0.133	0.91	S72A	0.35	OF	S72	
OSTEINB	9.08	2.27	0.134	0.53	S72A	0.33	ANO	S72	M 50 PC SIL
HILLCI	9.09	2.29	0.144	0.88	S72A	0.38	OF	S72	

ABBREV	NI	GA	GE	IR	REF	BW	STRC	REF	REMARKS
IVA									
NOVORY	9.1	2.45	0.20	0.90	W74A				
MART++	9.20	2.16	0.139	0.64	S72A	0.38	OF	S72	
SMITHL	9.23	2.38	0.133	0.89	S72A	---	ANO	S72	
DUCHES	9.32	2.19	0.126	0.42	S72A	0.35	OF	S72	
MOUNTA					B74A		OF		P DUCHESNE
NEWWES	9.36	2.40	0.139	0.55	S72A	0.42	OF	S72	
OCHINAU	9.54	2.08	0.112	0.12	S72A	0.35	OF	S72	
ODUEL54	10.39	1.93	0.111	0.64	S72A	0.35	OF	S72	
IVB									
TLACOT	15.82	0.195	0.031	20.	S72A	0.04	D	S72	
KOKOMO	15.88	0.193	0.032	31.	S72A	0.02	D	S72	
IQUIQU	16.03	0.170	0.051	28.	S72A	0.02	D	S72	
OCHINGA	16.18	0.181	0.082	3.6	S72A	0.01	D	S72	
HOBA++	16.56	0.192	0.047	27.	S72A	0.02	D	S72	
WEAVER	16.81	0.233	0.058	17.	S72A	0.006	D	S72	
CAPEOF	16.92	0.198	0.060	36.	S72A	0.03	D	S72	
TAWALL	17.06	0.248	0.068	16.	S72A	0.008	D	S72	
SKOOKU	17.13	0.272	0.057	15.	S72A	0.01	D	S72	
GALLEG	17.58	0.253	0.063	15.	S71A		D	S72	P TERNERA
WARBUR	17.80	0.244	0.064	13.	S72A		D	S72	
TERNER	18.13	0.261	0.056	16.	W74A	---	D	B74	
MESOSIDERITES									
CLOVER	6.10	11.0	42.	2.1	W73A		MES		
MOUNPA	6.10	8.9	44.9	2.0	W73A		MES		
CRABOR	7.02	13.3	49.5	2.9	W73A		MES		
MINCY+	7.2	13.1	52.4	2.5	W73A		MES		
BONDOC	7.3	15.6	48.	3.8	W73A		MES		
+VERAMI	7.32	14.8	56.	3.2	W73A		MES		
BUDULA	7.48	15.0	58.0	4.4	W73A		MES		
+LOWICZ	7.69	15.3	54.	3.8	W73A		MES		
MORRIS	7.82	13.2	57.	3.7	W73A		MES		
HAINHO	8.12	15.1	59.	3.7	W73A		MES		
VACAMU	8.8	9.6	42.8	2.2	W73A		MES		
DALGAR	8.8	15.5	56.0	4.2	W73A		MES		
+ESTHER	9.0	9.0	52.3	3.2	W73A		MES		
CHINGU	9.6	15.2	55.2	6.2	W73A		MES		
PINNAR	9.66	14.0	51.6	3.2	W73A		MES		
+PATWAR	9.9	9.7	39.3	2.1	W73A		MES		
PALLASITES									
+MARJAL	7.86	23.0	54.7	1.6	W74A		PAL		
AHUMAD	7.96	21.4	49.0	0.057	W69B		PAL		
KRASNJ	8.8	22.6	55.1	0.18	W69B		PAL		
GIROUX	9.	22.6	50.8	0.03	W74A		PAL		
SALTA+	9.4	21.4	48.6	0.07	W74A		PAL		
SOUTBE	9.62	21.2	44.3	0.055	W69B		PAL		
IMILAC	9.8	21.2	45.5	0.071	W69B		PAL		
ANTOFA	10.10	21.9	50.4	0.06	W74A		PAL		
THIELM	10.1	22.2	50.2	0.23	W74A		PAL		
ALBIPA	10.42	16.8	29.4	0.015	W69B		PAL		
HOPEWE	10.63	24.0	61.8	0.049	W69B		PAL		P BRENHAM

ABBREV	NI	GA	GE	IR	REF	BW	STRC	REF	REMARKS
PAL									
FINMAR	10.67	18.7	43.7	1.8	W69B		PAL		
ADMIRE	10.72	20.3	39.2	0.017	W69B		PAL		
NEWPOR	10.72	17.5	31.2	0.16	W69B		PAL		
BRENHA	11.1	26.2	70.8	0.041	W69B		PAL		
ANDERS	11.34	24.8	65.6	0.045	W69B		PAL		P BRENHAM
OGLORIE	12.04	13.2	10.7	0.014	W74A	0.85	OM	B74	
SANTFE					B74A		OM		P GLORIETA
SPRINW	12.3	14.1	31.9	0.062	W69B		PAL		
MOUNVE	12.88	23.7	52.0	0.14	W69B		PAL		
OITZAWI	14.7	5.73	85.9	15.0	W74A		PAL		G EAGLE STA
OEAGLES	15.37	4.54	75.3	10.	W74A		PAL		G ITZAWISIS
ANOMALOUS METAL-RICH METEORITES									
BUTLER	15.2	87.1	1970.	1.3	W70A	0.15	OPL	B74	
KENDAL	5.43	74.6	355.	1.7	W74A		ANO	B74	
LIVITE	6.64	45.3	250.	0.73	W74A	0.8	OM	B74	
SAIFRA	6.12	48.1	246.	0.11	W70A	2.7	OG	B74	
UNIONC	6.12	54.8	245.	2.1	W70A		OM	B74	G MT.DOOLIN
ARISPE	6.54	50.3	243.	9.7	W70A	2.9	OG	B74	
BENDEG	6.39	54.0	234.	0.20	W70A	1.8	OG	B74	G? CHIHUAHU
MOUNDO	6.26	52.0	234.	1.2	W70A	1-4	OGG		G UNION COU
SANTRO	6.63	50.6	222.	0.068	W70A	---	ANO	B74	G CHIHUAHUA
CHIHUA	6.68	52.7	212.	0.11	W70A		ANO	B74	G STA ROSA
MUNDRA	7.72	59.5	196.	0.87	W74A		ANO		G WATERVILL
WATERV	7.81	64.8	196.	0.30	W70A	1.1	ANO	W70	
REPEEV	14.3	11.6	193.	3.0	W74A				
CRUZDE	9.00	38.2	186.	5.9	W74A	0.48	OF	B74	
PREMIE	7.72	58.	180.		D72A		ANO		P MUNDRABIL
LOONGA	7.78	60.	179.		D72A		ANO		P MUNDRABIL
ELTON+	6.9	47.0	165.	0.053	W70A	0.55	ANO		
COROWA	13.13	10.1	159.	0.75	W69A		OPL		
NOCOLE	6.4	49.3	148.	8.2	W70A	0.8-2	ANO	B74	
MONAHA	10.60	8.9	127.	14.	W69A	0.05	OPL	B74	G DOROFEEVK
DOROFE	11.26	9.10	124.	23.	W74A	0.09	OPL	B74	G MONAHANS
MORRAD	19.54	46.3	119.	0.61	W69A	0.01	D	B74	
HORSEC	5.75	47.5	110.	2.5	W74A		ANO		G MOUNT EGE
MOUNEG	6.20	35.2	99.2	1.8	W74A		ANO		G HORSE CRE
DELRIO	11.34	9.19	98.6	19.	W74A	0.06	D	B74	G MONOHANS
REDFIE	6.91	39.3	95.4	0.82	W74A		ANO		
MURNPE	6.31	41.8	85.4	1.8	W74A		ANO	B74	
GLENOR	7.45	16.9	76.8	2.7	S73A		ANO	S73	
AUBURN	4.55	44.5	69.9	0.025	S73A		ANO		P TOMBIGBEE
GUNCRE	8.38	22.4	69.7	0.052	S73A	0.75	OM	S73	
TOMBIG	4.3	38.5	62.5	0.021	S73A	2-100	ANO	S73	G LA PRIMIT
KINGST	6.88	21.3	58.8	5.1	S73A	0.80	OM	S73	
HAMMON	8.07	26.2	58.4	0.098	S73A	0.60	OM	S73	
REEDCI	7.35	22.5	55.5	54.	S73A	1.8	OG	S73	
BELLSB	4.13	39.2	54.6	0.15	S73A	100	ANO	S73	G TOMBIGBEE
CHEBAN	8.80	21.8	52.5	0.11	S73A	2.5	OG	S73	
SOUTBY	17.8	20.0	45.0	28.	S73A	0.01	D	S73	G BABBS(TRO
BABBTR	17.7	18.6	41.0	35.	S73A	0.01	D	S73	G SOUTH BYR
TARAPH	4.35	33.6	38.6	0.05	S73A		ANO		P LA PRIMIT
ALGOMA	10.75	17.9	38.3	0.35	S73A	0.60	OM	S73	

ABBREV	NI	GA	GE	IR	REF	BW	STRC	REF	REMARKS
ANOMALOUS METAL-RICH METEORITES									
LAPRIM	4.90	33.3	37.3	0.039	S73A	100	ANO	S73	G TOMBIGBEE
NEWBAL	6.36	20.4	35.9	10.	S73A	1.0	ANO	S73	
EMSLAN	9.40	2.90	35.0	2.9	S73A	0.8	OM	S73	G MBOSI
BACUBI	9.62	17.7	31.9	4.9	S73A	0.08	OFF	S73	
VICTOR	11.8	15.3	31.4	0.022	S73A	0.22	OF	S73	
CACHIO	7.88	16.9	30.3	3.1	S73A	1.30	OM	S73	G MURFREESB
MURFRE	7.91	16.7	30.2	2.2	S73A	0.95	OM	S73	G CACHIY OM
LIMECR	29.1	15.5	28.5	1.1	S73A		D	S73	
MBOSI+	8.71	2.54	26.9	6.6	S73A	0.80	OM	S73	G EMSLAND
PIEDAD	7.51	15.1	25.7	11.	S73A	0.75	OM	S73	
VENTUR	10.07	14.0	25.0	0.18	S73A	0.4	OF	S73	
LAUREN	12.95	10.5	23.0	7.9	W71A	0.30	OF	B74	
LACAIL	9.11	13.7	21.5	9.7	W71A	1.10	OM	B74	
WASHIN	9.96	15.5	20.5	0.067	W71A		ANO	B74	
GARDEN	16.96	10.7	16.6	0.12	W71A	0.035	OPL		G GAY GULCH
GRANDR	9.26	17.9	13.8	14.	W71A	0.55	OM	B74	
COWRA+	12.94	73.5	12.4	14.0	W71A	0.07	OPL	B74	
ELQOSE	13.19	6.15	11.7	5.5	W71A		D		
SOPER+	5.70	9.71	10.8	0.011	W71A	---	ANO	B74	
GAYGUL	15.06	6.68	10.7	0.11	W71A	0.06	OPL	B74	G GARDEN HE
SANTCA	33.62	5.28	8.94	0.020	W71A	---	ANO	B74	
KOFA++	18.27	4.79	8.61	0.098	W71A	0.035	OPL	B74	G GAY GULCH
YARDEA	7.7	8.	8.		R72A				
TWINCI	30.06	4.54	7.42	0.015	W71A	0.005	D	B74	
MOUNMA	14.56	7.53	5.26	0.012	W71A	0.015	OPL	B74	
+SOROTI	12.88	14.1	5.22	0.060	W71A	0.13	ANO	B74	
ALIKAT	13.0	<3.	3.		R72A				
ILLIGU	11.68	2.80	2.76	5.3	W71A	0.04	D	B74	
CAMBRI	10.17	11.1	1.52	0.88	W71A	0.48	OF	B74	
PINON+	15.54	2.32	1.16	15.	W71A	0.015	D	B74	
NORDHE	11.64	0.550	0.644	11.	S72A	0.02	D	S72	
DERMBA	42.1	4.66	0.144	0.029	W74A		ANO		
SHINGL	16.95	2.06	0.130	2.6	S72A	0.015	D	S72	
YSLETA	7.62	0.144	0.12	7.0	S72A	---	ANO	S72	
DEEPSP	13.20	0.405	0.109	9.4	S72A	0.005	D	S72	
GUFFEY	9.94	0.146	0.082	5.0	S72A	0.015	D	S72	
RAFRUE	9.32	0.159	0.055	0.007	S72A	---	ANO	S72	
TUCSON	9.45	0.941	0.049	2.1	W74A	---	ANO	S72	
SANTSQ	7.48	0.573	0.040	4.0	S72A	---	ANO	S72	
BABBBL	11.8	0.203	0.029	1.7	S72A		D	B74	
+NGOURE	9.26	0.067	0.016	0.58	S72A	---	ANO	S72	
+NEDAGO	6.02	0.665	0.005	3.4	S72A	---	ANO	S72	

References, Appendix II D

B74A BUCHWALD, V. F.: Iron meteorites. Ariz. State Univ. Press, in press. (1974).

D72A DELAETER, J. R.: Meteoritics **7**, 285 (1972).

R72A REED, S. J. B.: Meteoritics **7**, 257 (1972).

S71A SCHAUDY, R., WASSON, J. T.: Chem. Erde **30**, 287 (1971).

S72A — SCHAUDY, R., WASSON, J.T., BUCHWALD, V.F.: Icarus **17**, 174 (1972).

S73A — SCOTT, E.R.D., WASSON, J.T., BUCHWALD, V.F.: Geochim. Cosmochim. Acta **37**, 1957 (1973).

W68A — WASSON, J.T., GOLDSTEIN, J.I.: Geochim. Cosmochim. Acta **32**, 329 (1968).

W69A — WASSON, J.T.: Geochim. Cosmochim. Acta **33**, 859 (1969).

W69B — WASSON, J.T., SEDWICK, S.P.: Nature **222**, 22 (1969).

W70A — WASSON, J.T.: Icarus **12**, 407 (1970).

W71A — WASSON, J.T., SCHAUDY, R.: Icarus **14**, 59 (1971).

W73A — WASSON, J.T., SCHAUDY, R., BILD, R.W., CHOU, C.-L.: Geochim. Cosmochim. Acta **38**, 135 (1974).

W74A — WASSON, J.T.: This work (1974).

Subject Index

Page numbers *in italics* indicate a photograph of meteorites or a glossary entry.

MIX
Papier aus verantwortungsvollen Quellen
Paper from responsible sources
FSC® C105338

If you have any concerns about our products,
you can contact us on
ProductSafety@springernature.com

In case Publisher is established outside the EU,
the EU authorized representative is:
Springer Nature Customer Service Center GmbH
Europaplatz 3, 69115 Heidelberg, Germany

Printed by Libri Plureos GmbH
in Hamburg, Germany